T0321224

MECHANICS OF FLUID DEFORMATIONS

Rigid Body Rotations and Plane Channel Flow Stability

MECHANICS OF FLUID DEFORMATIONS

Rigid Body Rotations and Plane Channel Flow Stability

Oleg V Troshkin

Russian Academy of Sciences, Russia

NEW JERSEY • LONDON • SINGAPORE • BEIJING • SHANGHAI • HONG KONG • TAIPEI • CHENNAI • TOKYO

Published by

World Scientific Publishing Co. Pte. Ltd.
5 Toh Tuck Link, Singapore 596224
USA office: 27 Warren Street, Suite 401-402, Hackensack, NJ 07601
UK office: 57 Shelton Street, Covent Garden, London WC2H 9HE

British Library Cataloguing-in-Publication Data
A catalogue record for this book is available from the British Library.

MECHANICS OF FLUID DEFORMATIONS
Rigid Body Rotations and Plane Channel Flow Stability

Copyright © 2021 by World Scientific Publishing Co. Pte. Ltd.

All rights reserved. This book, or parts thereof, may not be reproduced in any form or by any means, electronic or mechanical, including photocopying, recording or any information storage and retrieval system now known or to be invented, without written permission from the publisher.

For photocopying of material in this volume, please pay a copying fee through the Copyright Clearance Center, Inc., 222 Rosewood Drive, Danvers, MA 01923, USA. In this case permission to photocopy is not required from the publisher.

ISBN 978-981-123-051-6 (hardcover)
ISBN 978-981-123-052-3 (ebook for institutions)
ISBN 978-981-123-053-0 (ebook for individuals)

For any available supplementary material, please visit
https://www.worldscientific.com/worldscibooks/10.1142/12106#t=suppl

Printed in Singapore

In memory of my daughter Lyudmila.

Contents

Introduction

Apparently, in proper non-traditional forms [Troshkin, 1995], sometimes the new mimics the old and forgotten to give rise to the next new, and so on. Briefly, with the help of vector and double vector (or matrix) analysis, the theory of boundary value problems, quasicompact Lie algebras for Navier–Stokes equations, and the notion of fluid stability, the fundamentals of mechanics are studied herein to unite both rigid body rotations and fluid dynamics, which was initiated in 18th century for the Euler dynamic and hydrodynamic equations. Maybe doing so will help us to find some new answers to the following questions: Where do rotations come from? Why do they not fade out at infinity of the universe? Why do ordinary fluid vortices remain stable while other ones cause turbulence? Finally, are there any mechanical reasons behind the fact that our space started from one point and extended, contracted, and ended in another one?

Above all, in mechanics we shall deal with the initial first order *vector objects* of ordinary *orts* (orthonormal basis) $\mathbf{i}, \mathbf{j}, \mathbf{k}$ introduced, as is usual, with the Archimedean *screw* "$\times$" (cross product) and the Pythagorean *metric* "$\cdot$" (point product),

$$\mathbf{i} = \mathbf{j} \times \mathbf{k}, \quad \mathbf{j} = \mathbf{k} \times \mathbf{i}, \quad \mathbf{k} = \mathbf{i} \times \mathbf{j} \quad \text{and}$$
$$\mathbf{i} \cdot \mathbf{i} = \mathbf{j} \cdot \mathbf{j} = \mathbf{k} \cdot \mathbf{k} = |\mathbf{k}|^2 = 1,$$

to be immediately extended to proportions $x\mathbf{i}, y\mathbf{j}, z\mathbf{k}$, $-\infty < x, y, z < \infty$, and their sums in the *radius-vector* $\mathbf{r}$ (to be either arm, position, particle, or point) and its *velocity* $\mathbf{u}$,

$$\mathbf{r} = x\mathbf{i} + y\mathbf{j} + z\mathbf{k} \quad \text{and} \quad \mathbf{u} = u\mathbf{i} + v\mathbf{j} + w\mathbf{k}$$
$$= \mathbf{r}_t = \partial_t \mathbf{r} = \frac{\partial r}{\partial t} = \lim_{\varepsilon\to 0, \varepsilon>0} \frac{\mathbf{r}(t+\varepsilon) - \mathbf{r}(\varepsilon)}{\varepsilon},$$
$$|\mathbf{r}|^2|\mathbf{u}|^2 = |\mathbf{r}\times\mathbf{u}|^2 + (\mathbf{r}\cdot\mathbf{u})^2,$$

proving to be one of *Al-Khwarizmi said*, or, literally, of *algebra M*,

$$(x^2+y^2+z^2)(u^2+v^2+w^2) = (yw-zv)^2 + (zu-xw)^2$$
$$+ (xv-yw)^2 + (xu+yv+zw)^2,$$

all these constituents of the habitual surrounding space M will be accompanied by seemingly invisible *second order* vectors such as *diorts*, or *direct products* of orts $\mathbf{ii}, \mathbf{ij}, \mathbf{ji}, \ldots,$ with their *reflections* $(\mathbf{ii})_* = \mathbf{ii}$, $(\mathbf{ij})_* = \mathbf{ji}$, $(\mathbf{ji})_* = \mathbf{ij}, \ldots,$ and sums of proportions to be *divectors*, or *double vectors* reduced to the familiar *matrix unit*

$$\vec{\mathbf{e}} = \begin{pmatrix} 1 & 0 & 0 \\ 0 & 1 & 0 \\ 0 & 0 & 1 \end{pmatrix} = \mathbf{ii} + \mathbf{jj} + \mathbf{kk}$$

and general *matrices*,

$$\vec{\mathbf{p}} = \begin{pmatrix} p^{xx} & p^{xy} & p^{xz} \\ p^{yx} & p^{yy} & p^{yz} \\ p^{zx} & p^{zy} & p^{zz} \end{pmatrix} = \begin{pmatrix} p^{xx}\mathbf{ii} + p^{xy}\mathbf{ij} + p^{xz}\mathbf{ik} \\ +p^{yx}\mathbf{ji} + p^{yy}\mathbf{jj} + p^{yz}\mathbf{jk} \\ +p^{zx}\mathbf{ki} + p^{zy}\mathbf{kj} + p^{zz}\mathbf{kk} \end{pmatrix}$$
$$= \mathbf{p}^x\mathbf{i} + \mathbf{p}^y\mathbf{j} + \mathbf{p}^z\mathbf{k}$$

including their *conjugates*

$$\vec{\mathbf{p}}_* = (\mathbf{p}^x\mathbf{i} + \mathbf{p}^y\mathbf{j} + \mathbf{p}^z\mathbf{k})_* = \mathbf{i}\mathbf{p}^x + \mathbf{j}\mathbf{p}^y + \mathbf{k}\mathbf{p}^z$$
$$= \begin{pmatrix} p^{xx} & p^{yx} & p^{zx} \\ p^{xy} & p^{yy} & p^{zy} \\ p^{xz} & p^{yz} & p^{zz} \end{pmatrix} = \mathbf{p}^x_*\mathbf{i} + \mathbf{p}^y_*\mathbf{j} + \mathbf{p}^z_*\mathbf{k},$$

to produce an *element*

$$\vec{\mathbf{p}} \cdot d\mathbf{S} = d\mathbf{S} \cdot \vec{\mathbf{p}}_*$$

of the *contact force*

$$\int_{\partial V} \vec{\mathbf{p}} \cdot d\mathbf{S}$$

supplying by every *stress* $\vec{\mathbf{p}}$ for each *area vector* $d\mathbf{S}$ as for a common *external normal* $d\mathbf{S}$ to the boundary ∂V of a *finite volume* V (Chapter 3) and reducing then to *volume density*

$$\nabla \cdot \vec{\mathbf{p}}_* \quad \text{for } \nabla = \mathbf{i}\partial_x + \mathbf{j}\partial_y + \mathbf{k}\partial_z$$

by the *divergence theorem*

$$\int_{\partial V} \mathbf{p}^{x,y,z} \cdot d\mathbf{S} = \int_V \nabla \cdot \mathbf{p}^{x,y,z} dV,$$

of Lagrange (1762), Gauss (1813), Ostrogradsky (1826), and Green (1828), or its mere generalization

$$\int_{\partial V} \vec{\mathbf{p}} \cdot d\mathbf{S} = \int_V (\nabla \cdot \vec{\mathbf{p}}_*) dV \quad \text{when } \vec{\mathbf{p}}_{**} = \vec{\mathbf{p}},$$

following immediately from evident identities

$$\nabla \cdot \vec{\mathbf{p}} = \nabla \cdot \mathbf{p}^x\mathbf{i} + \nabla \cdot \mathbf{p}^y\mathbf{j} + \nabla \cdot \mathbf{p}^z\mathbf{k} = \nabla \cdot \mathbf{p}^x\mathbf{i} + \cdots = \mathbf{i}\nabla \cdot \mathbf{p}^x + \cdots$$

and

$$\int_V \nabla \cdot \vec{\mathbf{p}} dV = \mathbf{i} \int_V \nabla \cdot \mathbf{p}^x dV + \cdots$$
$$= \int_{\partial V} (\mathbf{i}\mathbf{p}^x + \cdots) \cdot d\mathbf{S} = \int_{\partial V} \vec{\mathbf{p}}_* \cdot d\mathbf{S}.$$

In turn, as in fluid dynamics, stresses $\vec{\mathbf{p}}$ may be delivered by the so-called *rates-of-strains* [Batchelor, 1967] to be *fluid deformations* of *velocity strain* and *flow gradient* (strain conjugation),

$$\mathbf{u_r} = \begin{pmatrix} u_x & u_y & u_z \\ v_x & v_y & v_z \\ w_x & w_y & w_z \end{pmatrix} = \mathbf{u}_x\mathbf{i} + \mathbf{u}_y\mathbf{j} + \mathbf{u}_z\mathbf{k} \quad \text{and}$$

$$\nabla\mathbf{u} = \begin{pmatrix} u_x & v_x & w_x \\ u_y & v_y & w_y \\ u_z & v_z & w_z \end{pmatrix} = \mathbf{i}\mathbf{u}_x + \mathbf{j}\mathbf{u}_y + \mathbf{k}\mathbf{u}_z,$$

with their *shear*

$$\vec{\tau} = \mathbf{u_r} + \nabla\mathbf{u} = \tau^x\mathbf{i} + \tau^y\mathbf{j} + \tau^z\mathbf{k} = (\mathbf{u}_x + \nabla u)\mathbf{i} + (\mathbf{u}_y + \nabla v)\mathbf{j} + (\mathbf{u}_z + \nabla w)\mathbf{k}$$

of double *deformation measure*

$$2D = 2D[\mathbf{u}] = \|\vec{\tau}\|^2 = |\tau^x|^2 + |\tau^y|^2 + |\tau^z|^2,$$

or the familiar *dissipation* [Abramovich, 1973],

$$D = 2u_x^2 + 2v_y^2 + 2w_z^2 + (v_x + u_y)^2 + (w_y + v_z)^2 + (u_z + w_x)^2,$$

consisting of scalar *heterogeneity*

$$A = \sqrt{(u_x - v_y)^2 + (v_y - w_z)^2 + (w_z - u_x)^2},$$

shift

$$B = \sqrt{(v_x + u_y)^2 + (w_y + v_z)^2 + (u_z + w_x)^2}$$

and *compressibility*

$$C = u_x + v_y + w_z = \nabla \cdot \mathbf{u} = Sp(\mathbf{u_r})$$

in the proper *identity of measure*

$$3D = 2A^2 + 3B^2 + 2C^2$$

(evidently derived, as in Section 2.3 of Chapter 2) and *torsion*

$$\vec{\omega} = \mathbf{u_r} - \nabla\mathbf{u} = \omega \times \vec{\mathbf{e}} = \vec{\mathbf{e}} \times \omega$$

of *vorticity*

$$\omega = \nabla \times \mathbf{u} = \omega^x \mathbf{i} + \omega^y \mathbf{j} + \omega^z \mathbf{k} = (w_y - v_z)\mathbf{i} + (u_z - w_x)\mathbf{j} + (v_x - u_y)\mathbf{k}$$

with *double enstrophy*

$$2|\omega|^2 = \|\vec{\omega}\|^2 = 2|\omega^x|^2 + 2|\omega^y|^2 + 2|\omega^z|^2$$

complementing the measure D, as in §1, by identities:

$$2\|\mathbf{u_r}\|^2 - D = 2\|\nabla \mathbf{u}\|^2 - D = |\omega|^2.$$

The latter would have to be grounded below in Part 1, devoted mainly to moving particles $\mathbf{r} = \mathbf{r}^t$ that are so close to each other that, together with their *smooth* (or infinitely differentiable) velocities $\mathbf{u} = \mathbf{u}(t, \mathbf{r})$, the flow strains $\mathbf{u_r}$ have to be taken into account rather than only the velocities of separate points in an emptiness whose *discrete matter* has been accepted initially in classical mechanics [Arnold, 1989] to be directly developed further into gases or liquids with the use of the Kelvin absolute temperature, the Maxwell velocity distribution, the Boltzmann kinetic equation [Libov, 1969], and the further *secondary* (or turbulent) averaging of the Newtonian dynamic law taken in the form of hydrodynamic equations [Monin and Yaglom, 1965].

However, instead of dealing with distinct particles in a vacuum, first of all we shall be interested in a *permanent continuum* of them to be a general *smooth substance* of particles initially infinitely close to each other. This one may be either a rigid body, a liquid, a gas, or even something else while being neither already a discrete matter like a sand or snow nor yet a *continuous medium* with its eternal *stress* $\vec{\mathbf{p}} = p\vec{\mathbf{e}}$ of scalar *pressure* $p = p(t, \mathbf{r})$ of molecular origin, which appear first only in Part 3.

The *moving volume* $V = V^t$ becomes the main object of interest in such a smooth medium. Its *kinematics* in Part 1 would depend on a strain $\mathbf{u_r}$, leading to rates of both its scalar element

$$dV = dxdydz = d\mathbf{r} \cdot d\mathbf{S},$$

and vector elements of length $d\mathbf{r}$ and area $d\mathbf{S}$, as above, the latter being the external normal to the boundary ∂V at the point $\mathbf{r}$ of any smooth part of ∂V:

$$\frac{d}{dt}dV = \left(\frac{d}{dt}d\mathbf{r}\right) \cdot d\mathbf{S} + d\mathbf{r} \cdot \frac{d}{dt}d\mathbf{S}, \quad \frac{d}{dt}d\mathbf{r} = \mathbf{u_r} \cdot d\mathbf{r},$$

$$\frac{d}{dt}d\mathbf{S} = (\nabla \cdot \mathbf{u})d\mathbf{S} - \nabla\mathbf{u} \cdot d\mathbf{S};$$

hence,

$$\frac{d}{dt}dV = (\nabla \cdot \mathbf{u})dV \quad (\text{since } (\mathbf{u_r} \cdot d\mathbf{r}) \cdot d\mathbf{S}$$
$$= d\mathbf{S} \cdot \mathbf{u_r} \cdot d\mathbf{r} = d\mathbf{r} \cdot \nabla\mathbf{u} \cdot d\mathbf{S}).$$

As this takes place, when taken together with Parts 2–4, the related constructions would then be aimed mainly at tracing the details of the *historical turn* that had occurred for ten years in the eighteenth century from the just obtained first fluid dynamics equations [Euler, 1757] to the *top* as an absolutely rigid body spinning around a fulcrum $\mathbf{r} = \mathbf{0}$ [Euler, 1765] that turned out to be a volume V undeformable, and therefore classical mechanics was forced to study its rotations among material points in a void [Arnold, 1989].

It seems that proper nuts and bolts would help us to understand the true mechanics concerning the formal analogy that is present between water fluids and top rotations that was detected in the former century [Moreau, 1959; Arnold, 1966] to lead eventually to the stability of the first of the two plane channel velocity profiles [Reynolds, 1883], profile P in Fig. 0.1, that had really proved to be stable for plane parallel flow disturbances [Troshkin, 1988a] as if it was a rotation of a top around its least principal axis of the proper greatest moment of inertia [Arnold, 1989], as in Fig. 0.2.

Moreover, both the velocity profiles P and S had turned out to be stable like the spinning coins on a table in Fig. 0.3.

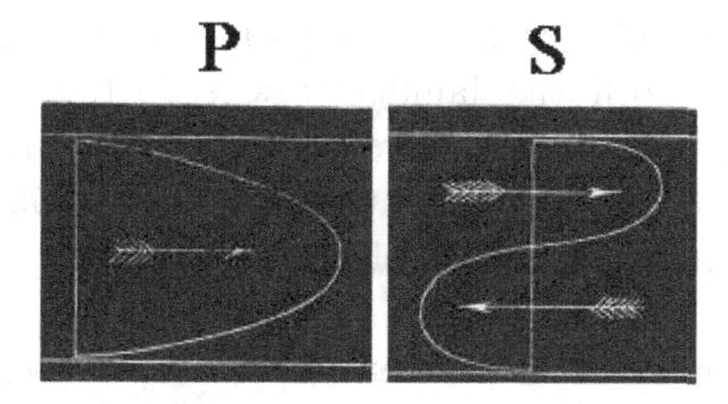

Figure 0.1. Parabolic and sinusoidal velocity profiles of Reynolds [1883].

Figure 0.2. Dan Burbank (on the left) and Anton Shcaplerov (on the right) validating the stability of a spinning can aboard ISS-30 (https://www.youtube.com).

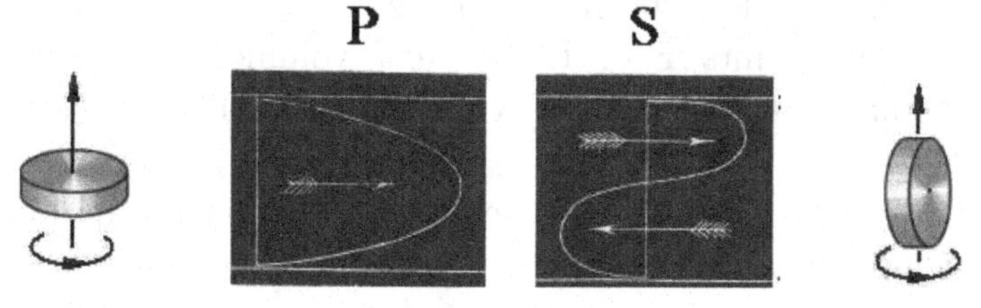

Figure 0.3. Velocity profiles [Reynolds, 1883] stable as rotations [Troshkin, 1988a].

Mechanically, stability of such a kind, while being far from exotic and supposed to be very first states of matter such as strongly symmetric and quark–gluon plasma where all conceivable mechanical conversations are generally possible, if any, would mean that ordinary flowing water looks like a spinning wheel.

Although incredible from commonsense, this may nevertheless be evidenced in both the familiar raw eggs that are converted into hard-boiled ones to reveal everyday the above mentioned from one mechanics to another [Moffatt, 2002] and similar well-known historical facts from the early science of forces, called *physics*, as follows.

Quite different from the habitual push-and-pull *body forces* $\mathbf{g}\rho dV$, with their *specific field* $\mathbf{g} = \mathbf{g}(t, \mathbf{r})$ prescribed and directly applied in points $\mathbf{r}$ to elements dV of either a rigid volume V, with a density $\rho = \rho_0$, or a fluid volume, with another $\rho > 0$, mechanics, be it classical or fluid, had not yet come into existence (287–212 BC) together with two such exactly measured actions as the Archimedean *buoyancy* $\mathbf{A} = -\int_V \mathbf{g}\rho dV$ reducing the immersed weight $\int_V \mathbf{g}\rho_0 dV$ by the dislodged one, $\int_V \mathbf{g}\rho dV$ (at the Earth's gravity $\mathbf{g} = -g\mathbf{j}$, $g = 9.8\,\mathrm{m/c^2}$), and the *torque* $\mathbf{r} \times \mathbf{g}\rho_0 dV$ applied to a fulcrum $\mathbf{0}$ through a point $\mathbf{r}$ as an arm $\mathbf{r}$.

As is well known, these two had been left for a long time to wait for the science of physical equations, called *mathematical physics* [Newton, 1687] with its *dynamic equilibrium*

$$\frac{d}{dt}\int_V \mathbf{u}\rho_0 dV = \int_V \mathbf{g}\rho_0 dV \quad \text{for } \mathbf{u} = \mathbf{u}(t), \text{ or } \mathbf{u}_\mathbf{r} = \vec{\mathbf{0}},$$

balancing the applied specific forces $\mathbf{g}$ with the acquired accelerations $\mathbf{u}_t$ in points $\mathbf{r}$ of those *rigid* volumes $V = V^t$ that move *translationally*, or for the same velocities $\mathbf{u}(t)$, without any strains $\mathbf{u}_\mathbf{r}$.

As to the *fluid* volumes V, before the Newtonian equilibrium, there was already a unification of buoyancy $\mathbf{A}$ for air and water ("We all are living immersed on the bottom of an air ocean", Torricelli, 1644); Pascal's explanation of $\mathbf{A}$ by a *pressure* p [Pascal, 1648] was later found to be of the same molecular origin as the Brownian motion [Brown, 1828] to make up an elementary*contact force* $d\mathbf{A} = -pd\mathbf{S}$ proportional to the vector element of area $d\mathbf{S}$ on boundary ∂V, with the necessary *press*

$$\vec{\mathbf{p}} = \vec{\mathbf{p}}_0 = p\vec{\mathbf{e}},$$

subsequently, thanks to the above-mentioned divergence theorem, resulting in the required buoyancy

$$\mathbf{A} = -\int_{\partial V} \vec{\mathbf{p}} \cdot d\mathbf{S} = -\int_{\partial V} p d\mathbf{S} = -\int_V (\nabla p) dV$$
$$= -\int_V (\nabla \cdot \vec{\mathbf{p}}) dV.$$

Then, there had come both matter-of-course *prolongation* of equilibrium in hand from translational solids to continuous media possessing the pressure stress $p\vec{\mathbf{e}}$, with the aim to keep *momentum balance* and *mass conservation* for any fluid volume $V = V^t$ [Euler, 1757],

$$\frac{d}{dt}\int_V \mathbf{u}\rho dV = \int_V \mathbf{g}\rho dV + \mathbf{A} \quad \text{for any}$$
$$\mathbf{u} = \mathbf{u}(t, \mathbf{r}), \quad \mathbf{u_r} \neq \vec{\mathbf{0}}, \quad \text{and} \quad \frac{d}{dt}\int_V \rho dV = 0,$$

or

$$(\rho\mathbf{u})_t + \nabla \cdot \vec{\mathbf{p}} = \rho\mathbf{g} \quad \text{and} \quad \rho_t + \nabla \cdot \rho\mathbf{u} = 0 \quad \text{for}$$

with the *first* fluid stress of the directly squared velocity $\mathbf{u}$ and pressure p,

$$\vec{\mathbf{p}} = \vec{\mathbf{p}}_1 = \rho\mathbf{u}\mathbf{u} + \vec{\mathbf{p}}_0 = \rho\mathbf{u}\mathbf{u} + p\vec{\mathbf{e}},$$

and an unexpected alternative *extension* of the momentum balance to a rigid body V not translating but *rotating* at an angular velocity

$$\boldsymbol{\Omega} = \boldsymbol{\Omega}(t) = \mathbf{i}\Omega^x + \mathbf{j}\Omega^y + \mathbf{k}\Omega^z,$$

around a fixed point $\mathbf{r} = \mathbf{0}$ to keep the *angular momentum equilibrium* [Euler, 1765],

$$\frac{d}{dt}\int_V \mathbf{r} \times \mathbf{u}\rho_0 dV = \int_V \mathbf{r} \times \mathbf{g}\rho_0 dV \quad \text{for } \mathbf{u} = \boldsymbol{\Omega} \times \mathbf{r}, \quad \text{or}$$

$$\mathbf{u_r} = \vec{\boldsymbol{\Omega}} = \begin{pmatrix} 0 & -\Omega^z & \Omega^y \\ \Omega^z & 0 & -\Omega^x \\ -\Omega^y & \Omega^x & 0 \end{pmatrix},$$

treated below in Part 2.

Meanwhile, after one century, just after initiating the internal *friction force*

$$\mathbf{B} = \int_{\partial V} \mu \vec{\mathbf{b}} \cdot d\mathbf{S} \quad \text{for } \vec{\mathbf{b}} = \vec{\tau} - \frac{2}{3}(\nabla \cdot \mathbf{u})\vec{\mathbf{e}},$$

by the dynamic *viscosity* μ consecutively in the cases of a one-dimensional flow $\mathbf{u} = u(t, y)\mathbf{i}$ near a no-slipping wall [Newton, 1687], then an incompressible fluid, $\nabla \cdot \mathbf{u} = 0$ [Navier, 1823], and, finally, a compressible gas, $\nabla \cdot \mathbf{u} \neq 0$ [Stokes, 1845], enforcing the momentum equilibrium by the *Navier–Stokes equations*

$$\frac{d}{dt}\int_V \mathbf{u}\rho dV = \int_V \mathbf{g}\rho dV + \mathbf{A} + \mathbf{B} \quad \text{and} \quad \rho_t + \nabla \cdot \rho\mathbf{u} = 0,$$

or

$$(\rho\mathbf{u})_t + \nabla \cdot \vec{\mathbf{p}} = \rho\mathbf{g} \quad \text{and} \quad \rho_t + \nabla \cdot \rho\mathbf{u} = 0;$$

already with the *second* fluid stress of velocity, pressure, and viscosity,

$$\vec{\mathbf{p}} = \vec{\mathbf{p}}_2 = \vec{\mathbf{p}}_1 - \mu\vec{\mathbf{b}} = \rho\mathbf{u}\mathbf{u} + p\vec{\mathbf{e}} - \mu\vec{\mathbf{b}},$$

a new turn had taken place, now, due to the *hydrodynamic instability* caused by small eddies viewed in flashes of light as spontaneously originating first in a tube [Reynolds, 1883], then in a duct [Comte-Bellot, 1969], and behind it [Hussain and Zaman, 1980], to collapse first a stationary *basic flow* $\mathbf{u}^{(0)} = \mathbf{P}$ at large *Reynolds numbers* $Re \sim 1/\mu$ [Reynolds, 1883] with the always *spatial* and *non-stationary* disturbing flows $\mathbf{u}$.

In other words, instead of the expected fading away, the *instabilities* as deviations $\mathbf{u} - \mathbf{u}^{(0)}$ of $\mathbf{u}$ from $\mathbf{u}^{(0)}$ somehow develop into *pulsations* $\mathbf{u}' = \mathbf{u} - \bar{\mathbf{u}}$ of the properly *averaged* velocity field

$$\bar{\mathbf{u}} = \lim_{T\to\infty} \frac{1}{T}\int_0^T \mathbf{u}(t+s, \mathbf{r})ds.$$

with a new fluid force

$$\mathbf{C} = -\int_{\partial V} \rho\overline{\mathbf{u}'\mathbf{u}'} \cdot d\mathbf{S}, \quad \mathbf{u}' = \mathbf{u} - \bar{\mathbf{u}},$$

found out first at $Re \to \infty$ [Thomson, 1887] and then for any $Re < \infty$ [Reynolds, 1894] to proceed in keeping the momentum equilibrium in the RANS,

$$\frac{d}{dt}\int_V \bar{\mathbf{u}}\rho dV = \int_V \mathbf{g}\rho dV + \bar{\mathbf{A}} + \bar{\mathbf{B}} + \mathbf{C} \quad \text{for } \rho = \bar{\rho} \quad \text{and}$$

$$\rho_t + \nabla \cdot \rho\bar{\mathbf{u}} = 0$$

(the Reynolds Averaged Navier–Stokes equations, or, to be more precise, the Kelvin–RANS), or

$$(\rho\bar{\mathbf{u}})_t + \nabla \cdot \bar{\vec{\mathbf{p}}} = \rho\mathbf{g} \quad \text{and} \quad \rho_t + \nabla \cdot \rho\bar{\mathbf{u}} = 0,$$

for the fluid stress of average velocity, pressure, viscosity, and one-point *correlation* $\overline{\mathbf{u}'\mathbf{u}'}$:

$$\bar{\vec{\mathbf{p}}} = \rho\bar{\mathbf{u}}\bar{\mathbf{u}} + \bar{p}\vec{\mathbf{e}} - \mu\bar{\vec{\mathbf{b}}} + \rho\overline{\mathbf{u}'\mathbf{u}'}.$$

However, in such a case, the momentum equilibrium has to be augmented additionally by the necessary *stress balance* [Thomson, 1887],

$$\overline{\mathbf{u}'\mathbf{u}'}_t + \bar{\mathbf{u}} \cdot \nabla\overline{\mathbf{u}'\mathbf{u}'} + \vec{\mathbf{G}} + \vec{\mathbf{R}} + \vec{\mathbf{D}} = \vec{\mathbf{0}}$$

with its [Frost and Moulden, 1977, Zubarev *et al.*, 1992] turbulent *generation*

$$\vec{\mathbf{G}} = \overline{\mathbf{u}'\mathbf{u}'} \cdot \nabla\bar{\mathbf{u}} + \bar{\mathbf{u}}_{\mathbf{r}} \cdot \overline{\mathbf{u}'\mathbf{u}'},$$

relaxation

$$\vec{\mathbf{R}} = -\frac{1}{\rho}\nabla \cdot (\overline{\vec{\mathbf{e}}p'\mathbf{u}' + p'\mathbf{u}'\vec{\mathbf{e}}}) - \frac{1}{\rho^2}(2\overline{p'\mathbf{u}'} \cdot \nabla\rho)\vec{\mathbf{e}},$$

and *diffusion*

$$\vec{\mathbf{D}} = -\frac{1}{\rho}\nabla \cdot \mu(\overline{\mathbf{u}'\vec{\tau}'} + \overline{\vec{\tau}'\mathbf{u}'}) + \frac{2\mu}{\rho^2}\overline{\vec{\tau}'\mathbf{u}'} \cdot \nabla\rho.$$

As a result, within an averagely resting *turbulent kernel* of zero averaged velocity $\bar{\mathbf{u}}^{(0)} = \mathbf{0}$, unchangeable density $\rho = 1$ (to be united), diagonal correlation matrix of homogeneous and isotropic turbulence, $\overline{\mathbf{u}'\mathbf{u}'}^{(0)} = c^2\vec{\mathbf{e}}$, $c = \text{const} > 0$, and constant pressure $p^{(0)} = \text{const}$, when disturbed slightly by the average fields of velocity $\bar{\mathbf{u}} = \xi$, symmetric correlation $\vec{\eta} = \overline{\mathbf{u}'\mathbf{u}'} - \overline{\mathbf{u}'\mathbf{u}'}^{(0)} = \vec{\eta}^*$, or $\overline{\mathbf{u}'\mathbf{u}'} =$

$c^2\vec{\mathbf{e}} + \vec{\eta}$, and specific pressure $\zeta = (\bar{p} - p^{(0)})/\rho$, or $\bar{p} = p^{(0)} + \rho\zeta$, for the proper exact but linearized generation

$$\vec{\mathbf{G}} = c^2(\nabla\xi + \xi_{\mathbf{r}}), \quad \xi_{\mathbf{r}} \cdot \vec{\eta} = \vec{\eta} \cdot \nabla\xi \to \vec{\mathbf{0}},$$

nonlinear *Rotta relaxation* [Rotta, 1972, formula 2.271] approximated as

$$\tau\vec{\mathbf{R}} = \overline{\mathbf{u}'\mathbf{u}'} - \frac{Sp(\overline{\mathbf{u}'\mathbf{u}'})}{3}\vec{\mathbf{e}} = \vec{\eta} - \frac{\eta}{3}\vec{\mathbf{e}}, \quad \tau = \text{const} > 0,$$
$$\eta = Sp(\vec{\eta}) = \eta^{xx} + \eta^{yy} + \eta^{zz}$$

and neglected diffusion

$$\vec{\mathbf{D}} = \vec{\mathbf{0}} \quad (\text{for either } \mu = 0 \text{ or } \mathrm{Re} \to \infty),$$

both momentum equilibrium and turbulent stress balance lead to the *linear turbulent system* [Troshkin, 1989, 1990a]

$$\xi_t + \nabla \cdot \vec{\eta} + \nabla\zeta = \mathbf{0}, \quad \nabla \cdot \xi = 0 \quad \textit{and}$$
$$\vec{\eta}_t + c^2(\nabla\xi + \xi_{\mathbf{r}}) + \frac{1}{\tau}\left(\vec{\eta} - \frac{\eta}{3}\vec{\mathbf{e}}\right) = \vec{\mathbf{0}}$$

with the necessary *pressure source*

$$-\Delta\zeta = -\nabla \cdot \nabla\zeta = \nabla \cdot (\nabla \cdot \vec{\eta} + \xi_t) = \nabla \cdot \nabla \cdot \vec{\eta} \quad (\nabla \cdot \xi = 0),$$

that assumes the *relaxation energy* η to be stationary,

$$\eta_t = Sp(\vec{\eta}_t) = -2c^2\nabla \cdot \xi - \frac{1}{\tau}Sp\left(\vec{\eta} - \frac{\eta}{3}\vec{\mathbf{e}}\right)$$
$$= -\frac{1}{\tau}\left(Sp(\vec{\eta}) - \frac{\eta}{3}Sp(\vec{\mathbf{e}})\right) = 0,$$

and in terms of the proper *tension*

$$\vec{\mathbf{N}} = \vec{\eta} - \frac{\eta}{3}\vec{\mathbf{e}},$$

it can be rewritten as

$$\xi_t + \nabla \cdot \vec{\mathbf{N}} + \nabla\left(\frac{\eta}{3} + \zeta\right) = \mathbf{0}, \quad \nabla \cdot \xi = 0, \quad \text{and}$$
$$\vec{\mathbf{N}}_t + c^2(\nabla\xi + \xi_{\mathbf{r}}) + \frac{1}{\tau}\vec{\mathbf{N}} = \vec{\mathbf{0}}$$

to reveal an electromagnetic structure in the mechanics of small disturbances $\xi, \vec{\eta}, \zeta$.

Really, with the familiar velocity and charge-to-mass constants

$$c = 299793000\frac{\text{m}}{\text{s}} \quad \text{and} \quad \frac{e}{m_e} = 1.75882 \cdot 10^{11}$$

$$\times \frac{C}{kg} = \frac{1}{\kappa}(C = 1\text{Coulomb}),$$

electrical conductivity

$$\sigma = 1/c\tau$$

and proper scalar and vector potentials, electrical and magnetic fields, current and charge densities,

$$\varphi = \kappa\left(\frac{\eta}{3} + \zeta\right) \quad \text{and} \quad \mathbf{A} = c\kappa\xi, \ \mathbf{E} = \kappa\nabla \cdot \vec{\mathbf{N}} \quad \text{and} \quad \mathbf{H} = \nabla \times \mathbf{A},$$

$$\mathbf{j} = \sigma\mathbf{E} \quad \text{and} \quad \delta = \nabla \cdot \mathbf{E} = -\Delta\varphi \quad (-\Delta\zeta = \nabla \cdot \nabla \cdot \vec{\eta}),$$

respectively, accounting for evident identities

$$-\nabla \times \nabla \times \mathbf{A} = \Delta\mathbf{A} = \nabla \cdot \nabla\mathbf{A} \quad \text{since } \nabla \cdot \mathbf{A} = 0,$$

both the curl $\nabla\times$ of the first equation and the divergence $\nabla\cdot$ of the second one in the above linear turbulent system immediately lead to the required Maxwell equations:

$$\frac{1}{c}\mathbf{H}_t + \nabla \times \mathbf{E} = \mathbf{0} \quad \text{and} \quad \frac{1}{c}\mathbf{E}_t - \nabla \times \mathbf{H} + \mathbf{j} = \mathbf{0}.$$

Thus, when started with buoyancy of press $p\vec{\mathbf{e}}$ and torque of *angular strain* $\mathbf{u}_\mathbf{r} = \vec{\boldsymbol{\Omega}}$, the mechanics of strains and stresses had managed into develop into instabilities $\mathbf{u} - \mathbf{u}^{(0)}$, leading to pulsations $\mathbf{u}'$ assuming both a *turbulence* $\overline{\mathbf{u}'\mathbf{u}'} \neq \vec{\mathbf{0}}$ and its cross-wise waves of small disturbances $\mathbf{E}, \mathbf{H}$.

Meanwhile, as a matter of fact, the latter does not exclude the alternative possibility that basic flow $\mathbf{u}^{(0)}$ is stable, and hence, *laminar* ($\overline{\mathbf{u}'\mathbf{u}'} = \vec{\mathbf{0}}$), whatever large Reynolds number $Re > 0$, provided that disturbing fields $\mathbf{u}$ are *plane-parallel*, or *2D*, and *solenoidal,*

$$\mathbf{u} = u\mathbf{i} + v\mathbf{j}, \quad \mathbf{u}_z = \mathbf{0}, \quad \text{and} \quad \nabla \cdot \mathbf{u} = u_x + v_y = 0,$$

including $\mathbf{u}^{(0)}$.

In terms of the rigid body stability in Fig. 0.3, the fluid stability in hand really becomes possible owing to a way [Arnold, 1966] paved from a continuous grope of ordinary rotations $SO(3)$ to an infinite-dimensional Lie grope $SDiff(V)$ of such invertible and smooth transformations of a bounded plane flow region V, with boundary ∂V and compact closure $\bar{V} = V + \partial V$, that preserve the volume elements $dV = dxdy$, as in the case of a 2D-flow $\mathbf{u}$ of an incompressible fluid ($\nabla \cdot \mathbf{u} = 0$) delivered, as $\mathbf{u} = \psi_y \mathbf{i} - \psi_x \mathbf{j}$, by a stream function $\psi = \psi(t, x, y)$ to be a *mass expenditure* per area and time (for $\rho = 1$) [Milne-Thomson, 1955] at least in a one-connected V [Hurwitz and Courant, 1922].

In such an event, one can introduce the relevant pre-Hilbert space of expenditures $M = C^\infty(t, \bar{V}) = \{\psi\}$ in which the subspace $M_0 = \{\varphi\} \subset M$ of zero boundary values $\varphi|_{\partial V} = 0$ is dense in M, respectively, to the natural topology of a *metric* $(\varphi, \psi) = \int_V \varphi\psi dV$ [Dezin, 1987].

Supplying M with a *screw* $[\varphi, \psi]$ as a commutator of Poisson brackets $[\varphi, \psi] = \varphi_y \psi_x - \varphi_x \psi_y$ and verifying the *volume invariance* [Troshkin, 1988a],

$$(\varphi, [\psi, \chi]) = (\chi, [\varphi, \psi]), \quad \varphi \in M_0, \quad \psi, \chi \in M$$

for either an ideal ($\mu = 0$) or viscous ($\mu = \text{const} > 0$) incompressible fluid ($\rho = 1$), we come to a *quasi-compact* Lie algebra M [Troshkin, 1995] of infinite-dimensional rotations as abstract angular velocities ψ supplied with angular momentums as scalar vorticities $\omega = v_x - u_y = -\Delta\psi$ to satisfy the *scalar vorticity equation*

$$\omega_t + [\psi, \omega] - \nu\Delta\omega = g^y_x - g^x_y \quad \text{for } \mathbf{g} = g^x\mathbf{i} + g^y\mathbf{j} \quad \text{and}$$
$$g^x_z = g^y_z = g^z = 0, \quad \nu = \frac{\mu}{\rho},$$

as a *dissipative top on M* algebraically similar to the ordinary top of *Euler's dynamic equations* [Arnold, 1989] equivalent to the above-mentioned angular momentum equilibrium [Euler, 1765].

Really, both the vector potential

$$\boldsymbol{\psi} = \psi\mathbf{k}, \quad \nabla \times \boldsymbol{\psi} = \mathbf{u}, \quad \nabla \cdot \boldsymbol{\psi} = 0(\psi_z = 0),$$

of stream function ψ, the curl

$$\boldsymbol{\omega} = \nabla \times \mathbf{u} = -\Delta\boldsymbol{\psi} = \omega\mathbf{k}, \quad \omega = v_x - u_y = -\Delta\psi = -\psi_{xx} - \psi_{yy},$$

of vorticity ω, and the rotor

$$\nabla \times (\boldsymbol{\omega} \times \mathbf{u}) = \mathbf{u} \cdot \nabla\boldsymbol{\omega} - \boldsymbol{\omega} \cdot \nabla\mathbf{u} = (u\omega_x + v\omega_y)\mathbf{k} = [\psi, \omega]\mathbf{k}$$
$$(\boldsymbol{\omega} \cdot \nabla\mathbf{u} = \omega\mathbf{u}_z = \mathbf{0})$$

of *vorticity acceleration*

$$\boldsymbol{\omega} \times \mathbf{u} = \mathbf{u} \cdot \nabla\mathbf{u} - \nabla\frac{|\mathbf{u}|^2}{2}$$

in the *Gromeka–Lamb form* of the Navier–Stokes equations for a viscous incompressible fluid,

$$\mathbf{u}_t + \boldsymbol{\omega} \times \mathbf{u} - \nu\Delta\mathbf{u} + \nabla\left(\frac{p}{\rho} + \frac{|\mathbf{u}|^2}{2}\right) = \mathbf{g} \quad \text{for}$$
$$\rho = \text{const} > 0, \quad \text{hence}, \nabla \cdot \mathbf{u} = 0,$$

produce a new screw to be a skew-symmetric *commutator* in M of Poisson brackets

$$[\psi, \omega] = \psi_y\omega_x - \psi_x\omega_y = -[\omega, \psi]$$

in the necessary *vector vorticity equation*

$$\boldsymbol{\omega}_t + \nabla \times (\boldsymbol{\omega} \times \mathbf{u}) - \nu\Delta\boldsymbol{\omega} = \nabla \times \mathbf{g} \quad \text{for}$$
$$\rho = const > 0, \quad \text{hence}, \quad \nabla \cdot \mathbf{u} = 0,$$

or the required scalar one as above.

Both *smooth uniqueness* in the class $C^\infty = C^\infty(\bar{V})$ and asymptotic stability delivered by the dissipative top to the one-dimensional basic flows $\mathbf{u}^{(0)} = \mathbf{P}, \mathbf{S}$ of a viscous fluid ($\nu > 0$) in a plane

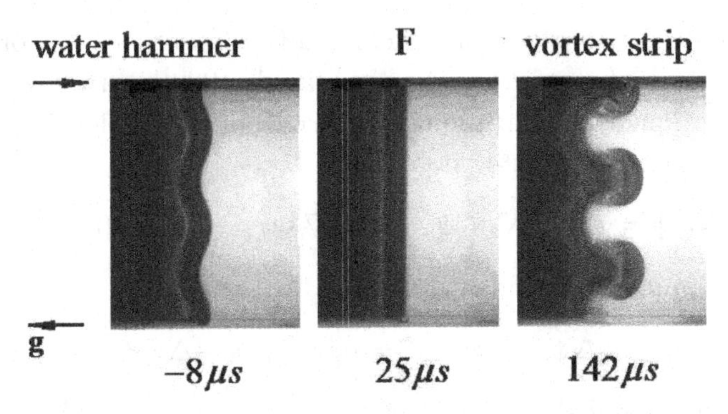

Figure 0.4. The first 150 microseconds (μs) after a tank filled with two liquids contacted by a sinusoidal surface and falling (from the right) in gravity **g** hits the floor (to the left) [Jacobs *et al.*, 1996].

periodic horizontal channel with no-slipping walls, as in Fig. 0.3, are complemented in Part 5 by the *analytical uniqueness* in the more narrow class $C^* = C^*(\bar{V})$ of functions $\psi \in C^\infty$ power-series expanded in the neighborhood of every point of $\bar{V}$ and *nonlinear* stability (maybe not asymptotic) given by the same but *degenerated* top ($\nu = 0$) to the following basic flows $\mathbf{u}^{(0)} = \mathbf{F}$ of the corresponding ideal fluid in a plane periodic vertical strip behind the shock front of the water hammer [Jacobs *et al.*, 1996; Troshkin, 2016], as in Fig. 0.4.

As to the importance to basic flows $\mathbf{u}^{(0)}$ of being unique, the analyticity of an ideal fluid together with the known uniqueness of analytical continuation [Hurwitz and Courant, 1922] replaces the viscosity of real ones, permitting the local disturbances to spread not only along the stream lines but also transversely to them and thereby providing the required uniqueness of analytical flow $\mathbf{u}^{(0)}$ [Troshkin, 1988b].

Alternatively, in the presence of a powerful vortex chamber in a region V of a unique analytical flow $\mathbf{u}^{(0)}$, with the violation of the maximum principle for the proper expenditure $\psi^{(0)} \in C^*$, an uncountable set of smooth non-analytical stationary flows $\tilde{\mathbf{u}}^{(0)}$ of $C^\infty \backslash C^*$ originates in V to satisfy the same boundary conditions and vorticity equation as $\mathbf{u}^{(0)}$ in the proper *smooth vortex catastrophe* [Troshkin, 1988b, 1989b].

The same cannot be said about *non-stationary* disturbances $\mathbf{u} \in M$ of $\mathbf{u}^{(0)}$ for $\nu = 0$ uniquely determined by arbitrary smooth initial solenoidal flows $\mathbf{u}|_{t=0} = \mathbf{u}_0$ submitted to the same boundary conditions as $\mathbf{u}^{(0)}$ assuming the vorticity prescribed in the inflow part of boundary ∂V [Yudovich, 1966].

Moreover, accounting for general near-boundary estimates [Agmon *et al.*, 1964], one may conclude that the required *non-linear disturbances* $\mathbf{u}$ uniquely determined by initial data $\mathbf{u}_0$ do exist in M at least in the case of either a plane periodic *channel* $K : -\infty < x < \infty$ of periodic cell $V_0 : 0 < x, y < l, h$ and horizontal period $l > 0$, with no-slipping walls $y = 0, h$, as in Fig. 0.1, for $\nu > 0$ [Ladyzhenskaya, 1969] or a plane periodic strip (a channel) $N : -\infty < y < \infty$ of the same periodic cell V_0 and vertical period $h > 0$, with an impenetrable wall on the right, $x = l$, and the inflow boundary parts to the left, $x = 0$, where the normal velocity component $u > 0$ and the vorticity $\omega = \omega^+$ are prescribed, as in Fig. 0.4, for $\nu = 0$ [Yudovich, 1966].

The same is true for the disturbances $\mathbf{u} \in M$ of two basic flows $\mathbf{u}^{(0)} = \mathbf{D}, \mathbf{E}$ added to vortex strip $\mathbf{u}^{(0)} = \mathbf{F}$ in Part 5 for completeness of stability considerations concerning the case $\nu = 0$. Both are treated in a periodic cell V_0 of K to become either a plane duct impenetrable both above and below ($y = 0, h$) or a backwater bounded from above, below, and on the right ($x = l$).

With get-at-able remedies of functional analysis and theory of boundary value problems including eigenvalues and eigenfunctions of spectral problems such as either

$$-\Delta\xi = \lambda\xi \quad \text{in} \quad V_0 \quad \text{for } \xi|_{\partial V_0} = 0, \quad \text{in } N \quad \text{for } \xi|_{x=0,l} = 0,$$

or

$$\Delta\Delta\eta = -\mu\Delta\eta \quad \text{in } K \quad \text{for } \eta, \eta_y|_{y=0,h} = 0,$$

to be either *principal* or *affiliated* moments and rotations, λ, ξ or μ, η, of the above-mentioned dissipative top for $\nu = 0$ or $\nu > 0$, respectively, the non-linear stability of $\mathbf{u}^{(0)} = \mathbf{D}, \mathbf{E}, \mathbf{F}, \mathbf{P}, \mathbf{S}$ is proved below in Part 5 to be *unconditional* (whatever $l, h > 0$) for $\mathbf{D}, \mathbf{E}, \mathbf{P}$, *long-wave* ($h >$ const > 0) for $\mathbf{F}$, and *short-wave* ($0 < l <$ const) for **S** or for either any or a sufficiently large *aspect ratio* $\alpha = h/l$, respectively, providing the required 2D-stability of both the viscous velocity profiles in Fig. 0.3 and the ideal mushroom vortices in Fig. 0.4.

Strains have to be taken into account at that in addition to velocities in the initial stability determination for isolated particles [Lyapunov, 1892], which can be justified by the following known stability origin in continuous media.

As is well known, not yet a hydrodynamic stability, the idea of either unstable or stable fluid motion had come into existence as both the *particular cause* of Reynolds [1883] on either collapse or conservation of the velocity profiles P and S in Fig. 0.1, and the *disturbed equilibrium* of Rayleigh [1883] on the deformation of a contact boundary between the upper heavy and the bottom light liquids (like in Fig. 0.4, to the left) coming in turn from the cirrus clouds of Jevons [Jevons, 1857] to later become the Rayleigh–Taylor instability in compressible media [Lewis, 1950; Taylor, 1950] and even literally ennobled further up to the RMI (the Richtmyer–Meshkov Instability) [Richtmyer, 1960; Meshkov, 1969] in such contacting dense and respectively rarefied noble gases as the left xenon (Xe) and the right argon (Ar) in Fig. 0.5, with the shock wave instead of the initial Earth's gravity, to come back into incompressible media, as above, already together with its detected physical reasons for gases [Lindl, 1995; Abarzhi, 2010; Mikhailov, 2011], revealed the phenomenon of contact inversion in liquids [Jacobs, 1996] and determined short-wave character [Aleshin *et al.*, 1988; Belotserkovskaya *et al.*, 2016;

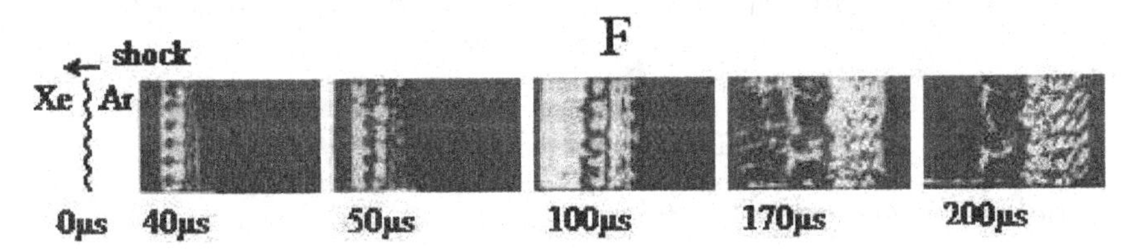

Figure 0.5. The Richtmayer–Meshkov instability [Richtmayer, 1960; Meshkov, 1969] for 200 microseconds (μs) beyond the shock wave front moving from argon (Ar) to xenon (Xe) and crossing a sinusoidally disturbed contact boundary.

Troshkin, 2016a] consistent with the mentioned long-wave stability of mushroom vortices behind the water–hammer front of the pressure jump [Joukowsky, 1900], as in the experiment of Fig. 0.4 [Troshkin, 2016a, 2017a].

Returning to the initial continuum, we see, thus, which familiar consequences the mass, momentum and energy conservation laws may lead to on including both stains and stresses in it. But when taken by itself as being deprived of physical actions, be they forces or torques, why not submit the smooth medium to somewhat that would provide it with the angular velocity $\mathbf{u} = \mathbf{\Omega}(t) \times \mathbf{r}$ in addition to the evident translational one $\mathbf{u} = \mathbf{u}(t)$ and independently of force or torque in the proper momentum or angular momentum equilibrium mentioned above, which have to be excluded therefore from relevant considerations restricted to velocities $\mathbf{u}(t, \mathbf{r})$ and strains $\mathbf{u}_\mathbf{r}$.

In accordance with the above measure identity, free of any forces and torques, translations $\mathbf{w} = \mathbf{w}(t)$ and rotations $\mathbf{w} = \mathbf{\Omega}(t) \times \mathbf{r}$ really realize the locally *least measure*

$$D[\mathbf{u}] \to \min = D[\mathbf{w}] = \frac{2}{3}(\nabla \cdot \mathbf{w})^2, \quad \text{at every } \mathbf{r},$$

of motion in Theorem 2.4.1 (from Section 2.4 of Chapter 2 of Part 1), both being incompressible ($\nabla \cdot \mathbf{w} = 0$) and providing zero internal friction force for any viscosity μ:

$$\mathbf{B} = \int_{\partial V} \mu \vec{\mathbf{b}} \cdot d\mathbf{S} = \mathbf{0} \quad \text{for } \vec{\mathbf{b}} = \vec{\tau} - \frac{2}{3}(\nabla \cdot \mathbf{u})\vec{\mathbf{e}} = \mathbf{u}_\mathbf{r} + \nabla \mathbf{u} = \vec{\mathbf{0}},$$
$$\mathbf{u} = \mathbf{w} = \mathbf{u}(t), \mathbf{\Omega}(t) \times \mathbf{r}.$$

At that, the only remaining linear independent least measure motion

$$\mathbf{w} = H(t)\mathbf{r} - \frac{\mathbf{r} \cdot \mathbf{r}}{2}\mathbf{\Theta}(t) + \mathbf{r} \cdot \mathbf{\Theta}(t)\mathbf{r}$$

in Theorem 2.4.1 proves to be compressible while corresponding unexpectedly to an astronomical question that arose in the last century that the universe is extending with the Hubble constant

$H(t) = H_0$ [Hubble, 1929] when presumably filled up with a hypothetical dark matter [Chwolson, 1924; Zwicky, 1937a, 1937b; Rubin, 1970] with its visible light part [Troshkin, 2017b] admitting both source and sink for a *braking vector* $\mathbf{\Theta}(t)$ discussed in Section 11.4 of Chapter 11 to illustrate another possibility of a smooth medium whose kinematics we have to study.

PART 1

Velocity Strain

Chapter 1

Vector Objects and Operations

1.1. The initial metric and screw

Perhaps, one may say that at the very beginning, formally (mathematically), the science of forces, or physics, looks like a *space* being of a simple and at the same time diverse *structure*, or an algebra of *elements* such as vectors (or arrows) including their immediate generalizations such as *double vectors*, or matrices, related to each other with such binary *operations* as point and cross products, which is the topic this section is devoted to.

To start with proper nuts and bolts, let us first remember the rule of *cyclic permutation* in which any initial record (entry, word, formula, identity, ratio, or sum) of three *ordered*, or counted elements (symbols, signs, numbers, or letters), as taken from their alphabets in Fig. 1.1, say, 123, is supplemented first by the second entry to from the *first permutation* 231 and then by the next (third) word, 312, so that the last (fourth) permutation has to return the initial (first) one, 123. In doing so, the second and third formulae in their *full cycle* (Fig. 1.1.1) will often be replaced by two points in *short records* $123, \ldots.$ and $a+\cdots$, $a+b+\cdots$, $a+b+c+\cdots$ to assume *full cycles* of permutations 123, 231, 312 and $a+b+c=s$, $2s$, $3s$, respectively.

The idea of surrounding *space* may be thought of as coming from either an ordinary *right frame* of origin $\mathbf{0}$ and orts $\mathbf{i}, \mathbf{j}$ and $\mathbf{k}$ as edges of the cube $\mathbf{0ijk}$ in Fig. 1.1.2, or a simple *algebra* of three initial vector elements $\mathbf{a}, \mathbf{b}, \mathbf{c} = \mathbf{i}, \mathbf{j}, \mathbf{k}$ and two operations of symmetric *metric*, or *convolution*, or *scalar* (point, dot) *product* $\mathbf{a} \cdot \mathbf{b} = \mathbf{b} \cdot \mathbf{a}$ of values $\mathbf{a} \cdot \mathbf{a} = 1$ and $\mathbf{a} \cdot \mathbf{b} = 0$ for $\mathbf{a} \neq \mathbf{b}$, and skew-symmetric *cross product*

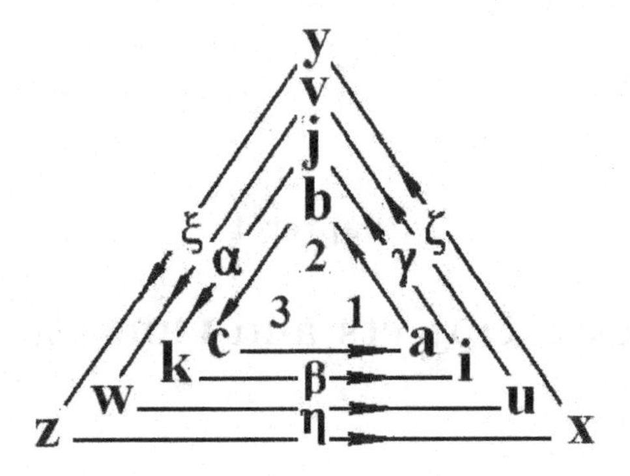

Figure 1.1.1. Full cycles of cyclic permutations.

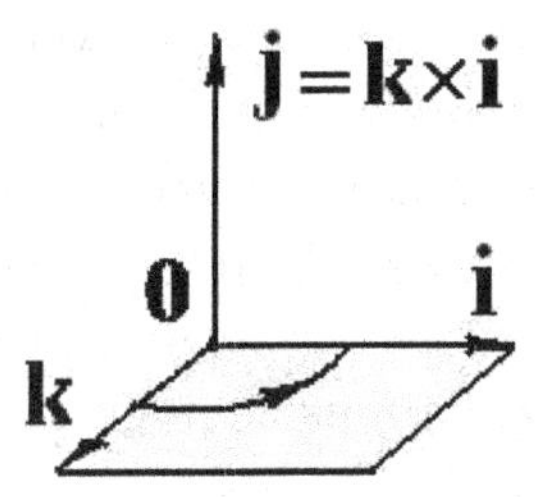

Figure 1.1.2. The right frame of origin and orts.

$\mathbf{a} \times \mathbf{b} = -\mathbf{b} \times \mathbf{a}$ of $\mathbf{a}$ and $\mathbf{b}$, or a *screw* $\mathbf{a} \times \mathbf{b}$ of *arm* $\mathbf{a}$ and *action* $\mathbf{b}$, or, finally, a *commutator* returning both origin, $\mathbf{i} \times \mathbf{i} = \cdots = \mathbf{0}$, and orts $\mathbf{i} = \mathbf{j} \times \mathbf{k}, \ldots$ and acting *by parts* as $\mathbf{a} \times (\mathbf{b} \times \mathbf{c}) = (\mathbf{a} \times \mathbf{b}) \times \mathbf{c} + \mathbf{b} \times (\mathbf{a} \times \mathbf{c})$ to satisfy the *Jacobi identity* $\mathbf{a} \times (\mathbf{b} \times \mathbf{c}) + \mathbf{b} \times (\mathbf{c} \times \mathbf{a}) + \mathbf{c} \times (\mathbf{a} \times \mathbf{b}) = \mathbf{0}$.

Proportions $x\mathbf{i}, y\mathbf{j}, z\mathbf{k}$, $-\infty < x, y, z < \infty$, of orts $\mathbf{i}, \mathbf{j}, \mathbf{k}$ form the *axes* of the Cartesian coordinate system $0xyz$, where every *particle* or *point* proves to be a vertex of a rectangular hexahedron $\mathbf{0}x\mathbf{i}y\mathbf{j}z\mathbf{k}$, or a sum of proportions $x\mathbf{i}, y\mathbf{j}, z\mathbf{k}$ to be a radius-vector $\mathbf{r} = x\mathbf{i} + y\mathbf{j} + z\mathbf{k}$ of *Pythagorean length* $|\mathbf{r}| = \sqrt{\mathbf{r} \cdot \mathbf{r}} = \sqrt{x^2 + y^2 + z^2}$ as in Fig. 1.1.3.

As is usual, the coordinate system $0xyz$ admits its parallel translations, so vectors $\mathbf{a}$ become *free* while forming a *linear space* of elements $\mathbf{a}$ as decomposed in sums $\mathbf{a} = a^x\mathbf{i} + \cdots$ of vector components $a^x\mathbf{i}, \ldots$ with scalar orthogonal projections $a^x = \mathbf{a} \cdot \mathbf{i}, \ldots$, as in Fig. 1.1.4.

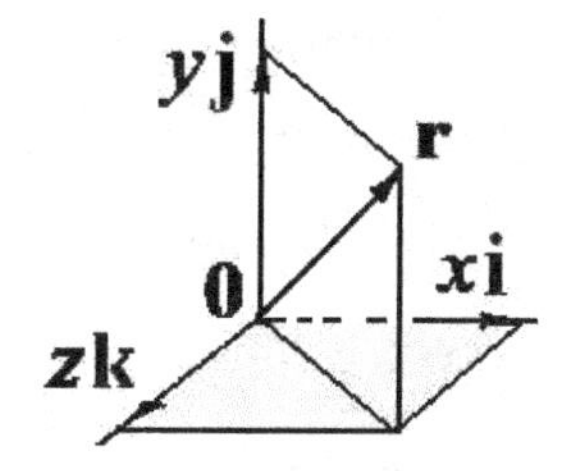

Figure 1.1.3. A Cartesian variant of the Pythagorean theorem.

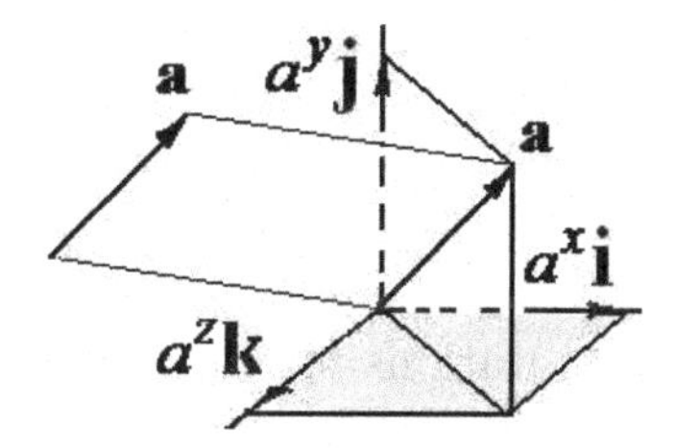

Figure 1.1.4. Orthogonal decomposition of a free vector.

The *metric* and *screw* are immediately transferred from orts to vectors

$$\mathbf{a} = a^x\mathbf{i}+a^y\mathbf{j}+a^z\mathbf{k} = a^i\mathbf{i}+a^j\mathbf{j}+a^k\mathbf{k}, \mathbf{b} = \mathbf{i}b^x+\cdots \quad \text{and} \quad \mathbf{c} = \mathbf{i}c^x+\cdots$$

as

$$\mathbf{a}\cdot\mathbf{b} = (a^x\mathbf{i}+a^y\mathbf{j}+a^z\mathbf{k})\cdot(\mathbf{i}b^x+\mathbf{j}b^y+\mathbf{k}b^z) = a^xb^x+a^yb^y+a^zb^z$$

and

$$\mathbf{b}\times\mathbf{c} = (b^x\mathbf{i}+b^y\mathbf{j}+b^z\mathbf{k})\times(\mathbf{i}c^x+\mathbf{j}c^y+\mathbf{k}c^z) = bc^x\mathbf{i}+bc^y\mathbf{j}+bc^z\mathbf{k}$$

with projections

$$bc^x = \begin{vmatrix} b^y & b^z \\ c^y & c^z \end{vmatrix} = b^yc^z - b^zc^y, \ldots,$$

so that, whatever

$$\mathbf{a} = a\mathbf{i}+b\mathbf{j}+c\mathbf{k} \quad \text{and} \quad \mathbf{u} = u\mathbf{i}+v\mathbf{j}+w\mathbf{k},$$

an *algebraic variant* of the Pythagorean theorem is valid:

$$(au + bv + cw)^2 + (bw - cv)^2 + (cu - aw)^2 + (av - bw)^2$$
$$= \left(a^2 + b^2 + c^2\right)\left(u^2 + v^2 + w^2\right),$$

or

$$(\mathbf{a} \cdot \mathbf{u})^2 + |\mathbf{a} \times \mathbf{u}|^2 = |\mathbf{a}|^2 \, |\mathbf{u}|^2 .$$

So does the action by parts

$$\mathbf{a} \times (\mathbf{b} \times \mathbf{c}) = (\mathbf{a} \times \mathbf{b}) \times \mathbf{c} + \mathbf{b} \times (\mathbf{a} \times \mathbf{c}),$$

or the Jacobi vector identity

$$\mathbf{a} \times (\mathbf{b} \times \mathbf{c}) + \mathbf{b} \times (\mathbf{c} \times \mathbf{a}) + \mathbf{c} \times (\mathbf{a} \times \mathbf{b}) = \mathbf{0}$$

as the sum of cyclic permutations for *triple products*

$$\mathbf{a} \times (\mathbf{b} \times \mathbf{c}) = \mathbf{ba} \cdot \mathbf{c} - \mathbf{ca} \cdot \mathbf{b}, \ldots, \tag{1.1.1}$$

or

$$a^y \left(b^x c^y - c^x b^y\right) - a^z \left(b^z c^x - c^z b^x\right)$$
$$= b^x \left(a^x c^x + a^y c^y + a^z c^z\right) - c^x \left(a^x b^x + a^y b^y + a^z b^z\right), \ldots .$$

Geometrically, the screw $\mathbf{b} \times \mathbf{c}$ is the *area vector* $\mathbf{S}$ of the parallelogram $\mathbf{0bc}$. So, the square of the triangular area $S_\Delta = |\mathbf{S}|/2$ equals the sum of squares of areas $X/2 = \overset{x}{bc}/2$, $Y/2 = \overset{y}{bc}/2$ and $Z/2 = \overset{z}{bc}/2$ as orthogonal projections of vector $\mathbf{S}/2$ on coordinate areas $0yz$, $0zx$, and $0xy$, respectively (*de Gua's theorem*) as in Fig. 1.1.5.

As a natural complement to metric, screw, and *linear operations* (of proportion and sum) for vectors, a new operation such as *derivative on* x is produced additionally for any *smooth* (infinitely differentiable) real-valued function $a = a(t, x, y, z)$:

$$a_x = \partial_x a = \frac{\partial a}{\partial x} = \lim_{h \to 0} \frac{a(t, x+h, y, z) - a(t, x, y, z)}{h} \quad (h > 0).$$

At that, by definition, the proportions of orts $\mathbf{i}, \mathbf{j}, \mathbf{k}$, with derivatives $\partial_x, \partial_y, \partial_z$ on the Cartesian coordinates x, y, z, respectively, when

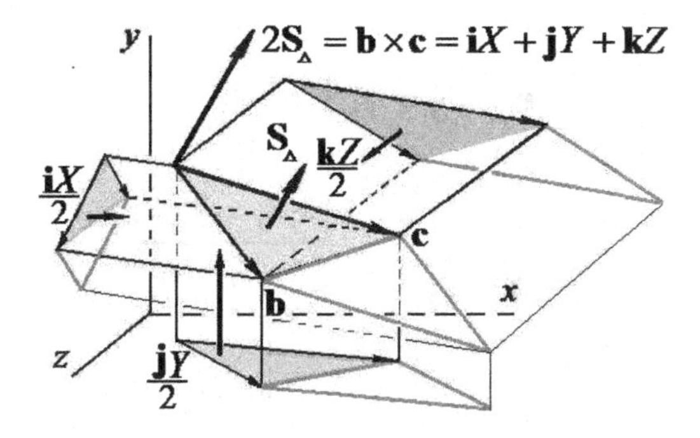

Figure 1.1.5. de Gua's variant of the Pythagorean theorem.

summarized, give a *coordinate strain* $\partial_{\mathbf{r}}$, or *gradient* ∇,

$$\partial_{\mathbf{r}} = \partial_x \mathbf{i} + \partial_y \mathbf{j} + \partial_z \mathbf{k} = \mathbf{i}\partial_x + \mathbf{j}\partial_y + \mathbf{k}\partial_z = \nabla,$$

so that

$$\mathbf{u}\cdot\nabla\boldsymbol{v} = u\boldsymbol{v}_x + \cdots = \boldsymbol{v}_x u + \cdots = \boldsymbol{v}_{\mathbf{r}}\cdot\mathbf{u} \text{ for vectors } \mathbf{u} = u\mathbf{i} + \cdots \text{ and } \boldsymbol{v}$$

and

$$a_{\mathbf{r}} = a_x \mathbf{i} + a_y \mathbf{j} + a_z \mathbf{k} = \mathbf{i}a_x + \mathbf{j}a_y + \mathbf{k}a_z = \nabla a \quad \text{for scalars } a.$$

These two are to be added to the above *vector algebra* to both differentiate *contra-variant derivatives*

$$\mathbf{a}_x = a_x \mathbf{i} + b_x \mathbf{j} + c_x \mathbf{k},$$

$$\mathbf{a}_y = a_y \mathbf{i} + \cdots \quad \text{and} \quad \mathbf{a}_z = a_z \mathbf{i} + \cdots$$

of vector $\mathbf{a} = a\mathbf{i} + b\mathbf{j} + c\mathbf{k}$ from *covariant gradients*

$$\nabla a, \quad \nabla b \quad \text{and} \quad \nabla c$$

of scalars a, b, c and make up the gradient *consistency* to coordinate frame $\mathbf{0}x\mathbf{i}y\mathbf{j}z\mathbf{k}$ (*and vice verse*) by means of the *correct orientation* of orts $\mathbf{i}, \mathbf{j}, \mathbf{k}$, as in Fig. 1.1.6, or the mutually *independence* of coordinates x, y, z that assumes $x_y = x_z = \cdots = 0$.

Thus, the set M of free vectors $\mathbf{a} = \mathbf{a}(t, x, y, z), \mathbf{b}, \mathbf{c}, \ldots$ in which the scalar components $a^{x,y,z}, b^{x,y,z}, c^{x,y,z}, \cdots$ are smooth real-valued functions of time $t \geq 0$ and coordinates $-\infty < x, y, z < \infty$ in a given

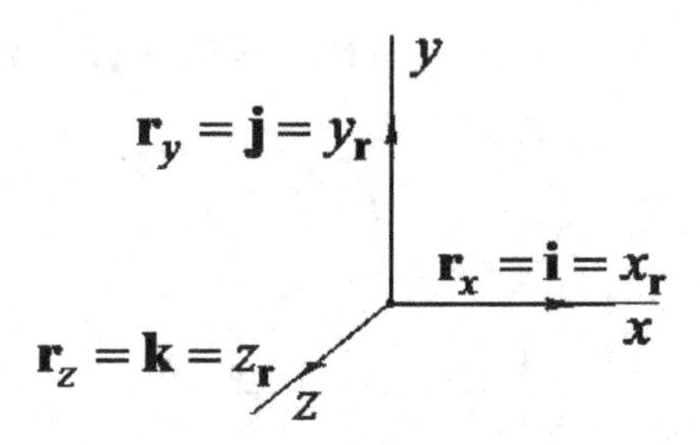

Figure 1.1.6. Correctly oriented orts.

reference frame $0x\mathbf{i}y\mathbf{j}z\mathbf{k}$ consistent with gradient ∇, proves to be a *Lie algebra* M of screw $\mathbf{a} \times \mathbf{b}$.

Moreover, respectively to the screw $\mathbf{a} \times \mathbf{b} \in M$ and the metric $-\infty < \mathbf{a} \cdot \mathbf{b} < \infty$, the Lie algebra M turns out to be *compact* in the sense that its *mixed product*, or *determinant*

$$J = \begin{vmatrix} a^x & b^x & c^x \\ a^y & b^y & c^y \\ a^z & b^z & c^z \end{vmatrix} = a^x \overset{x}{bc} + a^y \overset{y}{bc} + a^z \overset{z}{bc}$$

$$= \mathbf{a} \cdot \mathbf{b} \times \mathbf{c} \quad (\overset{x}{bc} = b^y c^z - b^z c^y, \cdots)$$

is invariant to cyclic permutation of multipliers $\mathbf{a}, \mathbf{b}, \mathbf{c}$,

$$\mathbf{a} \cdot \mathbf{b} \times \mathbf{c} = a^x b^y c^z - a^x b^z c^y + a^y b^z c^x - a^y b^x c^z + a^z b^x c^y - a^z b^y c^x$$
$$= b^x \overset{x}{ca} + b^y \overset{y}{ca} + b^z \overset{z}{ca} = \mathbf{b} \cdot \mathbf{c} \times \mathbf{a}$$

to be the *volume* of both the hexahedron $\mathbf{0abc}$ and its *conjugation* $\mathbf{0a_*b_*c_*}$:

$$\mathbf{a}_* \cdot \mathbf{b}_* \times \mathbf{c}_* = \begin{vmatrix} a^x & a^y & a^z \\ b^x & b^y & b^z \\ c^x & c^y & c^z \end{vmatrix} = a^x \overset{x}{bc} + \begin{pmatrix} b^x \overset{x}{ca} \\ +c^x \overset{x}{ab} \end{pmatrix}$$

$$= a^x \overset{x}{bc} + \begin{pmatrix} a^y \overset{y}{bc} \\ +a^z \overset{z}{bc} \end{pmatrix} = \mathbf{a} \cdot \mathbf{b} \times \mathbf{c} \tag{1.1.1}$$

Figure 1.1.7. A bust of Pythagoras of Samos in Capitoline Museum in Rome (on the left) and an image of Archimedes by Domenico Fetti, 1620, in Dresden Old Master Gallery (on the right).

since

$$\begin{pmatrix} b^x \overset{x}{ca} \\ +c^x \overset{x}{ab} \end{pmatrix} = \begin{pmatrix} b^x \left(c^y a^z - c^z a^y\right) \\ +c^x \left(a^y b^z - a^z b^y\right) \end{pmatrix}$$

$$= \begin{pmatrix} a^y \left(b^z c^x - b^x c^z\right) \\ +a^z \left(b^x c^y - b^y c^x\right) \end{pmatrix} = \begin{pmatrix} a^y \overset{y}{bc} \\ +a^z \overset{z}{bc} \end{pmatrix}.$$

As is well known, the *metric* $\mathbf{a} \cdot \mathbf{b}$ was first detected by the Pythagorean theorem (570–490 BC; Fig. 1.1.7) to give a norm, or a square root of the scalar square $\mathbf{c} \cdot \mathbf{c} = a^2 + b^2$ for plane vectors $\mathbf{c} = a\mathbf{i} + b\mathbf{j}$ to be corrected further with Bolyai–Lobachevskian geometry and Einstein's velocities about speed of light and massive bodies like the Sun. The *screw* $\mathbf{a} \times \mathbf{b}$ had come from the Archimedean *torque* $\mathbf{r} \times \mathbf{f}$ (287–212 BC; Fig. 1.1.7) of a *force* $\mathbf{f}$ acting on a *fulcrum* $\mathbf{0}$ through an *arm* $\mathbf{r}$ to develop subsequently into the Heisenberg commutator [Heisenberg, 1949].

All the developments of the screw and the metric would have hardly been possible without the well-known way from East to West although it remained invisible (being nevertheless not less

Figure 1.1.8. Muhammad ibn Musa al-Khwarizmi in a USSR postage stamp of 1983.

greater than the old Silk Route) paved with such words as "Dixit Algorismi" (literally, "Al-Khwarizmi said", Latin, Segovia, 1145, Robert of Chester) by Muhammad ibn Musa al-Khwarizmi (c. 783, Khiva, Khwaresm (Uzbekistan) — c. 850, Baghdad; Fig. 1.1.8) to become an *algebra* as a science of elements and operations [Kurosh, 1965, 1972] (the original Arabic "Al-Kitāb al-mukhtaṣar fīḥisāb al-jabr wal-muqābala", or "The Compendious Book on Calculation by Completion and Balancing" has been irretrievably lost).

1.2. Matrices as double vectors

Indirectly yet inevitably, with a table, or a *matrix* of volume J, the algebra M of the surrounding space had acquired such a new operation as a *direct product* **ij** of *action* **i** and *normal* **j** to be a *diort* **ij** (a double ort from the Greek "$\delta\iota\pi\lambda\varsigma$", i.e. "diplos") depicted in Fig. 1.2.1.
and *conjugation* of **ij**,

$$(\mathbf{ij})_* = \mathbf{ji}, \ldots \quad (\text{so that } (\mathbf{ij})_{**} = (\mathbf{ji})_* = \mathbf{ij}, \ldots).$$

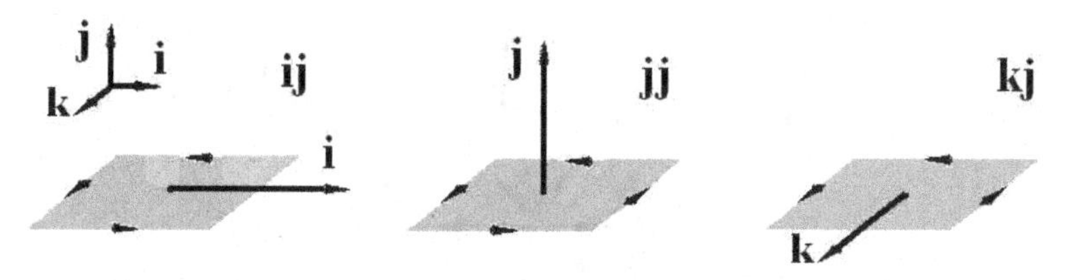

Figure 1.2.1. Diorts $\mathbf{ij}, \mathbf{jj}, \mathbf{kj}$ of three actions $\mathbf{i}, \mathbf{j}, \mathbf{k}$ and one normal $\mathbf{j} = \mathbf{k} \times \mathbf{i}$.

Nine diorts of three *normal* ones, **ii**, **jj**, **kk**, and six *tangential* ones, **ji**, **kj**, **ik**, **ij**, **jk**, **ki**, make up an additional linear basis in M related evidently to 3×3 –matrices of real components $-\infty < p^{xx}, p^{xy}, p^{yx}, \ldots < \infty$, or *divectors*

$$\vec{\mathbf{p}} = \begin{pmatrix} p^{xx} & p^{xy} & p^{xz} \\ p^{yx} & p^{yy} & p^{yz} \\ p^{zx} & p^{zy} & p^{zz} \end{pmatrix} = \begin{pmatrix} p^{xx}\mathbf{ii} + p^{xy}\mathbf{ij} + p^{xz}\mathbf{ik} \\ +p^{yx}\mathbf{ji} + p^{yy}\mathbf{jj} + p^{yz}\mathbf{jk} \\ +p^{zx}\mathbf{ki} + p^{zy}\mathbf{kj} + p^{zz}\mathbf{kk} \end{pmatrix}$$

$$= \mathbf{p}^x\mathbf{i} + \mathbf{p}^y\mathbf{j} + \mathbf{p}^z\mathbf{k} = \begin{pmatrix} \mathbf{i}\mathbf{p}^x_* \\ +\mathbf{j}\mathbf{p}^y_* \\ +\mathbf{k}\mathbf{p}^z_* \end{pmatrix},$$

consisting of *columns* $\mathbf{p}^{x,y,z}$ as *actions* and *rows* $\mathbf{p}^{*x,y,z}$ as *normals*,

$$\mathbf{p}^x = \begin{pmatrix} p^{xx}\mathbf{i} \\ +p^{yx}\mathbf{j} \\ +p^{zx}\mathbf{k} \end{pmatrix}, \ldots \quad \text{and} \quad \mathbf{p}^{*x} = p^{xx}\mathbf{i} + p^{xy}\mathbf{j} + p^{xz}\mathbf{k}, \ldots,$$

depicted in Fig. 1.2.2, and determine the same volume

$$J[\vec{\mathbf{p}}] = \mathbf{p}^x \cdot \mathbf{p}^y \times \mathbf{p}^z. \tag{1.2.1}$$

As we have already seen in (1.1.1), their *conjugations*

$$\vec{\mathbf{p}}^* = \vec{\mathbf{p}}_* = \begin{pmatrix} p^{xx} & p^{yx} & p^{zy} \\ p^{xy} & p^{yy} & p^{zy} \\ p^{xz} & p^{yz} & p^{zz} \end{pmatrix} = \mathbf{p}^{*x}\mathbf{i} + \mathbf{p}^{*y}\mathbf{j} + \mathbf{p}^{*z}\mathbf{k} = \mathbf{i}\mathbf{p}^x + \mathbf{j}\mathbf{p}^y + \mathbf{k}\mathbf{p}^z$$

change no volumes,

$$J[\vec{\mathbf{p}}_*] = \mathbf{p}^x_* \cdot \mathbf{p}^y_* \times \mathbf{p}^y_* = J[\vec{\mathbf{p}}], \tag{1.2.2}$$

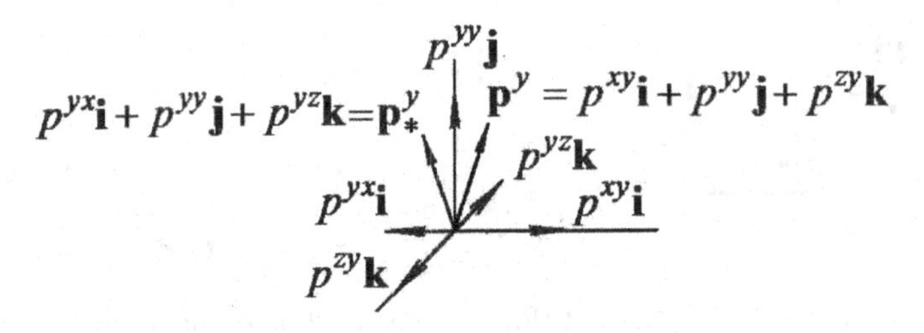

Figure 1.2.2. Action $\mathbf{p}^y$ and normal $\mathbf{p}^y_*$ of divector $\vec{\mathbf{p}}$.

and, when repeated, recover initial objects $\vec{\mathbf{p}}$ at that:

$$\vec{\mathbf{p}}^{**} = (\mathbf{i}\mathbf{p}^x + \mathbf{j}\mathbf{p}^y + \mathbf{k}\mathbf{p}^z)^* = \mathbf{p}^x\mathbf{i} + \mathbf{p}^y\mathbf{j} + \mathbf{p}^z\mathbf{k} = \vec{\mathbf{p}}.$$

For a *reducible* matrix

$$\begin{aligned}\mathbf{ab} &= (a^x\mathbf{i} + a^y\mathbf{j} + a^z\mathbf{k})(b^x\mathbf{i} + b^y\mathbf{j} + b^z\mathbf{k}) \\ &= a^x b^x\mathbf{ii} + a^x b^y\mathbf{ij} + a^x b^z\mathbf{ik} + \cdots,\end{aligned}$$

or a *parallelogram* **ab** of vectors **a** and **b**, the conjugation assumes merely the mutual replacement of direct multipliers:

$$\begin{aligned}(\mathbf{ab})^* &= a^x b^x\,(\mathbf{ii})^* + a^x b^y\,(\mathbf{ij})^* + a^x b^z\,(\mathbf{ik})^* + \cdots \\ &= b^x a^x\mathbf{ii} + a^x b^y\mathbf{ji} + a^x b^z\mathbf{ki} + \cdots = \mathbf{ba}.\end{aligned}$$

Reducible divectors are complemented alternatively by the *matrix unit*

$$\vec{\mathbf{e}} = \mathbf{ii} + \mathbf{jj} + \mathbf{kk} = \begin{pmatrix}1 & 0 & 0\\ 0 & 1 & 0\\ 0 & 0 & 1\end{pmatrix}, \quad \vec{\mathbf{e}}\cdot\mathbf{i} = \mathbf{i} = \mathbf{i}\cdot\vec{\mathbf{e}},$$

$$\vec{\mathbf{e}}\cdot\mathbf{ii} = \mathbf{ii} = \mathbf{ii}\cdot\vec{\mathbf{e}}, \quad \vec{\mathbf{e}}\cdot\mathbf{ij} = \mathbf{ij} = \mathbf{ij}\cdot\vec{\mathbf{e}}, \ldots,$$

that proves to be *irreducible*:

$$\vec{\mathbf{e}} \neq \mathbf{ab} \quad \text{whatever} \quad \mathbf{a}, \mathbf{b}.$$

Really, the assumption that $\vec{\mathbf{e}} = \mathbf{ab}$ for some $\mathbf{a}, \mathbf{b}$ would lead to absurdity:

$$a^x b^y = b^x a^y = \cdots = 0 \quad \text{for} \quad a^x b^x = \cdots = 1.$$

1.3. Metric and convolution

While extending vectors $\mathbf{a}, \mathbf{b}, \ldots$ to vector objects $\mathbf{a}, \mathbf{b}, \vec{\mathbf{p}}, \tilde{\mathbf{q}}, \ldots$, one may note that screws "$\times$" remain to be the same whereas metrics "$\cdot$" transform into *convolutions* as follows.

By definition, the convolution of two *vectors*

$$\mathbf{a} = a^x \mathbf{i} + \cdots = \mathbf{i} a^x + \cdots \quad \text{and} \quad \mathbf{b} = b^x \mathbf{i} + \cdots = \mathbf{i} b^x + \cdots$$

is their initial metric

$$\mathbf{a} \cdot \mathbf{b} = a^x b^x + \cdots = b^x a^x + \cdots = \mathbf{b} \cdot \mathbf{a},$$

with its *natural* (Pythagorean) *norm* $|\mathbf{a}| = \sqrt{\mathbf{a} \cdot \mathbf{a}} \geq 0$ and related *vector metric estimate*

$$\mathbf{a} \cdot \mathbf{b} \leq |\mathbf{a}|\,|\mathbf{b}|$$

immediately following from evident *algebraic relationships*:

$$-2\,|\mathbf{a}|\,|\mathbf{b}|\,(\mathbf{a} \cdot \mathbf{b} - |\mathbf{a}|\,|\mathbf{b}|) = (|\mathbf{a}|\,\mathbf{b} - \mathbf{a}\,|\mathbf{b}|) \cdot (|\mathbf{a}|\,\mathbf{b} - \mathbf{a}\,|\mathbf{b}|) \geq 0.$$

Further, the convolution of *matrix and vector*,

$$\begin{aligned} \vec{\mathbf{p}} &= \mathbf{p}^x \mathbf{i} + \mathbf{p}^y \mathbf{j} + \mathbf{p}^z \mathbf{k} \\ &= \mathbf{i} \mathbf{p}^x_* + \mathbf{j} \mathbf{p}^y_* + \mathbf{k} \mathbf{p}^z_* \quad \text{and} \quad \mathbf{a} = a^x \mathbf{i} + \cdots = \mathbf{i} a^x + \cdots, \end{aligned}$$

reduces to the common *action* (from the *left*) of $\vec{\mathbf{p}}$ (as a matrix) on $\mathbf{a}$ (as a column) that summarizes the proportions of columns taken with coefficients as vector components,

$$\vec{\mathbf{p}} \cdot \mathbf{a} = \mathbf{p}^x a^x + \mathbf{p}^y a^y + \mathbf{p}^z a^z,$$

or of familiar scalar multiplications $\mathbf{p}^x_* \cdot \mathbf{a}, \ldots$ of rows $\mathbf{p}^x_*, \ldots$ by a column $\mathbf{a}$:

$$\vec{\mathbf{p}} \cdot \mathbf{a} = \mathbf{i}\,(\mathbf{p}^x_* \cdot \mathbf{a}) + \mathbf{j}\,(\mathbf{p}^y_* \cdot \mathbf{a}) + \mathbf{k}\,(\mathbf{p}^z_* \cdot \mathbf{a})\,.$$

The corresponding *inverse action* (from the *right*) of $\vec{\mathbf{p}}$ on $\mathbf{a}$, or *the convolution of* $\mathbf{a}$ *and* $\vec{\mathbf{p}}$ as

$$\mathbf{a} \cdot \vec{\mathbf{p}} = (\mathbf{a} \cdot \mathbf{p}^x)\,\mathbf{i} + \mathbf{p}^y \cdot \mathbf{a}\mathbf{j} + \mathbf{p}^z \cdot \mathbf{a}\mathbf{k} = a^x \mathbf{p}^x_* + a^y \mathbf{p}^y_* + a^z \mathbf{p}^z_*$$

reveals the above-mentioned *conjugations*

$$\vec{\mathbf{p}}_* = \mathbf{p}_*^x\mathbf{i} + \mathbf{p}_*^x\mathbf{j} + \mathbf{p}_*^z\mathbf{k} = \mathbf{i}\mathbf{p}^x + \mathbf{j}\mathbf{p}^y + \mathbf{p}^z\mathbf{k} \quad \text{and} \quad \vec{\mathbf{p}}_{**} = \vec{\mathbf{p}}$$

such that

$$\vec{\mathbf{p}} \cdot \mathbf{a} = \mathbf{a} \cdot \vec{\mathbf{p}}_* \quad (\text{or} \quad \vec{\mathbf{p}}_* \cdot \mathbf{a} = \mathbf{a} \cdot \vec{\mathbf{p}}).$$

Finally, the *convolution* of divectors $\vec{\mathbf{p}}$ and $\vec{\mathbf{q}}$ reduces to the product of matrices $\vec{\mathbf{p}}, \vec{\mathbf{q}}$, or to their *row–column* multiplication

$$\vec{\mathbf{p}} \cdot \vec{\mathbf{q}} = \begin{pmatrix} \mathbf{i}\mathbf{p}_*^x \\ +\mathbf{j}\mathbf{p}_*^y \\ +\mathbf{k}\mathbf{p}_*^z \end{pmatrix} \cdot (\mathbf{q}^x\mathbf{i} + \mathbf{q}^y\mathbf{j} + \mathbf{q}^z\mathbf{k})$$

$$= \begin{pmatrix} (\mathbf{p}_*^x \cdot \mathbf{q}^x)\,\mathbf{ii} \\ +(\mathbf{p}_*^y \cdot \mathbf{q}^x)\,\mathbf{ji} \\ +(\mathbf{p}_*^z \cdot \mathbf{q}^x)\,\mathbf{ki} \end{pmatrix} + \begin{pmatrix} (\mathbf{p}_*^x \cdot \mathbf{q}^y)\,\mathbf{ij} \\ +(\mathbf{p}_*^y \cdot \mathbf{q}^y)\,\mathbf{jj} \\ +(\mathbf{p}_*^z \cdot \mathbf{q}^y)\,\mathbf{kj} \end{pmatrix} + \begin{pmatrix} (\mathbf{p}_*^x \cdot \mathbf{q}^z)\,\mathbf{ik} \\ +(\mathbf{p}_*^y \cdot \mathbf{q}^z)\,\mathbf{jk} \\ +(\mathbf{p}_*^z \cdot \mathbf{q}^z)\,\mathbf{kk} \end{pmatrix},$$

to be an alternative *column–row direct product* of the form

$$\vec{\mathbf{p}} \cdot \vec{\mathbf{q}} = (\mathbf{p}^x\mathbf{i} + \mathbf{p}^y\mathbf{j} + \mathbf{p}^z\mathbf{k}) \cdot \begin{pmatrix} \mathbf{i}\mathbf{q}_*^x \\ +\mathbf{j}\mathbf{q}_*^y \\ +\mathbf{k}\mathbf{q}_*^z \end{pmatrix} = \mathbf{p}^x\mathbf{q}_*^x + \mathbf{p}^y\mathbf{q}_*^y + \mathbf{p}^z\mathbf{q}_*^z.$$

The conjugation of convolution at that turns out to be that of conjugations taken in the reverse order:

$$\begin{aligned}(\vec{\mathbf{p}} \cdot \vec{\mathbf{q}})_* &= (\mathbf{p}^x\mathbf{q}_*^x + \cdots)_* = \mathbf{q}_*^x\mathbf{p}^x + \cdots \\ &= (\mathbf{q}_*^x\mathbf{i} + \cdots) \cdot (\mathbf{i}\mathbf{p}^x + \cdots) = \vec{\mathbf{q}}_* \cdot \vec{\mathbf{p}}_*.\end{aligned}$$

As this takes place, *the volume*

$$J\,[\vec{\mathbf{p}} \cdot \vec{\mathbf{q}}] = (\vec{\mathbf{p}} \cdot \mathbf{q}^x) \cdot ((\vec{\mathbf{p}} \cdot \mathbf{q}^y) \times (\vec{\mathbf{p}} \cdot \mathbf{q}^z))$$

of convolution

$$\vec{\mathbf{p}} \cdot \vec{\mathbf{q}} = \vec{\mathbf{p}} \cdot \mathbf{q}^x\mathbf{i} + \vec{\mathbf{p}} \cdot \mathbf{q}^y\mathbf{j} + \vec{\mathbf{p}} \cdot \mathbf{q}^z\mathbf{k}$$

reduces to the product of volumes of matrices convolved, $J\,[\vec{\mathbf{p}}]$ *and* $J\,[\vec{\mathbf{q}}]$.

Really, we have that

$$\vec{\mathbf{p}} \cdot \mathbf{q}^x = \mathbf{p}^x q^{xx} + \mathbf{p}^y q^{yx} + \mathbf{p}^z q^{zx}, \ldots,$$

so

$$(\vec{\mathbf{p}} \cdot \mathbf{q}^y) \times (\vec{\mathbf{p}} \cdot \mathbf{q}^z) = (\mathbf{p}^x q^{xy} + \mathbf{p}^y q^{yy} + \mathbf{p}^z q^{zy}) \times (\mathbf{p}^x q^{xz} + \mathbf{p}^y q^{yz} + \mathbf{p}^z q^{zz}),$$

or

$$(\vec{\mathbf{p}} \cdot \mathbf{q}^y) \times (\vec{\mathbf{p}} \cdot \mathbf{q}^z) = \mathbf{p}^y \times \mathbf{p}^z \begin{vmatrix} q^{yy} q^{yz} \\ q^{zy} q^{zz} \end{vmatrix} + \mathbf{p}^z \times \mathbf{p}^x \begin{vmatrix} q^{xz} q^{xy} \\ q^{zz} q^{zy} \end{vmatrix} + \mathbf{p}^x \times \mathbf{p}^y \begin{vmatrix} q^{xy} q^{xz} \\ q^{yy} q^{yz} \end{vmatrix},$$

which gives the volume

$$J[\vec{\mathbf{p}} \cdot \vec{\mathbf{q}}] = \mathbf{p}^x \cdot \mathbf{p}^y \times \mathbf{p}^z q^{xx} \begin{vmatrix} q^{yy} q^{yz} \\ q^{zy} q^{zz} \end{vmatrix} + \mathbf{p}^y \cdot \mathbf{p}^z \times \mathbf{p}^x q^{yx} \begin{vmatrix} q^{xz} q^{xy} \\ q^{zz} q^{zy} \end{vmatrix} + \mathbf{p}^z \cdot \mathbf{p}^x \times \mathbf{p}^y q^{zx} \begin{vmatrix} q^{xy} q^{xz} \\ q^{yy} q^{yz} \end{vmatrix}.$$

With the invariance

$$\mathbf{p}^x \cdot \mathbf{p}^y \times \mathbf{p}^z = \mathbf{p}^y \cdot \mathbf{p}^z \times \mathbf{p}^x = \mathbf{p}^z \cdot \mathbf{p}^x \times \mathbf{p}^y = J[\vec{\mathbf{p}}],$$

the volume reduces either to

$$J[\vec{\mathbf{p}} \cdot \tilde{\mathbf{q}}] = J[\vec{\mathbf{p}}] \left(q^{xx} \begin{vmatrix} q^{yy} q^{yz} \\ q^{zy} q^{zz} \end{vmatrix} + q^{yx} \begin{vmatrix} q^{xz} q^{xy} \\ q^{zz} q^{zy} \end{vmatrix} + q^{zx} \begin{vmatrix} q^{xy} q^{xz} \\ q^{yy} q^{yz} \end{vmatrix} \right),$$

or, since

$$q^{xx} \begin{vmatrix} q^{yy} q^{yz} \\ q^{zy} q^{zz} \end{vmatrix} + q^{yx} \begin{vmatrix} q^{xz} q^{xy} \\ q^{zz} q^{zy} \end{vmatrix} + q^{zx} \begin{vmatrix} q^{xy} q^{xz} \\ q^{yy} q^{yz} \end{vmatrix} = J[\vec{q}_*] = J[\vec{q}],$$

to the required product:

$$J[\vec{\mathbf{p}} \cdot \vec{\mathbf{q}}] = J[\vec{\mathbf{p}}]\, J[\vec{\mathbf{q}}]. \tag{1.3.1}$$

As is usual, to multiply *scalar* matrices $\vec{\mathbf{p}}$ and $\vec{\mathbf{q}}$, one needs to first turn them into 9-dimensional vectors, then multiply their identical components and, finally, sum up multiplications, which gives the corresponding *matrix metric* as a *double scalar* product

$$\vec{\mathbf{p}} \cdots \vec{\mathbf{q}} = \begin{pmatrix} p^{xx}q^{xx} \\ +p^{yx}q^{yx} \\ +p^{zx}q^{zx} \end{pmatrix} + \begin{pmatrix} p^{xy}q^{xy} \\ +p^{yy}q^{yy} \\ +p^{zy}q^{zy} \end{pmatrix} + \begin{pmatrix} p^{xz}q^{xz} \\ +p^{yz}q^{yz} \\ +p^{zz}q^{zz} \end{pmatrix}$$
$$= \mathbf{p}^x \cdot \mathbf{q}^x + \cdots = \mathbf{q}^x \cdot \mathbf{p}^x + \cdots, \tag{1.3.2}$$

the *matrix norm* as a square root of scalar square

$$\|\vec{\mathbf{p}}\|^2 = \vec{\mathbf{p}} \cdots \vec{\mathbf{p}} = |\mathbf{p}^x|^2 + |\mathbf{p}^y|^2 + |\mathbf{p}^z|^2, \tag{1.3.3}$$

and the *matrix metric estimate*

$$\vec{\mathbf{p}} \cdots \vec{\mathbf{q}} \leq \|\vec{\mathbf{p}}\| \, \|\vec{\mathbf{q}}\|$$

provided by the same algebraic relationships as above:

$$-2\,\|\vec{\mathbf{p}}\|\,\|\vec{\mathbf{q}}\|\,(\vec{\mathbf{p}} \cdots \vec{\mathbf{q}} - \|\vec{\mathbf{p}}\|\,\|\vec{\mathbf{q}}\|)$$
$$= (\|\vec{\mathbf{p}}\|\,\vec{\mathbf{q}} - \vec{\mathbf{p}}\,\|\vec{\mathbf{q}}\|) \cdots (\|\vec{\mathbf{p}}\|\,\vec{\mathbf{q}} - \vec{\mathbf{p}}\,\|\vec{\mathbf{q}}\|) \geq 0.$$

As reducible divectors, all nine diorts

$$\mathbf{ab} = \mathbf{a}^b\mathbf{b} = \mathbf{ii}, \mathbf{ij}, \mathbf{ik}, \ldots, \qquad \mathbf{a}, \mathbf{b} = \mathbf{i}, \mathbf{j}, \mathbf{k},$$

of actions $\mathbf{a} = \mathbf{a}^b$ and normals $\mathbf{b}$ making up metrics

$$\mathbf{ab} \cdots \mathbf{cd} = \mathbf{a}^b \cdot \mathbf{c}^d\mathbf{b} \cdot \mathbf{d}$$
$$= \mathbf{a} \cdot \mathbf{cb} \cdot \mathbf{d} = \begin{cases} 1, & \text{for} \quad \mathbf{a} = \mathbf{c} \quad \text{and} \quad \mathbf{b} = \mathbf{d}, \\ 0, & \text{othewise}, \end{cases}$$
$$\mathbf{c}, \mathbf{d} = \mathbf{i}, \mathbf{j}, \mathbf{k},$$

do form an *orthogonal basis* for matrices $\vec{\mathbf{p}}$ (as the sums of proportions of $\mathbf{ab}$), so

$$\begin{pmatrix}\mathbf{jj}\\ -\mathbf{kk}\end{pmatrix}\cdots\begin{pmatrix}\mathbf{jj}\\ -\mathbf{kk}\end{pmatrix}=2,\ldots,\begin{pmatrix}\mathbf{jj}\\ -\mathbf{kk}\end{pmatrix}\cdots\begin{pmatrix}\mathbf{kk}\\ -\mathbf{ii}\end{pmatrix}=-1,\ldots,$$

$$\begin{pmatrix}\mathbf{jj}\\ -\mathbf{kk}\end{pmatrix}\cdots\vec{\mathbf{e}}=\cdots=1-1=0,$$

$$\vec{\mathbf{e}}\cdots\mathbf{ii}=\cdots=1\quad\text{and}\quad\|\vec{e}\|^2=3.$$

1.4. Orthogonal decompositions

As a result, we come to the following series of immediately verified matrix *orthogonal resolutions*, or *decompositions* where every matrix

$$\vec{\mathbf{p}}=\begin{pmatrix}p^{xx} & p^{xy} & p^{xz}\\ p^{yx} & p^{yy} & p^{yz}\\ p^{zx} & p^{zy} & p^{zz}\end{pmatrix}(-\infty<p^{xx},\quad p^{xy},\cdots<\infty),$$

be it a *strain* (or a deformation) or a *stress* (or a force per unit area), is orthogonally resolved, or decomposed as

$$\vec{\mathbf{p}}=\frac{\vec{\omega}}{2}+\frac{\vec{\tau}}{2},\quad\text{or}\quad 2\vec{p}=\vec{\omega}+\vec{\tau},\quad\text{for}\quad\vec{\omega}\cdots\vec{\tau}=0,\tag{1.4.1}$$

into the *general torsion*

$$\vec{\omega}=\vec{p}-\vec{p}_*=\begin{pmatrix}0 & -\omega^z & \omega^y\\ \omega^z & 0 & -\omega^x\\ -\omega^y & \omega^x & 0\end{pmatrix}=\begin{pmatrix}-\mathbf{ij}\omega^z+\mathbf{ik}\omega^y\\ +\mathbf{ji}\omega^z-\mathbf{jk}\omega^x\\ -\mathbf{ki}\omega^y+\mathbf{kj}\omega^x\end{pmatrix},$$

$$\text{or}\quad\vec{\omega}=\begin{pmatrix}\mathbf{ii}\\ +\mathbf{jj}\\ +\mathbf{kk}\end{pmatrix}\times\begin{pmatrix}\mathbf{i}\omega^x+\\ \mathbf{j}\omega^y+\\ \mathbf{k}\omega^z\end{pmatrix}=\begin{pmatrix}\omega^x\mathbf{i}\\ +\omega^y\mathbf{j}\\ +\omega^z\mathbf{k}\end{pmatrix}\times\begin{pmatrix}\mathbf{ii}+\\ \mathbf{jj}+\\ \mathbf{kk}\end{pmatrix}$$

$$=\vec{\mathbf{e}}\times\omega=\omega\times\vec{\mathbf{e}},\tag{1.4.2}$$

of *rotation*

$$\omega=\omega^x\mathbf{i}+\cdots=-\mathbf{i}\,(\mathbf{j}\cdot\vec{\omega}\cdot\mathbf{k})-\cdots=(\mathbf{k}\cdot\vec{\omega}\cdot\mathbf{j})\,\mathbf{i}+\cdots,$$

with squared norms

$$\|\vec{\omega}\|^2 = 2|\omega|^2, \quad \text{or} \quad |\omega|^2 = \frac{1}{2}\|\vec{\omega}\|^2, \tag{1.4.3}$$

and the *shear*

$$\vec{\tau} = \vec{\mathbf{p}} + \vec{\mathbf{p}}_* = \begin{pmatrix} 2\sigma^x & \gamma^z & \gamma^y \\ \gamma^z & 2\sigma^y & \gamma^x \\ \gamma^y & \gamma^x & 2\sigma^z \end{pmatrix} = 2\vec{\sigma} + \vec{\gamma}, \quad \vec{\gamma} \cdots \vec{\sigma} = 0, \tag{1.4.4}$$

decomposed, in turn, into the *shear strain*, or the *shift*

$$\vec{\gamma} = \gamma^x \begin{pmatrix} \mathbf{kj} \\ +\mathbf{jk} \end{pmatrix} + \gamma^y \begin{pmatrix} \mathbf{ik} \\ +\mathbf{ki} \end{pmatrix} + \gamma^z \begin{pmatrix} \mathbf{ji} \\ +\mathbf{ij} \end{pmatrix}$$

$$= \gamma^x \mathbf{i} \times \begin{pmatrix} \mathbf{jj} \\ -\mathbf{kk} \end{pmatrix} + \gamma^y \mathbf{j} \times \begin{pmatrix} \mathbf{kk} \\ -\mathbf{ii} \end{pmatrix} + \gamma^z \mathbf{k} \times \begin{pmatrix} \mathbf{ii} \\ -\mathbf{jj} \end{pmatrix},$$

with its *vector* and *squared measure*,

$$\gamma = \gamma^x \mathbf{i} + \gamma^y \mathbf{j} + \gamma^z \mathbf{k} \quad \text{and} \quad B^2 = \frac{1}{2}\|\vec{\gamma}\|^2 = |\gamma|^2 = \gamma^{x2} + \gamma^{y2} + \gamma^{z2}, \tag{1.4.5}$$

respectively, and the *normal strain* $2\vec{\sigma}$ as the double *tension*

$$\vec{\sigma} = \sigma^x \mathbf{ii} + \sigma^y \mathbf{jj} + \sigma^z \mathbf{kk} = \vec{\varsigma} + \frac{C}{3}\vec{\mathbf{e}}, \quad Sp(\vec{\varsigma})$$

$$= \vec{\mathbf{e}} \cdots \vec{\varsigma} = 0, \quad \|\vec{e}\|^2 = \vec{\mathbf{e}} \cdots \vec{\mathbf{e}} = 3, \tag{1.4.6}$$

of vector

$$\sigma = \sigma^x \mathbf{i} + \sigma^y \mathbf{j} + \sigma^z \mathbf{k} \quad \text{with} \quad |\sigma| = \|\vec{\sigma}\|,$$

decomposed, finally, into the isotropic *press* $C\vec{\mathbf{e}}/3$ of *compressibility*

$$C = Sp(\vec{p}) = Sp(\vec{\sigma}) = \sigma^x + \sigma^y + \sigma^z, \tag{1.4.7}$$

and *heterogeneity*

$$\vec{\varsigma} = \frac{2\sigma^x - \sigma^y - \sigma^z}{3}\mathbf{ii} + \cdots = \mathbf{ii}\frac{\delta^z - \delta^y}{3} + \mathbf{jj}\frac{\delta^x - \delta^z}{3} + \mathbf{kk}\frac{\delta^y - \delta^x}{3}, \tag{1.4.8}$$

with proper vector δ and squared measure A^2,

$$\delta = \mathbf{i}\delta^x + \mathbf{j}\delta^y + \mathbf{k}\delta^z = \mathbf{i}(\sigma^y - \sigma^z) + \cdots \quad \text{and} \quad A^2 = |\delta|^2 = 3\,\|\vec{\varsigma}\|^2, \tag{1.4.9}$$

where

$$\delta^x + \delta^y + \delta^z = 0, \quad \text{hence,} \quad \varepsilon = \delta^z\delta^y + \delta^x\delta^z + \delta^y\delta^x$$
$$= -\delta^{z2} + \delta^y\delta^x,$$

so that

$$3\varepsilon = -|\delta|^2 + \varepsilon, \quad \text{or} \quad 9\,\|\vec{\varsigma}\|^2 = (\delta^z - \delta^y)^2 + \cdots = 2\,|\delta|^2 - 2\varepsilon = 3\,|\delta|^2.$$

Thus, due to (1.4.1) and (1.4.3), the squared norm $\|\vec{\mathbf{p}}\|^2$ of the vector object (matrix) $\vec{\mathbf{p}}$, as of a *structured* action (a divector), is still composed of two squares as of *general enstrophy* $|\omega|^2$ and *half dissipation* $\|\vec{\tau}/\|^2$ to be squared norms of *rotation* ω and *half shear deformation* $\vec{\tau}/2$ in the following Pythagorean *matrix theorem*:

$$\|\vec{\mathbf{p}}\|^2 = |\omega|^2 + \|\vec{\tau}/2\|^2. \tag{1.4.10}$$

However, unlike an ordinary vector, the second component in (1.4.10) has got its own structure (1.4.4)–(1.4.9) in the following.

Theorem 1.4.1 (on deformation measure). *The shear deformation measure, or dissipation*

$$D = \frac{1}{2}\,\|\vec{\tau}\|^2 = 2\,\|\vec{\sigma}\|^2 + \frac{1}{2}\,\|\vec{\gamma}\|^2 \tag{1.4.11}$$

consists of measures of heterogeneity A, shift B, and compressibility C in the proper ABC–resolution:

$$D = \frac{2}{3}A^2 + B^2 + \frac{2}{3}C^2 \quad \textit{for} \quad A^2 = \|\vec{\sigma}\|^2, B^2$$
$$= \|\vec{\gamma}\|^2 \quad \textit{and} \quad C = Sp(\vec{\sigma}). \tag{1.4.12}$$

Proof. Really, we have that

$$D = 2\,|\sigma|^2 + |\gamma|^2 = 2\,\|\vec{\varsigma}\|^2 + 2\frac{C^2}{9}\,\|\vec{\mathbf{e}}\|^2 + B^2 = \frac{2A^2}{3} + \frac{2C^2}{3} + B^2,$$

which proves the required statement. □

1.5. Area of the contact force

Meanwhile, as a stress, any matrix

$$\vec{\mathbf{p}} = \mathbf{p}^x\mathbf{i} + \mathbf{p}^y\mathbf{j} + \mathbf{p}^z\mathbf{k}$$

proves to be a *structure* $\vec{\mathbf{p}}$ of the *contact force* $\mathbf{f} = \vec{\mathbf{p}} \cdot \mathbf{S}$ applied to the *oriented area*

$$\mathbf{S} = \mathbf{b} \times \mathbf{c} = \mathbf{i}X + \mathbf{j}Y + \mathbf{k}Z,$$

$$X = \mathbf{i} \cdot \mathbf{b} \times \mathbf{c} = \overset{x}{bc} = \begin{vmatrix} b^y & c^y \\ b^z & b^z \end{vmatrix} = b^y c^z - b^z c^y, \ldots,$$

of the parallelogram, or a reducible divector $\vec{\mathbf{S}} = \mathbf{bc}$ of vectors $\mathbf{b}$ and $\mathbf{c}$.

At that, whatever the force $\mathbf{f} = \vec{\mathbf{p}} \cdot \mathbf{S}$ of a *non-degenerate* structure $\vec{\mathbf{p}}$, $J[\vec{\mathbf{p}}] \neq 0$, the components X, Y, Z of the oriented area $\mathbf{S}$ are readily found from the corresponding *area equation*

$$\mathbf{p}^x X + \mathbf{p}^y Y + \mathbf{p}^z Z = \mathbf{f}, \quad \text{or} \quad \vec{\mathbf{p}} \cdot \mathbf{S} = \mathbf{f},$$

$$\text{for } J = J[\vec{\mathbf{p}}] = \mathbf{p}^x \cdot \mathbf{p}^y \times \mathbf{p}^z \neq 0,$$

after convolving it with $\mathbf{p}^y \times \mathbf{p}^z$ and crossing with $\mathbf{p}^z$ and $\mathbf{p}^y$,

$$JX = \mathbf{f} \cdot \mathbf{p}^y \times \mathbf{p}^z, \quad \mathbf{p}^x \times \mathbf{p}^z X + \mathbf{p}^y \times \mathbf{p}^z Y = \mathbf{f} \times \mathbf{p}^z \quad \text{and}$$

$$\mathbf{p}^y \times \mathbf{p}^x X + \mathbf{p}^y \times \mathbf{p}^z Z = \mathbf{p}^y \times \mathbf{f},$$

to lead to the well-known *Cramer's formulae*

$$X = \frac{1}{J}\mathbf{f} \cdot \mathbf{p}^y \times \mathbf{p}^z, \quad Y = \frac{1}{J}\mathbf{p}^x \cdot \mathbf{f} \times \mathbf{p}^z, \quad \text{and} \quad Z = \frac{1}{J}\mathbf{p}^x \cdot \mathbf{p}^y \times \mathbf{f},$$

as illustrated in Fig. 1.5.1.

For example, a contact force $\mathbf{f} = p\vec{\mathbf{e}} \cdot \mathbf{S} = p\mathbf{S}$ of Pascal [1648] having the structure of a matrix unity $\vec{\mathbf{e}}$ and being created at any moment t and in every internal point (x, y, z) by the scalar pressure $p = p(t, x, y, z)$ found subsequently to be of molecular origin evidenced with Brownian motion [Brown, 1828] into such a

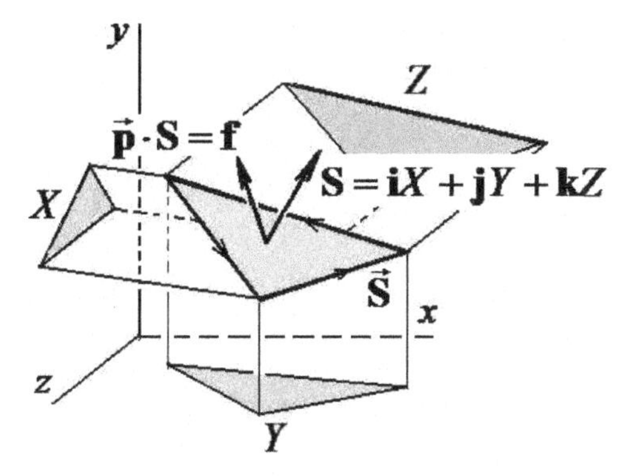

Figure 1.5.1. The area vector $\mathbf{S}$ of the contact force $\mathbf{f}$ with a non-degenerate structure $\vec{\mathbf{p}}$.

continuous medium as either air, water, or mercury, is *orthogonally* applied to a rectangular $\vec{\mathbf{S}} = \mathbf{bc}$ as

$$\vec{\mathbf{S}} \cdot \mathbf{S} = \mathbf{S} \cdot \vec{\mathbf{S}} = \vec{0}(\mathbf{S} = \mathbf{b} \times \mathbf{c}).$$

Accounting for the evident *cross conjugation*

$$(\mathbf{a} \times \mathbf{b})_* = \mathbf{a}_* \times \mathbf{b}_*, \tag{1.5.1}$$

convolutions $\vec{\mathbf{p}}^{-1} \cdot \vec{\mathbf{p}}$ and $\vec{\mathbf{p}} \cdot \vec{\mathbf{p}}^{-1}$ with the *inverse* matrix

$$\begin{aligned} \vec{\mathbf{p}}^{-1} &= \mathbf{i}\frac{\mathbf{p}^y \times \mathbf{p}^z}{J} + \mathbf{j}\frac{\mathbf{p}^z \times \mathbf{p}^x}{J} + \mathbf{k}\frac{\mathbf{p}^x \times \mathbf{p}^y}{J} \\ &= \frac{\mathbf{p}_*^y \times \mathbf{p}_*^z}{J}\mathbf{i} + \frac{\mathbf{p}_*^z \times \mathbf{p}_*^x}{J}\mathbf{j} + \frac{\mathbf{p}_*^x \times \mathbf{p}_*^y}{J}\mathbf{k} \end{aligned} \tag{1.5.2}$$

(delivered by Cramer's formulae) are reduced to the matrix unit, namely,

$$\begin{aligned} \tilde{\mathbf{p}}^{-1} \cdot \tilde{\mathbf{p}} = \left(\frac{\mathbf{p}_*^y \times \mathbf{p}_*^z}{J}\mathbf{i} + \frac{\mathbf{p}_*^z \times \mathbf{p}_*^x}{J}\mathbf{j} + \frac{\mathbf{p}_*^x \times \mathbf{p}_*^y}{J}\mathbf{k} \right) \\ \cdot (\mathbf{i}\mathbf{p}_*^x + \mathbf{j}\mathbf{p}_*^y + \mathbf{k}\mathbf{p}_*^z) = \tilde{\mathbf{e}} \end{aligned}$$

and

$$\vec{\mathbf{p}} \cdot \vec{\mathbf{p}}^{-1} = (\mathbf{p}^x\mathbf{i} + \mathbf{p}^y\mathbf{j} + \mathbf{p}^z\mathbf{k}) \cdot \left(\mathbf{i}\frac{\mathbf{p}^y \times \mathbf{p}^z}{J} + \mathbf{j}\frac{\mathbf{p}^z \times \mathbf{p}^x}{J} + \mathbf{k}\frac{\mathbf{p}^x \times \mathbf{p}^y}{J} \right) = \vec{\mathbf{e}}.$$

Chapter 2

Continuum Motion

2.1. A smooth substance

Meanwhile, as a part of the analysis of infinitesimals, fluid dynamics starts from classical mechanics as follows.

Given a time $t \geq 0$ and a set of arbitrarily closed *particles* (points)

$$\mathbf{r} = x\mathbf{i} + y\mathbf{j} + z\mathbf{k},$$

a *smooth* (infinitely differentiable) velocity vector field, or *flow*

$$\mathbf{u} = \mathbf{u}(t, \mathbf{r}) = u\mathbf{i} + v\mathbf{j} + w\mathbf{k},$$

and resolving the *Cauchy initial problem*

$$\frac{d\mathbf{r}'}{dt} = \left.\frac{d\mathbf{r}'}{dt'}\right|_{t'=t} = \lim_{t'-t \to +0} \frac{\mathbf{r}' - \mathbf{r}}{t' - t} = \mathbf{u}\left(t', \mathbf{r}'\right)$$

$$\text{for} \quad t' > t \quad \text{and} \quad \left.\mathbf{r}'\right|_{t'=t} = \mathbf{r}$$

at least for sufficiently small $0 < t' - t < \varepsilon = const$ [Pontryagin, 1962], we obtain a *smooth substance*, or a *continuum* of particles moving along *trajectories*

$$\mathbf{r}' = x'\mathbf{i} + y'\mathbf{j} + z'\mathbf{k} = \mathbf{r}^{t'} = \mathbf{r} + \int_t^{t'} \mathbf{u}(s, \mathbf{r}^s)\, ds,$$

$$\text{or} \quad \mathbf{r} = \mathbf{r}^t = \mathbf{r}' + \int_{t'}^{t} \mathbf{u}(s, \mathbf{r}^s)\, ds),$$

visibly forward, from the past position $\mathbf{r}$ to the present one $\mathbf{r}'$, or *invisibly* backward, from the future $\mathbf{r}'$ to the present $\mathbf{r}$, respectively.

In doing so, infinitely close particles of the corresponding *general smooth medium* or *substance* deform the coordinate cube **Oijk** splitting common orts

$$\mathbf{i} = \mathbf{r}_x, \quad \mathbf{j} = \mathbf{r}_y, \quad \mathbf{k} = \mathbf{r}_z$$

of the congruent cubes **rijk** and **r′ijk** into both *direct* and *inverse edges*,

$$\mathbf{i}^+ = \mathbf{r}'_x = x'_x\mathbf{i} + y'_x\mathbf{j} + z'_x\mathbf{k}, \quad \mathbf{j}^+ = \mathbf{r}'_y, \quad \mathbf{k}^+ = \mathbf{r}'_z \tag{2.1.1}$$

and

$$\mathbf{i}^- = x_{\mathbf{r}'} = \mathbf{i}x_{x'} + \mathbf{j}x_{y'} + \mathbf{k}x_{z'}, \quad \mathbf{j}^- = y_{\mathbf{r}'}, \quad \mathbf{k}^- = z_{\mathbf{r}'}, \tag{2.1.2}$$

$\mathbf{i}^- = x_{\mathbf{r}'} = \mathbf{i}x_{x'} + \mathbf{j}x_{y'} + \mathbf{k}x_{z'}$, $\mathbf{j}^- = y_{\mathbf{r}'}$, $\mathbf{k}^- = z_{\mathbf{r}'}$, of hexahedrons $\mathbf{r}'\mathbf{i}^+\mathbf{j}^+\mathbf{k}^+$ and $\mathbf{r}\mathbf{i}^-\mathbf{j}^-\mathbf{k}^-$, with origins $\mathbf{r}'$ and $\mathbf{r}$, respectively, the first being translated to the second, as shown in Fig. 2.1.1.

Direct and inverse edges, $\mathbf{i}^+, \mathbf{j}^+, \mathbf{k}^+$ and $\mathbf{i}^-, \mathbf{j}^-, \mathbf{k}^-$, prove to be *dual*,

$$\mathbf{i}^+ \cdot \mathbf{i}^- = \mathbf{j}^+ \cdot \mathbf{j}^- = \mathbf{k}^+ \cdot \mathbf{k}^- = 1, \quad \mathbf{i}^+ \cdot \mathbf{j}^- = \mathbf{i}^+ \cdot \mathbf{k}^- = \cdots = 0,$$

and *consistent*,

$$\mathbf{i}^+\mathbf{i}^- + \mathbf{j}^+\mathbf{j}^- + \mathbf{k}^+\mathbf{k}^- = \mathbf{ii} + \mathbf{jj} + \mathbf{kk} = \mathbf{i}^-\mathbf{i}^+ + \mathbf{j}^-\mathbf{j}^+ + \mathbf{k}^-\mathbf{k}^+,$$

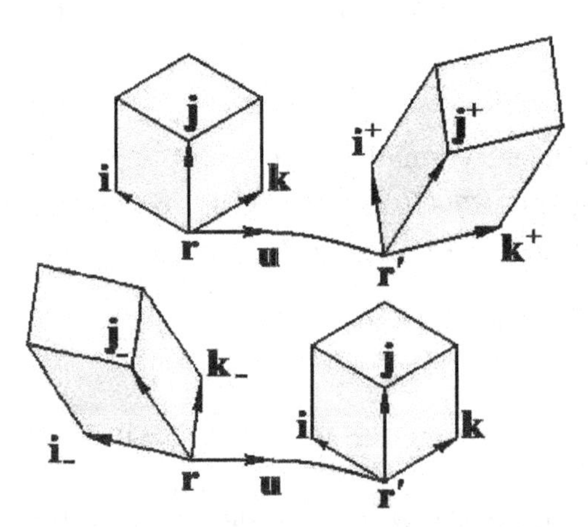

Figure 2.1.1. Flow deformations of the cubes **rijk** and **r′ijk** into hexahedrons $\mathbf{r}'\mathbf{i}^+\mathbf{j}^+\mathbf{k}^+$ and $\mathbf{r}\mathbf{i}_-\mathbf{j}_-\mathbf{k}_-$.

which can be verified immediately:

$$\mathbf{i}^+ \cdot \mathbf{i}^- = \mathbf{i}^- \cdot \mathbf{i}^+ = x_{\mathbf{r}'} \cdot \mathbf{r}'_x \cdot = x_{x'}x'_x + x_{y'}y'_x + x_{z'}z'_x = x_x = 1, \ldots,$$

further,

$$\mathbf{i}^+ \cdot \mathbf{j}^- = \mathbf{j}^- \cdot \mathbf{i}^+ = y_{\mathbf{r}'} \cdot \mathbf{r}'_x \cdot = y_{x'}x'_x + y_{y'}y'_x + y_{z'}z'_x = y_x = 0, \ldots,$$

and finally,

$$\begin{aligned}\mathbf{i}^+\mathbf{i}^- + \cdots &= \mathbf{r}'_x x_{\mathbf{r}'} + \cdots = \mathbf{i}\left(x'_x x_{\mathbf{r}'} + x'_y y_{\mathbf{r}'} + x'_z z_{\mathbf{r}'}\right) + \cdots \\ &= \mathbf{i}x'_{\mathbf{r}'} + \cdots = \mathbf{ii} + \cdots\end{aligned}$$

and

$$\begin{aligned}\mathbf{i}^-\mathbf{i}^+ + \cdots &= x_{\mathbf{r}'}\mathbf{r}'_x + \cdots = \left(x_{\mathbf{r}'}x'_x + y_{\mathbf{r}'}x'_y + z_{\mathbf{r}'}x'_z\right)\mathbf{i} + \cdots \\ &= x'_{\mathbf{r}'}\mathbf{i} + \cdots = \mathbf{ii} + \cdots\end{aligned}$$

In fact, at least for sufficiently small $t'-t > 0$, when the *coefficient of compressibility*

$$J = \mathbf{i}^+ \cdot \mathbf{j}^+ \times \mathbf{k}^+ = \mathbf{r}'_x \cdot \mathbf{r}'_y \times \mathbf{r}'_z = \frac{\partial\left(x', y', z'\right)}{\partial\left(x, y, z\right)} = J_{\mathbf{r}}^{\mathbf{r}'}$$

retains the sign

$$J|_{t'>t} > 0 \quad \text{for} \quad J|_{t'=t} = J_{\mathbf{r}}^{\mathbf{r}} = \mathbf{i} \cdot \mathbf{j} \times \mathbf{k} = 1,$$

hence,

$$\mathbf{i}^- \cdot \mathbf{j}^- \times \mathbf{k}^- = J_{\mathbf{r}'}^{\mathbf{r}} = \frac{1}{J},$$

the *inverse edges* prove to be *proportions of* the *direct normals,*

$$\mathbf{i}^- = \frac{\mathbf{j}^+ \times \mathbf{k}^+}{J}, \quad \mathbf{j}^- = \frac{\mathbf{k}^+ \times \mathbf{i}^+}{J}, \quad \mathbf{k}^- = \frac{\mathbf{i}^+ \times \mathbf{j}^+}{J}, \quad J = J_{\mathbf{r}}^{\mathbf{r}'}, \tag{2.1.3}$$

and *vise versa:*

$$\mathbf{i}^+ = J\mathbf{j}^- \times \mathbf{k}^-, \quad \mathbf{j}^+ = J\mathbf{k}^- \times \mathbf{i}^-, \quad \mathbf{k}^+ = J\mathbf{i}^- \times \mathbf{j}^-.$$

Really, in this case, vectors $\mathbf{i}^-, \mathbf{j}^-, \mathbf{k}^-$ are *linear independent*, i.e.

$$\alpha\mathbf{i}^- + \beta\mathbf{j}^- + \gamma\mathbf{k}^- = \mathbf{0} \text{ implies } \left(\alpha\mathbf{i}^- + \beta\mathbf{j}^- + \gamma\mathbf{k}^-\right)\cdot\mathbf{j}^- \times \mathbf{k}^- = \alpha\mathbf{i}^- \cdot \mathbf{j}^- \times \mathbf{k}^- = 0,$$

or $\alpha = 0$. This is the case with $\beta = \gamma = 0$.

Therefore,

$$\mathbf{j}^+ \times \mathbf{k}^+ = \alpha_1^-\mathbf{i}^- + \beta_1^-\mathbf{j}^- + \gamma_1^-\mathbf{k}^-, \quad \mathbf{k}^+ \times \mathbf{i}^+ = \alpha_2^-\mathbf{i}^- + \beta_2^-\mathbf{j}^- + \gamma_2^-\mathbf{k}^-$$

and

$$\mathbf{j}^+ \times \mathbf{k}^+ = \alpha_3^-\mathbf{i}^- + \beta_2^-\mathbf{j}^- + \gamma_2^-\mathbf{k}^-,$$

and the above duality implies that:

$$J = \mathbf{i}^+ \cdot \mathbf{j}^+ \times \mathbf{k}^+ = \alpha_1^-\mathbf{i}^+ \cdot \mathbf{i}^- + \beta_1^-\mathbf{i}^+ \cdot \mathbf{j}^- + \gamma_1^-\mathbf{i}^+ \cdot \mathbf{k}^- = \alpha_1^-,$$
$$0 = \mathbf{j}^+ \cdot \mathbf{j}^+ \times \mathbf{k}^+ = \alpha_1^-\mathbf{j}^+ \cdot \mathbf{i}^- + \beta_1^-\mathbf{j}^+ \cdot \mathbf{j}^- + \gamma_1^-\mathbf{j}^+ \cdot \mathbf{k}^- = \beta_1^-,$$
$$0 = \mathbf{k}^+ \cdot \mathbf{j}^+ \times \mathbf{k}^+ = \alpha_1^-\mathbf{k}^+ \cdot \mathbf{i}^- + \beta_1^-\mathbf{k}^+ \cdot \mathbf{j}^- + \gamma_1^-\mathbf{k}^+ \cdot \mathbf{k}^- = \gamma_1^-.$$

Similarly, $J = \beta_2^- = \gamma_3^-$ and $0 = \alpha_2^- = \gamma_2^- = \alpha_3^- = \beta_3^-$. This is the case for normals $\mathbf{i}^+, \mathbf{j}^+, \mathbf{k}^+$, which completes the poof of (2.1.3).

2.2. Metric of deformation

The motions $\mathbf{r}'\mathbf{i}^+\mathbf{j}^+\mathbf{k}^+$ and $\mathbf{r}\mathbf{i}^-\mathbf{j}^-\mathbf{k}^-$ of cubes $\mathbf{rijk}$ and $\mathbf{r}'\mathbf{ijk}$ of a smooth substance correspond to the following *direct* and *inverse cube deformations*:

$$\vec{\mathbf{e}}^+ = \mathbf{i}^+\mathbf{i} + \mathbf{j}^+\mathbf{j} + \mathbf{k}^+\mathbf{k} = \mathbf{r}'_x\mathbf{i} + \mathbf{r}'_y\mathbf{j} + \mathbf{r}'_z\mathbf{k}_\mathbf{r} = \mathbf{r}'_\mathbf{r} \tag{2.2.1}$$

and

$$\vec{\mathbf{e}}^- = \mathbf{i}\mathbf{i}^- + \mathbf{j}\mathbf{j}^- + \mathbf{k}\mathbf{k}^- = \mathbf{i}x_{\mathbf{r}'} + \mathbf{j}y_{\mathbf{r}'} + \mathbf{k}z_{\mathbf{r}'} = \mathbf{r}_{\mathbf{r}'}, \tag{2.2.2}$$

of the unit matrix as a *cube structure*

$$\vec{\mathbf{e}} = \mathbf{ii} + \mathbf{jj} + \mathbf{kk} = \mathbf{r}_x\mathbf{i} + \mathbf{r}_y\mathbf{j} + \mathbf{r}_z\mathbf{k} = \mathbf{i}x_\mathbf{r} + \mathbf{j}y_\mathbf{r} + \mathbf{k}z_\mathbf{r} = \mathbf{r}_\mathbf{r},$$

with *conjugations* as conjugate deformations

$$\vec{\mathbf{e}}_*^+ = \mathbf{i}\mathbf{i}^+ + \mathbf{j}\mathbf{j}^+ + \mathbf{k}\mathbf{k}^+ \quad \text{and} \quad \vec{\mathbf{e}}_*^- = \mathbf{i}^-\mathbf{i} + \mathbf{j}^-\mathbf{j} + \mathbf{k}^-\mathbf{k}, \tag{2.2.3}$$

so that, using both the above duality and consistency of edges, we have

$$\vec{\mathbf{e}}^{+} \cdot \vec{\mathbf{e}}^{-} = \vec{\mathbf{e}} = \vec{\mathbf{e}}^{-} \cdot \vec{\mathbf{e}}^{+}.$$

The convolution of the cube conjugate and direct deformations will be referred to as the *deformation metric*, namely, the *direct metric*

$$\vec{\mathbf{g}} = \vec{\mathbf{g}}^{+} = \vec{\mathbf{e}}_{*}^{+} \cdot \vec{\mathbf{e}}^{+} = \left(\mathbf{i}\mathbf{i}^{+} + \mathbf{j}\mathbf{j}^{+} + \mathbf{k}\mathbf{k}^{+}\right) \cdot \left(\mathbf{i}^{+}\mathbf{i} + \mathbf{j}^{+}\mathbf{j} + \mathbf{k}^{+}\mathbf{k}\right)$$
$$= \begin{pmatrix} \mathbf{i}^{+} \cdot \mathbf{i}^{+} & \mathbf{i}^{+} \cdot \mathbf{j}^{+} & \mathbf{i}^{+} \cdot \mathbf{k}^{+} \\ \mathbf{j}^{+} \cdot \mathbf{i}^{+} & \mathbf{j}^{+} \cdot \mathbf{j}^{+} & \mathbf{j}^{+} \cdot \mathbf{k}^{+} \\ \mathbf{k}^{+} \cdot \mathbf{i}^{+} & \mathbf{k}^{+} \cdot \mathbf{j}^{+} & \mathbf{k}^{+} \cdot \mathbf{k}^{+} \end{pmatrix} = \vec{\mathbf{g}}_{*}$$

and the *inverse metric*

$$\vec{\mathbf{g}}^{-} = \vec{\mathbf{e}}^{-} \cdot \vec{\mathbf{e}}_{*}^{-} = \left(\mathbf{i}\mathbf{i}^{-} + \mathbf{j}\mathbf{j}^{-} + \mathbf{k}\mathbf{k}^{-}\right) \cdot \left(\mathbf{i}^{-}\mathbf{i} + \mathbf{j}^{-}\mathbf{j} + \mathbf{k}^{-}\mathbf{k}\right)$$
$$= \begin{pmatrix} \mathbf{i}^{-} \cdot \mathbf{i}^{-} & \mathbf{i}^{-} \cdot \mathbf{j}^{-} & \mathbf{i}^{-} \cdot \mathbf{k}^{-} \\ \mathbf{j}^{-} \cdot \mathbf{i}^{-} & \mathbf{j}^{-} \cdot \mathbf{j}^{-} & \mathbf{j}^{-} \cdot \mathbf{k}^{-} \\ \mathbf{k}^{-} \cdot \mathbf{i}^{-} & \mathbf{k}^{-} \cdot \mathbf{j}^{-} & \mathbf{k}^{-} \cdot \mathbf{k}^{-} \end{pmatrix} = \vec{\mathbf{g}}^{-1}$$

such that

$$\vec{\mathbf{g}}^{-} \cdot \vec{\mathbf{g}} = \vec{\mathbf{e}}^{-} \cdot \vec{\mathbf{e}}_{*}^{-} \cdot \vec{\mathbf{e}}_{*}^{+} \cdot \vec{\mathbf{e}}^{+} = \vec{\mathbf{e}}^{-} \cdot \left(\vec{\mathbf{e}}^{+} \cdot \vec{\mathbf{e}}^{-}\right)_{*} \cdot \vec{\mathbf{e}}^{+} = \vec{\mathbf{e}}^{-} \cdot \vec{\mathbf{e}} \cdot \vec{\mathbf{e}}^{+} = \vec{\mathbf{e}}^{-} \cdot \vec{\mathbf{e}}^{+} = \vec{\mathbf{e}}$$

and

$$\vec{\mathbf{g}} \cdot \vec{\mathbf{g}}^{-} = \vec{\mathbf{e}}_{*}^{+} \cdot \vec{\mathbf{e}}^{+} \cdot \vec{\mathbf{e}}^{-} \cdot \vec{\mathbf{e}}_{*}^{-} = \vec{\mathbf{e}}_{*}^{+} \cdot \vec{\mathbf{e}} \cdot \vec{\mathbf{e}}_{*}^{-} = \vec{\mathbf{e}}_{*}^{+} \cdot \vec{\mathbf{e}}_{*}^{-} = \left(\vec{\mathbf{e}}^{-} \cdot \vec{\mathbf{e}}^{+}\right)_{*} = \vec{\mathbf{e}}_{*} = \vec{\mathbf{e}}.$$

The corresponding *coordinate deformations* of vectors $\mathbf{a}$ as

$$\mathbf{a}^{+} = \mathbf{a} \cdot \vec{\mathbf{e}}_{*}^{+} = \mathbf{i}a^{+} + \mathbf{j}b^{+} + \mathbf{k}c^{+}, \quad a^{+} = \mathbf{i}^{+} \cdot \mathbf{a},$$
$$b^{+} = \mathbf{j}^{+} \cdot \mathbf{a}, \ \ c^{+} = \mathbf{k}^{+} \cdot \mathbf{a},$$

or

$$\mathbf{a}^{-} = \mathbf{a} \cdot \vec{\mathbf{e}}^{-} = \mathbf{i}a^{-} + \mathbf{j}b^{-} + \mathbf{k}c^{-}, \quad a^{-} = \mathbf{i}^{-} \cdot \mathbf{a},$$
$$b^{-} = \mathbf{j}^{-} \cdot \mathbf{a}, \ \ c^{-} = \mathbf{k}^{-} \cdot \mathbf{a},$$

keep unchanged the statement of Pythagorean theorem

$$|\mathbf{a}^{\pm}|^2 = (a^{\pm})^2 + (b^{\pm})^2 + (c^{\pm})^2$$

while merely *transporting* the components of $\mathbf{a} = \mathbf{i}a + \mathbf{j}b + \mathbf{k}c$ by the edges $\mathbf{i}^+, \mathbf{j}^+, \mathbf{k}^+$ or $\mathbf{i}^-, \mathbf{j}^-, \mathbf{k}^-$ as

$$\mathbf{a}^+ = \vec{\mathbf{e}}^+ \cdot \mathbf{a} = \mathbf{i}^+ a + \mathbf{j}^+ b + \mathbf{k}^+ c \quad \text{or}$$
$$\mathbf{a}^- = \vec{\mathbf{e}}_*^- \cdot \mathbf{a} = \mathbf{i}^- a + \mathbf{j}^- b + \mathbf{k}^- c,$$

respectively.

For example, every point of a spinning rigid body V fixed at $\mathbf{r} = \mathbf{0}$ rotates, transforming into

$$\mathbf{r}^+ = \vec{\mathbf{e}}^+ \cdot \mathbf{r} = x\mathbf{i}^+ + y\mathbf{j}^+ + z\mathbf{k}^+ \quad \text{for}$$
$$\mathbf{r} = \vec{\mathbf{e}} \cdot \mathbf{r} = x\mathbf{i} + y\mathbf{j} + z\mathbf{k}. \tag{2.2.4}$$

Transformations $\mathbf{a}^{\pm}$ are complemented by *metric deformations*

$$\tilde{\mathbf{a}} = \tilde{\mathbf{a}}^+ = \vec{\mathbf{g}} \cdot \mathbf{a} = \mathbf{a} \cdot \vec{\mathbf{g}} (\vec{\mathbf{g}} = \vec{\mathbf{g}}^+) \quad \text{and}$$
$$\tilde{\mathbf{a}}^- = \vec{\mathbf{g}}^- \cdot \mathbf{a} = \mathbf{a} \cdot \vec{\mathbf{g}}^-$$

to trace *distortions*

$$\mathbf{a}^+ \cdot \mathbf{b}^+ = \mathbf{a} \cdot \vec{\mathbf{e}}_*^+ \cdot \vec{\mathbf{e}}^+ \cdot \mathbf{b} = \mathbf{a} \cdot \vec{\mathbf{g}} \cdot \mathbf{b} \quad \text{or}$$
$$\mathbf{a}^- \cdot \mathbf{b}^- = \mathbf{a} \cdot \vec{\mathbf{e}}^- \cdot \vec{\mathbf{e}}_*^- \cdot \mathbf{b} = \mathbf{a} \cdot \vec{\mathbf{g}}^- \cdot \mathbf{b}$$

of *metric* $\mathbf{a} \cdot \mathbf{b}$.

At that, both *coordinate* and *metric* deformations, $\vec{\mathbf{e}}^{\pm}$, $\mathbf{a}^{\pm}$ and $\vec{\mathbf{g}}^{\pm}$, $\tilde{\mathbf{a}}^{\pm}$, may be *compressible* ones as multiplied additionally by the degree J^σ of *medium compressibility* J:

$$J^\sigma \vec{\mathbf{e}}^{\pm}, J^\sigma \mathbf{a}^{\pm} \quad \text{and} \quad J^\sigma \vec{\mathbf{g}}^{\pm}, J^\sigma \tilde{\mathbf{a}}^{\pm}, \quad \sigma = \text{const},$$

respectively,

2.3. Rates of orts and deformation measure resolution

Let us turn to *deviations* of orts,

$$\mathbf{i}^{+} - \mathbf{i} = \mathbf{r}'_x - \mathbf{r}_x = \int_t^{t+\varepsilon} \mathbf{u}_x\left(s, \mathbf{r}^s\right) ds = \varepsilon \mathbf{u}_x\left(t, \mathbf{r}\right) + \mathbf{o}\left(\varepsilon\right), \ldots,$$
$$\lim_{\varepsilon\to+0} \left(\mathbf{o}\left(\varepsilon\right)/\varepsilon\right) = \mathbf{0},$$

and

$$\mathbf{i}^{-} - \mathbf{i} = x_{\mathbf{r}'} - x_{\mathbf{r}} = \int_{t+\varepsilon}^{t} u_{\mathbf{r}}(s, \mathbf{r}^s) ds = -\varepsilon u_{\mathbf{r}}(t, \mathbf{r}) + \mathbf{o}(\varepsilon), \ldots$$

as occurring at *basic rates*

$$\mathbf{i}_t^{+} = \mathbf{i}_{t'}^{+}\big|_{t'=t} = \lim_{t'-t=\varepsilon\to+0} \frac{\mathbf{r}'_x - \mathbf{r}_x}{\varepsilon} = \mathbf{u}_x, \quad \mathbf{j}_t^{+} = \mathbf{u}_y, \quad \mathbf{k}_t^{+} = \mathbf{u}_z \tag{2.3.1}$$

and

$$\mathbf{i}_t^{-} = -u_{\mathbf{r}}, \quad \mathbf{j}_t^{-} = -v_{\mathbf{r}}, \quad \mathbf{k}_t^{-} = -w_{\mathbf{r}}, \tag{2.3.2}$$

to form the mutually conjugate divectors

$$u_{\mathbf{r}} = (\nabla \mathbf{u})^* = (\nabla \mathbf{u})_* \quad \text{and} \quad \nabla \mathbf{u} = (u_{\mathbf{r}})^* = (u_{\mathbf{r}})_*$$

of *flow* (or *fluid*) *deformation*, or *strain*

$$u_{\mathbf{r}} = \begin{pmatrix} u_x & u_y & u_z \\ v_x & v_y & v_z \\ w_x & w_y & w_z \end{pmatrix} = \mathbf{u}_x\mathbf{i} + \mathbf{u}_y\mathbf{j} + \mathbf{u}_z\mathbf{k} = \mathbf{i}\nabla u + \mathbf{j}\nabla v + \mathbf{k}\nabla w \tag{2.3.3}$$

and *flow gradient*, or *co-strain*

$$\nabla \mathbf{u} = \begin{pmatrix} u_x & v_x & w_x \\ u_y & v_y & w_y \\ u_z & v_z & w_z \end{pmatrix} = \nabla u\mathbf{i} + \nabla v\mathbf{j} + \nabla w\mathbf{k} = \mathbf{i}\mathbf{u}_x + \mathbf{j}\mathbf{u}_y + \mathbf{k}\mathbf{u}_z = u_{\mathbf{r}}^*. \tag{2.3.4}$$

The fluid deformations are revealed already by a rigid body moving not *translationally*, with

$$\mathbf{u} = \mathbf{u}(t) \quad \text{and} \quad \mathbf{u_r} = \nabla \mathbf{u} = \vec{\mathbf{0}},$$

but *rotationally*, with angular velocity $\boldsymbol{\Omega}(t) = \Omega^x \mathbf{i} + \Omega^y \mathbf{j} + \Omega^z \mathbf{k}$, as

$$\mathbf{u} = \boldsymbol{\Omega}(t) \times \mathbf{r}, \quad \mathbf{u_r} = \begin{pmatrix} 0 & -\Omega^z & \Omega^y \\ \Omega^z & 0 & -\Omega^x \\ -\Omega^y & \Omega^x & 0 \end{pmatrix}$$
$$= \boldsymbol{\Omega} \times \vec{\mathbf{e}} = \vec{\mathbf{e}} \times \boldsymbol{\Omega} \quad \text{and} \quad \nabla \mathbf{u} = -\mathbf{u_r}.$$

Consequently, *deformation rates* of cube structures $\vec{\mathbf{e}}^{\pm}$, their conjugations $\vec{\mathbf{e}}_*^{\pm}$, and *compressibility of medium* $J = \mathbf{i}^+ \cdot \mathbf{j}^+ \times \mathbf{k}^+$ are

$$\vec{\mathbf{e}}_t^+ = \mathbf{i}_t^+ \mathbf{i} + \mathbf{j}_t^+ \mathbf{j} + \mathbf{k}_t^+ \mathbf{k} = \mathbf{u_r}, \quad \vec{\mathbf{e}}_t^- = \mathbf{i}\mathbf{i}_t^- + \mathbf{j}\mathbf{j}_t^- + \mathbf{k}\mathbf{k}_t^- = -\mathbf{u_r} \tag{2.3.5}$$

with

$$\vec{\mathbf{e}}_{*t}^+ = \mathbf{i}\mathbf{i}_t^+ + \mathbf{j}\mathbf{j}_t^+ + \mathbf{k}\mathbf{k}_t^+ = \nabla \mathbf{u}, \quad \vec{\mathbf{e}}_{*t}^- = \mathbf{i}_t^- \mathbf{i} + \mathbf{j}_t^- \mathbf{j} + \mathbf{k}_t^- \mathbf{k} = -\nabla \mathbf{u}, \tag{2.3.6}$$

and

$$J_t = \mathbf{i}_t^+ \cdot \mathbf{j} \times \mathbf{k} + \mathbf{i} \cdot \mathbf{j}_t^+ \times \mathbf{k} + \mathbf{i} \cdot \mathbf{j} \times \mathbf{k}_t^+ = \nabla \cdot \mathbf{u} = C, \tag{2.3.7}$$

where

$$C = \nabla \cdot \mathbf{u} = u_x + v_y + w_z = \mathbf{u}_x \cdot \mathbf{i} + \mathbf{u}_y \cdot \mathbf{j} + \mathbf{u}_z \cdot \mathbf{k}$$
$$= \mathbf{i}_t^+ \cdot \mathbf{i} + \mathbf{j}_t^+ \cdot \mathbf{j} + \mathbf{k}_t^+ \cdot \mathbf{k}$$

is the *compressibility of flow* $\mathbf{u}$.

The rate of motion of metric $\vec{\mathbf{g}} = \vec{\mathbf{g}}^+ = \vec{\mathbf{e}}_*^+ \cdot \vec{\mathbf{e}}^+$ gives the *full shear*

$$\begin{aligned}\vec{\tau} = \mathbf{g}_t &= (\vec{\mathbf{e}}_*^+ \cdot \vec{\mathbf{e}}^+)_{t'}|_{t'=t} = \vec{\mathbf{e}}_{*t}^+ \cdot (\vec{\mathbf{e}}^+|_{t'=t}) + (\vec{\mathbf{e}}_*^+|_{t'=t}) \cdot \vec{\mathbf{e}}_t^+ \\ &= \nabla\mathbf{u} \cdot \vec{\mathbf{e}} + \vec{\mathbf{e}} \cdot \mathbf{u_r} = \mathbf{u_r} + \nabla\mathbf{u},\end{aligned}$$

or

$$\begin{aligned}\vec{\tau} = \mathbf{g}_t &= \mathbf{u_r} + \nabla\mathbf{u} = (\mathbf{u}_x + \nabla u)\mathbf{i} + \cdots \\ &= \begin{pmatrix} 2u_x & v_x + u_y & u_z + w_x \\ v_x + u_y & 2v_y & w_y + v_z \\ u_z + w_x & w_y + v_z & 2w_z \end{pmatrix}\end{aligned} \tag{2.3.8}$$

(as in § 1.4), with the opposite rate

$$\vec{\mathbf{g}}_t^- = \vec{\mathbf{e}}_{*t}^- \cdot (\vec{\mathbf{e}}^-|_{t'=t}) + (\vec{\mathbf{e}}_*^-|_{t'=t}) \cdot \vec{\mathbf{e}}_t^- = -\nabla\mathbf{u} \cdot \vec{\mathbf{e}} - \vec{\mathbf{e}} \cdot \mathbf{u_r} = -\vec{\tau}. \tag{2.3.9}$$

As in (1.4.11), the half squared local norm of fluid shear,

$$\begin{aligned}D = \frac{\|\vec{\tau}\|^2}{2} = \frac{|\mathbf{u}_x + \nabla u|^2}{2} + \cdots = 2u_x^2 + 2v_y^2 + 2w_z^2 + (v_x + u_y)^2 \\ + (w_y + v_z)^2 + (u_z + w_x)^2,\end{aligned}$$

or the *deformation measure*, reduces to the *norms* of compressibility C, *heterogeneity*

$$A = \sqrt{(u_x - v_y)^2 + (v_y - w_z)^2 + (w_z - u_x)^2}$$

and *shift shear*, or *shift*

$$B = \sqrt{(v_x + u_y)^2 + (w_y + v_z)^2 + (u_z + w_x)^2}$$

in the evident equalities

$$3D - 3B^2 = 6\left(u_x^2 + v_y^2 + w_z^2\right) = 2C^2 + 2A^2,$$

or in the ABC-measure resolution (1.4.12), or in the *identity of measure*

$$D = 2u_x^2 + 2v_y^2 + 2w_z^2 + (v_x + u_y)^2 + (w_y + v_z)^2 + (u_z + w_x)^2$$
$$= \frac{2}{3}A^2 + B^2 + \frac{2}{3}C^2$$

to complement the orthogonal decomposition (1.4.1), or

$$\mathbf{u_r} = \frac{\vec{\tau}}{2} + \frac{\vec{\omega}}{2}, \quad \nabla\mathbf{u} = \frac{\vec{\tau}}{2} - \frac{\vec{\omega}}{2}, \quad \vec{\tau} = \vec{\gamma} + 2\sigma, \quad \vec{\sigma} = \frac{C}{3}\vec{\mathbf{e}} + \vec{\varsigma},$$
$$C = \nabla \cdot \mathbf{u}, \quad \text{for} \quad \vec{\tau}\cdots\vec{\omega} = \vec{\gamma}\cdots\vec{\omega} = \vec{\gamma}\cdots\vec{\sigma} = \vec{\sigma}\cdots\vec{\omega} = \vec{\mathbf{e}}\cdots\vec{\varsigma} = 0, \tag{2.3.10}$$

as taken with *torsion*

$$\vec{\omega} = \mathbf{u_r} - \nabla\mathbf{u} = (\mathbf{u}_x - \nabla u)\,\mathbf{i} + \cdots = \begin{pmatrix} 0 & -\omega^z & \omega^y \\ \omega^z & 0 & -\omega^x \\ -\omega^y & \omega^x & 0 \end{pmatrix}$$
$$= \omega \times \vec{\mathbf{e}} = \vec{\mathbf{e}} \times \omega \tag{2.3.11}$$

of *vorticity*

$$\omega = \nabla \times \mathbf{u} = \mathbf{i}\,(w_y - v_z) + \mathbf{j}\,(u_z - w_x) + \mathbf{k}\,(v_x - u_y) \tag{2.3.12}$$

to be rotation

$$\omega = (\mathbf{k}\cdot\vec{\omega}\cdot\mathbf{j})\,\mathbf{i} + (\mathbf{i}\cdot\vec{\omega}\cdot\mathbf{k})\,\mathbf{j} + + (\mathbf{j}\cdot\vec{\omega}\cdot\mathbf{i})\,\mathbf{k} = \mathbf{i}\omega^x + \mathbf{j}\omega^y + \mathbf{k}\omega^z.$$

2.4. Least measure motions

The above identity of measure delivers both the *least measure* as the two-thirds of squared compressibility

$$2C^2/3 \leq D \tag{2.4.1}$$

and its realizations in the following.

Theorem 2.4.1 (principle of least deformation). *The least local measure* $2C^2/3$ *is realized by a strain* $\mathbf{u_r}$ *that is homogeneous,*

$$A = 0, \quad or \quad u_x = v_y = w_z, \tag{2.4.2}$$

and without shear,

$$B = 0, \quad or \quad v_x + u_y = w_y + v_z = u_z + w_x = 0, \tag{2.4.3}$$

or by the flow

$$\mathbf{u} = \mathbf{w}\,(t, \mathbf{r}) = \mathbf{w}^{(i)} + \mathbf{w}^{(ii)} + \mathbf{w}^{(iii)}, \tag{2.4.4}$$

of the translation $\mathbf{w}^{(i)}$ *without strain,* $\mathbf{w}_\mathbf{r}^{(i)} = \vec{0}$, *the rotation* $\mathbf{w}^{(ii)}$ *of angular velocity* $\boldsymbol{\Omega} = \boldsymbol{\Omega}(t)$ *and the central expansion (or contraction)* $\mathbf{w}^{(iii)}$, *with a press coefficient* $H = H(t) \geq 0$ *(or* $H \leq 0$, *respectively) and braking vector* $\boldsymbol{\Theta} = \boldsymbol{\Theta}(t)$,

$$\begin{aligned} \mathbf{w}^{(i)} &= \mathbf{U}(t), \quad \mathbf{w}^{(ii)} = \boldsymbol{\Omega}(t) \times \mathbf{r} \quad \textit{and} \\ \mathbf{w}^{(iii)} &= H(t)\mathbf{r} - \frac{\mathbf{r}\cdot\mathbf{r}}{2}\boldsymbol{\Theta}(t) + \mathbf{r}\cdot\boldsymbol{\Theta}(t)\mathbf{r}, \end{aligned} \tag{2.4.5}$$

with their strains

$$\mathbf{w}_\mathbf{r}^{(ii)} = \vec{\boldsymbol{\Omega}} = \boldsymbol{\Omega} \times \vec{\mathbf{e}} \quad \textit{and} \quad \mathbf{w}_\mathbf{r}^{(iii)} = (H + \mathbf{r}\cdot\boldsymbol{\Theta})\vec{\mathbf{e}} - \boldsymbol{\Theta}\mathbf{r} + \mathbf{r}\boldsymbol{\Theta},$$

and gradients

$$\nabla\mathbf{w}^{(ii)} = -\vec{\boldsymbol{\Omega}} \quad \textit{and} \quad \nabla\mathbf{w}^{(iii)} = (H + \mathbf{r}\cdot\boldsymbol{\Theta})\vec{\mathbf{e}} - \mathbf{r}\boldsymbol{\Theta} + \boldsymbol{\Theta}\mathbf{r},$$

torsions

$$\vec{\omega}^{(ii)} = 2\vec{\boldsymbol{\Omega}} \quad \textit{and} \quad \vec{\omega}^{(iii)} = 2(\mathbf{r}\boldsymbol{\Theta} - \boldsymbol{\Theta}\mathbf{r}),$$

of vorticities

$$\omega^{(ii)} = 2\boldsymbol{\Omega} \quad \textit{and} \quad \omega^{(iii)} = \boldsymbol{\Theta} \times \mathbf{r} - \mathbf{r} \times \boldsymbol{\Theta},$$

shears as zero and non-zero presses

$$\vec{\tau}^{(ii)} = \vec{0} \quad \textit{and} \quad \vec{\tau}^{(iii)} = 2\vec{\sigma}^{(iii)} = \frac{2C}{3}\vec{\mathbf{e}} \quad (\vec{\gamma}^{(iii)} = \vec{\varsigma}^{(iii)} = \vec{0})$$

of compressibilities

$$C^{(ii)} = 0 \quad \textit{and} \quad C^{(iii)} = C = 3(H + \mathbf{r}\cdot\boldsymbol{\Theta}),$$

respectively, for any transfer $\mathbf{r} \to \mathbf{r} - \mathbf{r}_0$ *(of origin* $\mathbf{r} = \mathbf{0}$ *in an arbitrary point* $\mathbf{r}_0$*).*

Proof. Really, conditions (2.3.9) and (2.4.1) assume that equations (2.4.2) and (2.4.3) hold true for smooth u, v, w after differentiating,

$$v_{zx} + u_{yz} = w_{xy} + v_{zx} = u_{yz} + w_{xy} = 0, \quad \text{or} \quad u_{yz} = v_{zx} = w_{xy},$$

so we have zero mixed derivatives

$$u_{yz} = \frac{1}{2}(v_{zx} + u_{yz}) = 0, \quad v_{zx} = \frac{1}{2}(w_{xy} + v_{zx}) = 0,$$
$$w_{xy} = \frac{1}{2}(u_{yz} + w_{xy}) = 0,$$

repeated derivatives

$$u_{yy} = -v_{yx} = -u_{xx} = -w_{zx} = u_{zz},$$
$$-v_{zz} = w_{zy} = v_{yy} = u_{xy} = -v_{xx},$$
$$-w_{xx} = u_{xz} = w_{zz} = v_{yz} = -w_{yy},$$

and derivatives of the third order

$$u_{yyy} = u_{zzy} = (u_{yz})_z = 0,$$

further,

$$u_{yyx} = u_{zzx} = -u_{xxx} \quad u \quad u_{yyx} = w_{yyz} = -v_{yzz} = -u_{xzz} = u_{xxx},$$

hence,

$$u_{yyx} = u_{yyz} = u_{xxx} = 0,$$

and, finally,

$$u_{yxx} = w_{xyz} = (w_{xy})_z = 0,$$
$$u_{zzz} = -w_{xzz} = -v_{xyz} = -(v_{zx})_y = 0$$
$$u \quad u_{zxx} = v_{zyx} = (v_{zx})_y = 0,$$

so

$$u_{yyy} = u_{yyx} = u_{yyz} = u_{yxx} = u_{zzz} = u_{zxx} = u_{xxx} = 0,$$

including analogous derivatives for v, w:

$$v_{zzz} = v_{zzy} = v_{zzx} = v_{zyy} = v_{xxx} = v_{xyy} = v_{yyy} = 0,$$
$$w_{xxx} = w_{xxz} = w_{xxy} = w_{xzz} = w_{yyy} = w_{yzz} = w_{zzz} = 0.$$

As a consequence,

$$u_{yyy} = 0 \quad \text{implies} \quad u = y^2 a(t,x,z) + yb(t,x,z) + c(t,x,z),$$
$$u_{yyx} = u_{yyz} = 0 \quad \text{implies} \quad a_x = a_z = 0, \quad \text{or} \quad a = a_1(t),$$
$$u_{yz} = 0 \quad \text{implies} \quad b_z = 0, \quad \text{or} \quad b = b(t,x),$$
$$u_{yxx} = 0 \quad \text{implies} \quad b_{xx} = 0, \quad \text{or} \quad b(t,x) = xb_1(t) + e_1(t),$$
$$u_{zzz} = 0 \quad \text{implies} \quad c_{zzz} = 0, \quad \text{or}$$
$$c(t,x,z) = z^2 d(t,x) + ze(t,x) + f(t,x),$$
$$u_{zz} = u_{yy} \quad \text{implies} \quad c_{zz} = u_{zz} = u_{yy}, \quad \text{or} \quad d(t,x) = a_1(t),$$
$$u_{zxx} = 0 \quad \text{implies} \quad e_{xx} = 0, \quad \text{or} \quad e(t,x) = xc_1(t) + f_1(t),$$
$$u_{xxx} = 0 \quad \text{implies} \quad f_{xxx} = 0, \quad \text{or} \quad f(t,x) = x^2 g(t) + xd_1(t) + U_1(t),$$
$$u_{xx} = -u_{zz} \quad \text{implies} \quad f_{xx} = c_{xx} = -u_{zz}, \quad \text{or} \quad g = -a_1(t).$$

Evidently, the same is true for v and w, so

$$\begin{aligned} u &= a_1(t)(y^2 + z^2 - x^2) + b_1(t)xy + c_1(t)xz + d_1(t)x \\ &\quad + e_1(t)y + f_1(t)z + U_1(t), \\ v &= a_2(t)(z^2 + x^2 - y^2) + b_2(t)yz + c_2(t)yx + d_2(t)x \\ &\quad + e_2(t)y + f_2(t)z + U_2(t), \\ w &= a_3(t)(x^2 + y^2 - z^2) + b_3(t)zx + c_3(t)zy + d_3(t)x \\ &\quad + e_3(t)y + f_3(t)z + U_3(t). \end{aligned}$$

Repeated substitution of u, v, w in (2.4.2) and (2.4.3) yields:

$$b_3 = c_2 = -2a_1 = \Theta_1, \quad b_1 = c_3 = -2a_2 = \Theta_2,$$
$$b_2 = c_1 = -2a_3 = \Theta_3, \quad d_1 = e_2 = f_3 = H(A = 0),$$

and

$$d_2 = -e_1 = \Omega_3, \quad e_3 = -f_2 = \Omega_1, \quad f_1 = -d_3 = \Omega_2 (B = 0).$$

As a result, we have

$$\begin{aligned} u &= U_1 + \Omega_2 z - \Omega_3 y + Hx - \frac{\Theta_1}{2}(x^2 + y^2 + z^2) \\ &\quad + (\Theta_1 x + \Theta_2 y + \Theta_3 z)x, \\ v &= U_2 + \Omega_3 x - \Omega_1 z + Hy - \frac{\Theta_2}{2}(x^2 + y^2 + z^2) \\ &\quad + (\Theta_1 x + \Theta_2 y + \Theta_3 z)y, \\ w &= U_3 + \Omega_1 y - \Omega_2 x + Hz - \frac{\Theta_3}{2}(x^2 + y^2 + z^2) \\ &\quad + (\Theta_1 x + \Theta_2 y + \Theta_3 z)z. \end{aligned}$$

Putting

$$\begin{aligned} &\mathbf{U} = U_1\mathbf{i} + U_2\mathbf{j} + U_3\mathbf{k}, \quad \boldsymbol{\Omega} = \Omega_1\mathbf{i} + \Omega_2\mathbf{j} + \Omega_3\mathbf{k}, \\ &\boldsymbol{\Theta} = \Theta_1\mathbf{i} + \Theta_2\mathbf{j} + \Theta_3\mathbf{k}, \end{aligned}$$

we come to (2.4.4) and (2.4.5), which with (2.4.4) and evident strains

$$\begin{aligned} \mathbf{w}_{\mathbf{r}}^{(\mathrm{i})} &= \tilde{\mathbf{0}}, \quad \mathbf{w}_{\mathbf{r}}^{(\mathrm{ii})} = \boldsymbol{\Omega} \times \mathbf{r}_x\mathbf{i} + \boldsymbol{\Omega} \times \mathbf{r}_y\mathbf{j} + \boldsymbol{\Omega} \times \mathbf{r}_z\mathbf{k} \\ &= \boldsymbol{\Omega} \times \tilde{\mathbf{e}} = \tilde{\boldsymbol{\Omega}} \\ \text{and} \quad \mathbf{w}_{\mathbf{r}}^{(\mathrm{iii})} &= \mathbf{w}_x^{(\mathrm{iii})}\mathbf{i} + \cdots = H\mathbf{ii} - \boldsymbol{\Theta} x\mathbf{i} + \mathbf{r}\Theta^x\mathbf{i} + \mathbf{r} \cdot \boldsymbol{\Theta}\mathbf{ii} + \cdots \\ &= (H + \mathbf{r} \cdot \boldsymbol{\Theta})\,\tilde{\mathbf{e}} - \boldsymbol{\Theta}\mathbf{r} + \mathbf{r}\boldsymbol{\Theta}, \end{aligned}$$

completes the proof. □

2.5. Alternative measure resolution

Now, let us take orthogonal decompositions (1.4.1), (1.4.4), and (1.4.6) for real flow *strain* $\vec{\mathbf{p}} = \mathbf{u}_{\mathbf{r}}$, *gradient* $\vec{\mathbf{p}}_* = \nabla\mathbf{u}$, *torsion*

$$\vec{\omega} = \begin{pmatrix} 0 & -\omega^z & \omega^y \\ \omega^z & 0 & -\omega^x \\ -\omega^y & \omega^x & 0 \end{pmatrix} = \vec{\mathbf{e}} \times \omega = \omega \times \vec{\mathbf{e}}$$

of *vorticity*

$$\omega = \nabla \times \mathbf{u} = \mathbf{i}\omega^x + \mathbf{j}\omega^y + \mathbf{k}\omega^z$$
$$= \mathbf{i}\,(w_y - v_z) + \mathbf{j}\,(u_z - w_x) + \mathbf{k}\,(v_x - u_y), \qquad (2.5.1)$$

shear $\vec{\mathbf{p}} + \vec{\mathbf{p}}_* = \vec{\tau}$ of *shift*

$$\vec{\gamma} = \begin{pmatrix} 0 & \gamma^z & \gamma^y \\ \gamma^z & 0 & \gamma^x \\ \gamma^y & \gamma^x & 0 \end{pmatrix} = \begin{pmatrix} 0 & v_x + u_y & u_z + w_x \\ v_x + u_y & 0 & w_y + v_z \\ u_z + w_x & w_y + v_z & 0 \end{pmatrix}, \quad \text{or}$$

$$\gamma = \begin{pmatrix} (w_y + v_z)\,\mathbf{i} + \\ (u_z + w_x)\,\mathbf{j} + \\ (v_x + u_y)\,\mathbf{k} \end{pmatrix},$$

and double *tension*

$$\vec{\sigma} = \begin{pmatrix} u_x & 0 & 0 \\ 0 & v_y & 0 \\ 0 & 0 & w_z \end{pmatrix} = \vec{\varsigma} + \frac{C}{3}\vec{\mathbf{e}}$$

with its *heterogeneity*

$$\vec{\varsigma} = \frac{1}{3}\begin{pmatrix} 2u_x - v_y - w_z & 0 & 0 \\ 0 & 2v_y - w_z - u_x & 0 \\ 0 & 0 & 2w_z - u_x - v_y \end{pmatrix}, \quad \text{or}$$

$$\delta = \begin{pmatrix} (v_y - w_z)\,\mathbf{i} + \\ (w_z - u_x)\,\mathbf{j} + \\ (u_x - v_y)\,\mathbf{k} \end{pmatrix},$$

and *homogeneous press* $C\vec{\mathbf{e}}/3$ of *compressibity*

$$C = \nabla \cdot \mathbf{u} = u_x + v_y + w_z.$$

Accounting for *enstrophy*

$$|\omega|^2 = \frac{\|\vec{\omega}\|^2}{2},$$

we have, from (1.4.1)–(1.4.6), that

$$|\mathbf{u_r}|^2 = |\nabla \mathbf{u}|^2 = \frac{D + |\omega|^2}{2}, \quad D = \frac{\|\vec{\tau}\|^2}{2} = 2\,\|\vec{\sigma}\|^2 + \frac{\|\vec{\gamma}\|^2}{2},$$
$$\|\vec{\gamma}\|^2 = 2\,|\gamma|^2,$$
$$\|\vec{\sigma}\|^2 = \frac{(\nabla \cdot \mathbf{u})^2}{3} + \|\vec{\varsigma}\|^2 \quad \text{and} \quad \|\vec{\varsigma}\|^2 = \frac{|\delta|^2}{3},$$

which eventually give us the identity of measure to be a general matrix resolution (1.4.12).

Besides, in the case of flow deformation $\mathbf{u_r}$, we have an alternative resolution:

Theorem 2.5.1 (on flow measure). *In addition to* (1.4.12), *the flow deformation measure, or dissipation*

$$D = \frac{\|\vec{\tau}\|^2}{2} = |\omega|^2 + 2C^2 + \nabla \cdot 2\mathbf{D} \quad \textit{for}$$
$$\mathbf{D} = \mathbf{u} \cdot \nabla \mathbf{u} - C\mathbf{u}, \quad C = \nabla \cdot \mathbf{u}, \tag{2.5.2}$$

or with the deformation flow $\mathbf{D}$.

Proof. Really, we have that

$$\begin{aligned} D &= 2u_x^2 + 2v_y^2 + 2w_z^2 + (v_x + u_y)^2 + (w_y + v_z)^2 + (u_z + w_x)^2 \\ &= 2(u_x + v_y + w_z)^2 - 4u_x v_y - 4v_y w_z - 4w_z u_x + (v_x - u_y)^2 \\ &\quad + 4v_x u_y + (w_y - v_z)^2 + 4w_y v_z + (u_z - w_x)^2 + 4u_z w_x, \end{aligned}$$

or

$$D - 2\,(\nabla \cdot \mathbf{u})^2 - |\omega|^2 = -2\chi,$$

where

$$\chi = 2\begin{pmatrix} u_x v_y \\ -u_y v_x \end{pmatrix} + 2\begin{pmatrix} v_y w_z \\ -v_z w_y \end{pmatrix} + 2\begin{pmatrix} w_z u_x \\ -w_x u_z \end{pmatrix}.$$

At the same time,

$$2\begin{pmatrix} u_x v_y \\ -u_y v_x \end{pmatrix} = \begin{pmatrix} u v_y \\ -u_y v \end{pmatrix}_x + \begin{pmatrix} u_x v \\ -u v_x \end{pmatrix}_y,$$

$$2\begin{pmatrix} v_y w_z \\ -v_z w_y \end{pmatrix} = \begin{pmatrix} v w_z \\ -v_z w \end{pmatrix}_y + \begin{pmatrix} v_y w \\ -v w_y \end{pmatrix}_z$$

$$\text{and} \quad 2\begin{pmatrix} w_z u_x \\ -w_x u_z \end{pmatrix} = \begin{pmatrix} w u_x \\ -w_x u \end{pmatrix}_z + \begin{pmatrix} w_z u \\ -w u_z \end{pmatrix}_x.$$

So,

$$\chi = 2\begin{pmatrix} u_x v_y \\ -u_y v_x \end{pmatrix} + 2\begin{pmatrix} v_y w_z \\ -v_z w_y \end{pmatrix} + 2\begin{pmatrix} w_z u_x \\ -w_x u_z \end{pmatrix}$$

$$= \begin{pmatrix} u v_y - u_y v \\ +w_z u - w u_z \end{pmatrix}_x + \begin{pmatrix} v w_z - v_z w \\ +u_x v - u v_x \end{pmatrix}_y + \begin{pmatrix} w u_x - w_x u \\ +v_y w - v w_y \end{pmatrix}_z,$$

or

$$\chi = \begin{pmatrix} u u_x - u u_x \\ +u v_y - v u_y \\ +u w_z - w u_z \end{pmatrix}_x + \begin{pmatrix} v u_x - u v_x \\ +v v_y - v v_y \\ +v w_z - w v_z \end{pmatrix}_y + \begin{pmatrix} w u_x - u w_x \\ +w v_y - v w_y \\ +w w_z - w w_z \end{pmatrix}_z,$$

or

$$\chi = (u\nabla \cdot \mathbf{u} - \mathbf{u} \cdot \nabla u)_x + (v\nabla \cdot \mathbf{u} - \mathbf{u} \cdot \nabla v)_y$$
$$+ (w\nabla \cdot \mathbf{u} - \mathbf{u} \cdot \nabla w)_z = -\nabla \cdot \mathbf{D},$$

which concludes the proof. □

2.6. Bulk acceleration

Let us take again the *absolute* (motionless) frame of reference related to the unit cube **0ijk**, with points, orts, and *matrix structure*

$$\mathbf{r} = x\mathbf{i} + y\mathbf{j} + z\mathbf{k}, \quad \mathbf{r}_x = \mathbf{i}, \quad \mathbf{r}_y = \mathbf{j}, \quad \mathbf{r}_z = \mathbf{k}, \quad \text{and}$$
$$\vec{\mathbf{e}} = \mathbf{r}_\mathbf{r} = \mathbf{ii} + \mathbf{jj} + \mathbf{kk},$$

respectively, and suppose that the *local flow*

$$\mathbf{u} = u\mathbf{i} + v\mathbf{j} + w\mathbf{k} = \mathbf{u}\,(t, \mathbf{r})$$

is given, with which we can determine the *displacement*

$$\mathbf{r}' = \mathbf{r}^{t'} = \mathbf{r} + \int_t^{t'} \mathbf{u}\,(s, \mathbf{r}^s)\, ds$$

of a *moving particle* (point) $\mathbf{r} = \mathbf{r}^t$, then compute its *internal* (as local, or relative) *velocity*

$$\mathbf{r}_t = \lim_{t' \to t+0} \frac{\mathbf{r}' - \mathbf{r}}{t' - t} = \mathbf{u}\,(t, \mathbf{r}),$$

and, finally, find out its *rigid* (as solid-state, or translational) *acceleration*

$$\mathbf{u}_t = \partial_t \mathbf{u} = \frac{\partial \mathbf{u}}{\partial t} = \lim_{t' \to t+0} \frac{\mathbf{u}\,(t', \mathbf{r}) - \mathbf{u}\,(t, \mathbf{r})}{t' - t} \quad \text{as if}$$

$$\mathbf{u}_\mathbf{r} = \mathbf{u}_x \mathbf{i} + \mathbf{u}_y \mathbf{j} + \mathbf{u}_z \mathbf{k} = \vec{\mathbf{0}},$$

fluid (or *convective*) *acceleration*

$$\delta_t \mathbf{u} = \frac{\delta \mathbf{u}}{\delta t} = \lim_{t' \to t+0} \frac{\mathbf{u}\,(t, \mathbf{r}') - \mathbf{u}\,(t, \mathbf{r})}{t' - t} = \mathbf{u} \cdot \nabla \mathbf{u} = \mathbf{u}_\mathbf{r} \cdot \mathbf{u},$$

and *full acceleration* reduced to preceding ones:

$$d_t \mathbf{u} = \frac{d\mathbf{u}}{dt} = \lim_{t' - t \to +0} \frac{\mathbf{u}\,(t', \mathbf{r}') - \mathbf{u}\,(t, \mathbf{r})}{t' - t} = \mathbf{u}_t + \delta_t \mathbf{u}.$$

Besides, let the *bulk flow*

$$\mathbf{w} = \mathbf{w}\,(t, \mathbf{r})$$

be given to transfer particles $\mathbf{r} = \mathbf{r}^+ = \mathbf{r}^{+t}$ in new positions

$$\mathbf{r}'^+ = \mathbf{r}'^{+t'} = \mathbf{r} + \int_t^{t'} \mathbf{w}\left(s, \mathbf{r}'^{+s}\right) ds,$$

with velocities

$$\mathbf{r}_t^+ = \mathbf{w}\left(t, \mathbf{r}^+\right),$$

and to deform orts of the cube $\mathbf{0ijk}$ into edges

$$\mathbf{i}^+ = \mathbf{r}'^+_x = \mathbf{i} + \int_t^{t'} \mathbf{w}_x ds,$$

$$\mathbf{j}^+ = \mathbf{r}'^+_y = \mathbf{j} + \int_t^{t'} \mathbf{w}_y ds,$$

$$\mathbf{k}^+ = \mathbf{r}'^+_z = \mathbf{k} + \int_t^{t'} \mathbf{w}_z ds,$$

of the moving cube, or hexahedron $\mathbf{0i^+j^+k^+}$, with *rates* (velocities)

$$\mathbf{i}^+_t = \mathbf{w}_x, \quad \mathbf{j}^+_t = \mathbf{w}_y, \quad \mathbf{k}^+_t = \mathbf{w}_z,$$

so that the matrix structure $\vec{\mathbf{e}}$ of $\mathbf{0ijk}$ takes the form of the structure

$$\vec{\mathbf{e}}^+ = \mathbf{i}^+\mathbf{i} + \mathbf{j}^+\mathbf{j} + \mathbf{k}^+\mathbf{k}, \quad \vec{\mathbf{e}}^+\big|_{t'=t} = \vec{\mathbf{e}},$$

of $\mathbf{0i^+j^+k^+}$, with the rate

$$\vec{\mathbf{e}}^+_t = \lim_{t' \to t+0} \frac{\vec{\mathbf{e}}^+ - \vec{\mathbf{e}}}{t' - t} = \mathbf{i}^+_t\mathbf{i} + \mathbf{j}^+_t\mathbf{j} + \mathbf{k}^+_t\mathbf{k} = \mathbf{w}_x\mathbf{i} + \mathbf{w}_y\mathbf{j} + \mathbf{w}_z\mathbf{k} = \mathbf{w}_\mathbf{r}.$$

As this takes place, the *moving* frame of reference related to $\mathbf{0i^+j^+k^+}$ transfers the radius-vector $\mathbf{r}$,

$$\vec{\mathbf{e}}^+ \cdot \mathbf{r} = \mathbf{i}^+ x + \mathbf{j}^+ y + \mathbf{k}^+ z, \quad \mathbf{r} = x\mathbf{i} + y\mathbf{j} + z\mathbf{k} = \vec{\mathbf{e}} \cdot \mathbf{r} = \mathbf{r} \cdot \vec{\mathbf{e}},$$

with the rate

$$\mathbf{w}_\mathbf{r} \cdot \mathbf{r} = \vec{\mathbf{e}}^+_t \cdot \mathbf{r} = \lim_{t' \to t+0} \frac{\tilde{\mathbf{e}}^+ \cdot \mathbf{r} - \mathbf{r}}{t' - t}$$

that complements the internal, or local velocity $\mathbf{u}$ up to the *external* (or absolute) velocity as

$$\boldsymbol{v} = \lim_{t' \to t+0} \frac{\vec{\mathbf{e}}^+ \cdot \mathbf{r}' - \mathbf{r}}{t' - t} = \left(\vec{\mathbf{e}}^+ \cdot \mathbf{r}\right)_t = \vec{\mathbf{e}}^+\big|_{t'=t} \cdot \mathbf{r}_t + \vec{\mathbf{e}}^+_t \cdot \mathbf{r} = \mathbf{u} + \mathbf{w}_\mathbf{r} \cdot \mathbf{r}.$$

As applied to a particle $\mathbf{r}$, the specific external force turns out by Newton to the *full external acceleration*

$$d^+_t \boldsymbol{v} = \frac{d^+ \boldsymbol{v}}{dt} = \lim_{t' \to t+0} \frac{\vec{\mathbf{e}}^+ \cdot \boldsymbol{v}\left(t', \vec{\mathbf{e}}^+ \cdot \mathbf{r}'\right) - \boldsymbol{v}\left(t, \mathbf{r}\right)}{t' - t},$$

that reduces in turn to the sum

$$d_t^+ v = v_t^+ + \delta_t^+ v,$$

of rigid and fluid components as

$$v_t^+ = \lim_{t' \to t+0} \frac{\vec{\mathbf{e}}^+ \cdot v\left(t', \vec{\mathbf{e}}^+ \cdot \mathbf{r}\right) - v\left(t, \mathbf{r}\right)}{t' - t}$$

$$= \lim_{t' \to t+0} \frac{\vec{\mathbf{e}}^+ \cdot \boldsymbol{v}\left(t', \vec{\mathbf{e}}^+ \cdot \mathbf{r}'\right) - \boldsymbol{v}\left(t, \mathbf{r}'\right)}{t' - t}$$

and

$$\delta_t^+ \boldsymbol{v} = \lim_{t' \to t+0} \frac{\boldsymbol{v}(t, \vec{\mathbf{e}}^+ \cdot \mathbf{r}') - \boldsymbol{v}(t, \mathbf{r})}{t' - t},$$

respectively, computed as vectors:

$$\boldsymbol{v}_t^+ = \mathbf{u}_t + 2\mathbf{w}_\mathbf{r} \cdot \mathbf{u} + 2\mathbf{w}_\mathbf{r} \cdot (\mathbf{w}_\mathbf{r} \cdot \mathbf{r}) + \mathbf{w}_{t\mathbf{r}} \cdot \mathbf{r} + (\mathbf{w}_{\mathbf{rr}} \cdot \mathbf{r}) \cdot (\mathbf{w}_\mathbf{r} \cdot \mathbf{r})$$

and

$$\delta_t^+ \boldsymbol{v} = \mathbf{u}_\mathbf{r} \cdot (\mathbf{u} + 2\mathbf{w}_\mathbf{r} \cdot \mathbf{r}) + (\mathbf{w}_\mathbf{r} + \mathbf{w}_{\mathbf{rr}} \cdot \mathbf{r}) \cdot \mathbf{u} + 2\left(\mathbf{w}_\mathbf{r} + \mathbf{w}_{\mathbf{rr}} \cdot \mathbf{r}\right) \cdot (\mathbf{w}_\mathbf{r} \cdot \mathbf{r}).$$

Together with the external acceleration $d_t^+ \boldsymbol{v}$, the *specific* (here, taken per unit mass) *inertial force*, or the *bulk acceleration*

$$d_t \mathbf{u} - d_t^+ \boldsymbol{v} = \mathbf{u}_t - \boldsymbol{v}_t^+ + \delta_t \mathbf{u} - \delta_t^+ \boldsymbol{v},$$

comes into existence in the moving frame of reference as the sum of the *rigid* one,

$$\mathbf{u}_t - \boldsymbol{v}_t^+ = -2\mathbf{w}_\mathbf{r} \cdot \mathbf{u} - 2\mathbf{w}_\mathbf{r} \cdot (\mathbf{w}_\mathbf{r} \cdot \mathbf{r}) - \mathbf{w}_{t\mathbf{r}} \cdot \mathbf{r} - (\mathbf{w}_{\mathbf{rr}} \cdot \mathbf{r}) \cdot (\mathbf{w}_\mathbf{r} \cdot \mathbf{r}),$$

acting on a particle of rigid or fluid medium, and *fluid* complement

$$\delta_t \mathbf{u} - \delta_t^+ v = -2\mathbf{u}_\mathbf{r} \cdot (\mathbf{w}_\mathbf{r} \cdot \mathbf{r}) - (\mathbf{w}_\mathbf{r} + \mathbf{w}_{\mathbf{rr}} \cdot \mathbf{r}) \cdot$$
$$\mathbf{u} - 2\left(\mathbf{w}_\mathbf{r} + \mathbf{w}_{\mathbf{rr}} \cdot \mathbf{r}\right) \cdot (\mathbf{w}_\mathbf{r} \cdot \mathbf{r}),$$

acting on a particle of fluid medium, such that

$$\mathbf{u}_t - \boldsymbol{v}_t^+ - \left(\delta_t \mathbf{u} - \delta_t^+ \boldsymbol{v}\right) = (\mathbf{w}_{\mathbf{rr}} \cdot \mathbf{r} - \mathbf{w}_\mathbf{r}) \cdot \mathbf{u} + 2\mathbf{u}_\mathbf{r} \cdot (\mathbf{w}_\mathbf{r} \cdot \mathbf{r})$$
$$- \mathbf{w}_{t\mathbf{r}} \cdot \mathbf{r} + (\mathbf{w}_{\mathbf{rr}} \cdot \mathbf{r}) \cdot (\mathbf{w}_\mathbf{r} \cdot \mathbf{r}).$$

For example, the rotation with constant angular velocity

$$\mathbf{w} = \mathbf{w}^{(\mathrm{ii})} = \boldsymbol{\Omega} \times \mathbf{r}, \quad \boldsymbol{\Omega} = \mathbf{const} \quad (\mathbf{w}_{\mathbf{r}t} = \mathbf{w}_{\mathbf{rr}} \cdot \mathbf{r} = \vec{\mathbf{0}}),$$

hence, with the deformation matrix $\mathbf{w_r} = \boldsymbol{\Omega} \times \vec{\mathbf{e}}$, transfer velocity $\mathbf{w_r} \cdot \mathbf{r} = \boldsymbol{\Omega} \times \mathbf{r}$, *Coriolis* and *centrifugal accelerations*, $2\mathbf{w_r} \cdot \mathbf{r} = 2\boldsymbol{\Omega} \times \mathbf{r}$ and

$$\begin{aligned} 2\mathbf{w_r} \cdot (\mathbf{w_r} \cdot \mathbf{r}) &= 2\boldsymbol{\Omega} \times (\boldsymbol{\Omega} \times \mathbf{r}) = 2\,(\boldsymbol{\Omega}\boldsymbol{\Omega} \cdot \mathbf{r} - \boldsymbol{\Omega} \cdot \boldsymbol{\Omega}\mathbf{r}) \\ &= \nabla\left((\boldsymbol{\Omega} \cdot \mathbf{r})^2 - |\boldsymbol{\Omega}|^2\,|\mathbf{r}|^2\right) = -\nabla\,|\boldsymbol{\Omega} \times \mathbf{r}|^2, \end{aligned}$$

are detected by specific inertial forces,

$$\begin{aligned} &\mathbf{u}_t - \boldsymbol{v}_t^+ = 2\mathbf{u} \times \boldsymbol{\Omega} + \nabla\,|\boldsymbol{\Omega} \times \mathbf{r}|^2 \\ &\quad \text{or} \quad d_t\mathbf{u} - d_t^+\boldsymbol{v} = 3\mathbf{u} \times \boldsymbol{\Omega} + 2\nabla\,|\boldsymbol{\Omega} \times \mathbf{r}|^2 - 2\,(\boldsymbol{\Omega} \times \mathbf{r}) \cdot \nabla\mathbf{u}, \end{aligned}$$

acting on rigid or fluid particles, respectively.

Finally, for the stationary central expansion

$$\mathbf{w} = \mathbf{w}^{(\mathrm{iii})} = H\mathbf{r} - \frac{1}{2}\mathbf{r} \cdot \mathbf{r}\boldsymbol{\Theta} + \mathbf{r} \cdot \boldsymbol{\Theta}\mathbf{r}, \quad H_t = 0, \quad \boldsymbol{\Theta}_t = \mathbf{0},$$

taking into account identities

$$\begin{aligned} &\mathbf{w_r} = (H + \mathbf{r} \cdot \boldsymbol{\Theta})\,\vec{\mathbf{e}} + \mathbf{r}\boldsymbol{\Theta} - \boldsymbol{\Theta}\mathbf{r} \\ \text{and} \quad &\mathbf{w_{rr}} = (\mathbf{w_r})_\mathbf{r} = \vec{\mathbf{e}}\Theta - \Theta\vec{\mathbf{e}} + \mathbf{i}\boldsymbol{\Theta}\mathbf{i} + \mathbf{j}\boldsymbol{\Theta}\mathbf{j} + \mathbf{k}\boldsymbol{\Theta}\mathbf{k}, \end{aligned}$$

we find that

$$\begin{aligned} \mathbf{u}_t - \boldsymbol{v}_t^+ &= -2\,(H + \boldsymbol{\Theta} \cdot \mathbf{r})\,\mathbf{u} + 2\,(\boldsymbol{\Theta} \times \mathbf{r}) \times \mathbf{u} + \mathbf{r} \cdot \mathbf{r}\,(5H + 6\boldsymbol{\Theta} \cdot \mathbf{r})\,\boldsymbol{\Theta} \\ &\quad - 2\left(\left(H + \frac{5}{2}\boldsymbol{\Theta} \cdot \mathbf{r}\right)^2 - \frac{6\,|\boldsymbol{\Theta}|^2 + (\boldsymbol{\Theta} \cdot \mathbf{r})^2}{4}\right)\mathbf{r} \end{aligned}$$

and

$$\begin{aligned} \mathbf{u}_t - \boldsymbol{v}_t^+ - \left(\delta_t\mathbf{u} - \delta_t^+\boldsymbol{v}\right) &= -H\mathbf{u} + 2\,((H + 2\boldsymbol{\Theta} \cdot \mathbf{r})\,\mathbf{r} - \mathbf{r} \cdot \mathbf{r}\boldsymbol{\Theta}) \cdot \nabla\mathbf{u} \\ &\quad + \left(2H\boldsymbol{\Theta} \cdot \mathbf{r} - |\boldsymbol{\Theta}|^2 + 4\,(\boldsymbol{\Theta} \cdot \mathbf{r})^2\right)\mathbf{r} - \mathbf{r} \cdot \mathbf{r}\,(H + 2\boldsymbol{\Theta} \cdot \mathbf{r})\,\boldsymbol{\Theta}. \end{aligned}$$

Chapter 3

Flows of Elements

3.1. Kinematic structure of a moving body point

Apart from algebraic beginnings of screw "$\times$" and metric "$\cdot$" (from § 1), space constituents of particles $\mathbf{r}$ and contact forces $\mathbf{f}$ (of structures $\vec{\mathbf{p}}$), and kinematic generatrices of velocity $\mathbf{u}$ and strain $\mathbf{u_r}$ (from Chapter 2), a smooth medium has got the evident infinitesimal *geometric elements* of vector elements of length $d\mathbf{r}$ and area $d\mathbf{S}$ and scalar element of volume dV in proper *flows* as changing Riemannian sums, or integrals taken over deforming length, area, and volume of a substance in hand, be it a rigid body translated, such as a vehicle, or rotated, such as a spinning wheel, a gas mixture compressed or extended in any internal combustion engine, an incompressible fluid flowing in a river or falling in a waterfall, or, finally, an invisible dark matter keeping together all the remaining light parts of the universe while pulling the parts to each other with gravity rather than pushing them away as Archimedean buoyancy.

The integral length, area and volume flow rates reduce at that to the following rates of elements, $d\mathbf{r}$, $d\mathbf{S}$ and dV.

When treated *instantly*, at a moment $t = t_0 \geq 0$, and *locally*, in the neighborhood $|\mathbf{r} - \mathbf{r}_0| < \varepsilon$ of the internal or boundary point $\mathbf{r}_0$, every fluid volume V as a *domain* (an open and connected set) proves to be a *body* whose particles

$$\mathbf{r} = \mathbf{i}x + \cdots = \mathbf{i}x\,(a, b, c) + \cdots = \mathbf{r}\,(\mathbf{a}),$$

by definition, *smoothly*, or infinitely differentiable, and *reversibly*, or with smooth *inverse images* $\mathbf{a} = \mathbf{a}\,(\mathbf{r})$, depend on the *vector*

$\mathbf{a} = \mathbf{i}a + \cdots$ of parameters a, b, c given as independent variables $a_- < a < a_+, \ldots$ for $a_\pm, \ldots = \text{const}$.

Consequently, every *body point* $\mathbf{r} = \mathbf{r}(\mathbf{a})$ has got its own *differential structure*

$$\mathbf{r}_\mathbf{a} = \mathbf{r}_a\mathbf{i} + \mathbf{r}_b\mathbf{j} + \mathbf{r}_c\mathbf{k} = \begin{pmatrix} x_a\mathbf{i} \\ +y_a\mathbf{j} \\ +z_a\mathbf{k} \end{pmatrix}\mathbf{i} + \begin{pmatrix} x_b\mathbf{i} \\ +y_b\mathbf{j} \\ +z_b\mathbf{k} \end{pmatrix}\mathbf{j} + \begin{pmatrix} x_c\mathbf{i} \\ +y_c\mathbf{j} \\ +z_c\mathbf{k} \end{pmatrix}\mathbf{k}, \tag{3.1.1}$$

or

$$\mathbf{r}_\mathbf{a} = \mathbf{i}x_\mathbf{a} + \mathbf{j}y_\mathbf{a} + \mathbf{k}z_\mathbf{a} = \mathbf{i}\begin{pmatrix} x_a\mathbf{i} \\ +x_b\mathbf{j} \\ +x_c\mathbf{k} \end{pmatrix} + \mathbf{j}\begin{pmatrix} y_a\mathbf{i} \\ +y_b\mathbf{j} \\ +y_c\mathbf{k} \end{pmatrix} + \mathbf{k}\begin{pmatrix} z_a\mathbf{i} \\ +z_b\mathbf{j} \\ +z_c\mathbf{k} \end{pmatrix}, \tag{3.1.2}$$

with coordinate *body vectors* $\mathbf{r}_a$, $\mathbf{r}_b$, $\mathbf{r}_c$ or *gradients* $x_\mathbf{a}$, $y_\mathbf{a}$, $z_\mathbf{a}$, respectively.

Further, given a smooth inverse image

$$\mathbf{a} = \mathbf{i}a + \cdots = \mathbf{i}a(x, y, z) + \cdots = \mathbf{a}(\mathbf{r}),$$

of body point $\mathbf{r} = \mathbf{r}(\mathbf{a})$, we can find directly the proper *inverse structure* as either

$$\mathbf{r}_\mathbf{a}^{-1} = \mathbf{a}_\mathbf{r} = \mathbf{a}_x\mathbf{i} + \mathbf{a}_y\mathbf{j} + \mathbf{a}_z\mathbf{k} = \begin{pmatrix} a_x\mathbf{i} \\ +b_x\mathbf{j} \\ +c_z\mathbf{k} \end{pmatrix}\mathbf{i} + \begin{pmatrix} a_y\mathbf{i} \\ +b_y\mathbf{j} \\ +c_y\mathbf{k} \end{pmatrix}\mathbf{j} + \begin{pmatrix} a_z\mathbf{i} \\ +b_z\mathbf{j} \\ +c_z\mathbf{k} \end{pmatrix}\mathbf{k}$$

with $\mathbf{a}_x$, $\mathbf{a}_y$, $\mathbf{a}_z$, or

$$\mathbf{r}_\mathbf{a}^{-1} = \mathbf{a}_\mathbf{r} = \mathbf{i}a_\mathbf{r} + \mathbf{j}b_\mathbf{r} + \mathbf{k}c_\mathbf{r} = \mathbf{i}\begin{pmatrix} a_x\mathbf{i} \\ +a_y\mathbf{j} \\ +a_z\mathbf{k} \end{pmatrix} + \mathbf{j}\begin{pmatrix} b_x\mathbf{i} \\ +b_y\mathbf{j} \\ +b_z\mathbf{k} \end{pmatrix} + \mathbf{k}\begin{pmatrix} c_x\mathbf{i} \\ +c_y\mathbf{j} \\ +c_z\mathbf{k} \end{pmatrix}$$

with $a_\mathbf{r}, b_\mathbf{r}, c_\mathbf{r}$.

Then

$$\mathbf{a_r}\cdot\mathbf{r_a}=\begin{pmatrix}a_x\mathbf{i}\\+b_x\mathbf{j}\\+c_z\mathbf{k}\end{pmatrix}\begin{pmatrix}x_a\mathbf{i}\\+x_b\mathbf{j}\\+x_c\mathbf{k}\end{pmatrix}+\begin{pmatrix}a_y\mathbf{i}\\+b_y\mathbf{j}\\+c_y\mathbf{k}\end{pmatrix}\begin{pmatrix}y_a\mathbf{i}\\+y_b\mathbf{j}\\+y_c\mathbf{k}\end{pmatrix}$$

$$+\begin{pmatrix}a_z\mathbf{i}\\+b_z\mathbf{j}\\+c_z\mathbf{k}\end{pmatrix}\begin{pmatrix}z_a\mathbf{i}\\+z_b\mathbf{j}\\+z_c\mathbf{k}\end{pmatrix}$$

and

$$\mathbf{r_a}\cdot\mathbf{a_r}=\begin{pmatrix}x_a\mathbf{i}\\+y_a\mathbf{j}\\+z_a\mathbf{k}\end{pmatrix}\begin{pmatrix}a_x\mathbf{i}\\+a_y\mathbf{j}\\+a_z\mathbf{k}\end{pmatrix}+\begin{pmatrix}x_b\mathbf{i}\\+y_b\mathbf{j}\\+z_b\mathbf{k}\end{pmatrix}\begin{pmatrix}b_x\mathbf{i}\\+b_y\mathbf{j}\\+b_z\mathbf{k}\end{pmatrix}$$

$$+\begin{pmatrix}x_c\mathbf{i}\\+y_c\mathbf{j}\\+z_c\mathbf{k}\end{pmatrix}\begin{pmatrix}c_x\mathbf{i}\\+c_y\mathbf{j}\\+c_z\mathbf{k}\end{pmatrix},$$

so

$$\mathbf{a_r}\cdot\mathbf{r_a}=\mathbf{a}_x x_\mathbf{a}+\mathbf{a}_y y_\mathbf{a}+\mathbf{a}_z z_\mathbf{a}=\mathbf{a_a}=\vec{\mathbf{e}},\quad\text{hence,}\quad J^\mathbf{a}_\mathbf{r}J^\mathbf{r}_\mathbf{a}=J\left[\vec{\mathbf{e}}\right]=1$$

(as follows from (1.3.1)), and

$$\mathbf{r_a}\cdot\mathbf{a_r}=\mathbf{r}_a a_\mathbf{r}+\mathbf{r}_b b_\mathbf{r}+\mathbf{r}_c c_\mathbf{r}=\mathbf{r_r}=\vec{\mathbf{e}},$$

respectively.

Finally, with the *volume factor*

$$J^\mathbf{r}_\mathbf{a}=\mathbf{r}_a\cdot\mathbf{r}_b\times\mathbf{r}_c>0\quad\text{such that}\quad J^\mathbf{a}_\mathbf{r}J^\mathbf{r}_\mathbf{a}=1\quad(\text{since }\mathbf{a_r}\cdot\mathbf{r_a}=\vec{\mathbf{e}}),\tag{3.1.3}$$

and formulas (1.5.1), and *vector areas* $\mathbf{r}_b\times\mathbf{r}_c,\ldots$, we obtain

$$\mathbf{r_a}^{-1}=\mathbf{i}\frac{\mathbf{r}_b\times\mathbf{r}_c}{J^\mathbf{r}_\mathbf{a}}+\mathbf{j}\frac{\mathbf{r}_c\times\mathbf{r}_a}{J^\mathbf{r}_\mathbf{a}}+\mathbf{k}\frac{\mathbf{r}_a\times\mathbf{r}_b}{J^\mathbf{r}_\mathbf{a}}=\mathbf{i}a_\mathbf{r}+\mathbf{j}b_\mathbf{r}+\mathbf{k}c_\mathbf{r}$$

and may draw a conclusion:

Theorem 3.1.1 (on invariance failure). *Body vector areas*

$$\mathbf{r}_b \times \mathbf{r}_c = J_{\mathbf{a}}^{\mathbf{r}} a_{\mathbf{r}}, \quad \mathbf{r}_c \times \mathbf{r}_a = J_{\mathbf{a}}^{\mathbf{r}} b_{\mathbf{r}} \quad \textit{and} \quad \mathbf{r}_a \times \mathbf{r}_b = J_{\mathbf{a}}^{\mathbf{r}} c_{\mathbf{r}} \tag{3.1.4}$$

are neither covariant nor contravariant.

3.2. Rates of length, area, and volume elements

Thus, together with structure $\mathbf{r_a}$ and *independent differentials* as differences $da = a - a_0 \neq 0, \ldots$, every point $\mathbf{r}_0 = \mathbf{r}(\mathbf{a}_0)$ acquires the *geometric elements*

$$\mathrm{da} = d\mathbf{r}, d\mathbf{S}, dV = \mathrm{da}(t, \mathbf{r})$$

of vector *length* and *area*

$$d\mathbf{r} = \mathbf{r}_a da \quad \text{and} \quad d\mathbf{S} = \mathbf{r}_b \times \mathbf{r}_c dbdc, \ldots,$$

and scalar *volume*

$$dV = d\mathbf{r} \cdot d\mathbf{S} = J dadbdc \quad \text{with } J = J_{\mathbf{a}}^{\mathbf{r}} = \mathbf{r}_a \cdot \mathbf{r}_b \times \mathbf{r}_c > 0,$$

as shown in Fig. 3.2.1.

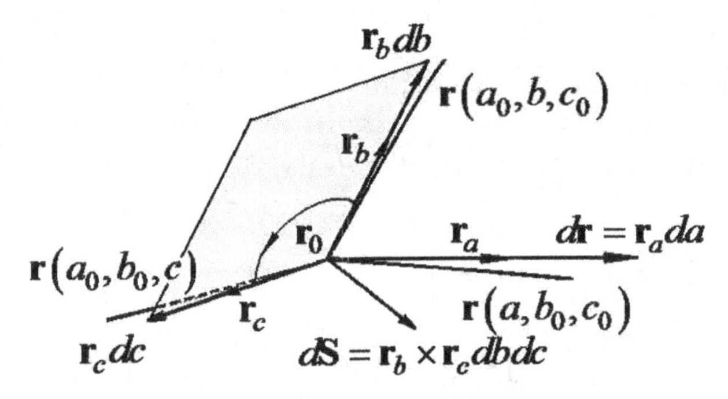

Figure 3.2.1. Vectors of length $d\mathbf{r}$ and area $d\mathbf{S}$ as forming a hexahedron $\mathbf{r}_0 d\mathbf{r}_a d\mathbf{r}_b d\mathbf{r}_c$ of volume $dV = d\mathbf{r} \cdot d\mathbf{S}$.

Given the *velocity flow* $\mathbf{u}(t, \mathbf{r})$, geometric elements da are deformed as

$$da\prime = \mathrm{da}\left(t', \mathbf{r}'\right), \quad \mathbf{r}' = \mathbf{r}^{t'} = \mathbf{r} + \int_t^{t'} \mathbf{u}\left(s, \mathbf{r}^s\right) ds, \quad t' = t + \varepsilon$$

with rates

$$\frac{d}{dt}\mathrm{da} = \frac{d}{dt'}\mathrm{da}'\bigg|_{t'=t} = \lim_{\varepsilon \to +0} \frac{\mathrm{da}' - \mathrm{da}}{\varepsilon}.$$

For example, for

$$d\mathbf{r}' = \mathbf{r}'_a da,$$

identities (2.1.1) and (2.3.1) imply that

$$\mathbf{r}'_a = x'_a\mathbf{i} + y'_a\mathbf{j} + z'_a\mathbf{k} = x_a\mathbf{r}'_x + y_a\mathbf{r}'_y + z_a\mathbf{r}'_z = x_a\mathbf{i}^+ + y_a\mathbf{j}^+ + z_a\mathbf{k}^+,$$

or

$$d\mathbf{r}' = d\mathbf{r} \cdot \mathbf{i}\mathbf{i}^+ + d\mathbf{r} \cdot \mathbf{j}\mathbf{j}^+ + d\mathbf{r} \cdot \mathbf{k}\mathbf{k}^+ = \mathbf{e}^+ \cdot d\mathbf{r},$$

so

$$d\mathbf{r}' - d\mathbf{r} = (\mathbf{r}'_a - \mathbf{r}_a)da = x_a(\mathbf{i}^+ - \mathbf{i}) + y_a(\mathbf{j}^+ - \mathbf{j}) + z_a(\mathbf{k}^+ - \mathbf{k}),$$

and

$$\frac{d}{dt}d\mathbf{r} = \mathbf{i}_t^+ x_a da + \mathbf{j}_t^+ y_a da + \mathbf{k}_t^+ z_a da = \mathbf{u_r} \cdot d\mathbf{r}. \tag{3.2.1}$$

The same follows from (2.2.1) and (2.3.5):

$$\mathbf{r}'_a = \mathbf{r}'_x x_a + \mathbf{r}'_y y_a + \mathbf{r}'_z z_a = \mathbf{r}'_\mathbf{r} \cdot \mathbf{r} = \vec{\mathbf{e}}^+ \cdot \mathbf{r}, \quad \text{and}$$

$$\frac{d}{dt}d\mathbf{r} = \vec{\mathbf{e}}_t^+ \cdot d\mathbf{r} = \mathbf{u_r} \cdot d\mathbf{r}.$$

Further, since

$$\mathbf{r'_a} = \mathbf{r'}_a\mathbf{i} + \mathbf{r'}_b\mathbf{j} + \mathbf{r'}_c\mathbf{k} = \begin{pmatrix} \mathbf{r'}_x x_a \\ +\mathbf{r'}_y y_a \\ +\mathbf{r'}_z z_a \end{pmatrix} \mathbf{i} + \begin{pmatrix} \mathbf{r'}_x x_b \\ +\mathbf{r'}_y y_b \\ +\mathbf{r'}_z z_b \end{pmatrix} \mathbf{j}$$

$$+ \begin{pmatrix} \mathbf{r'}_x x_c \\ +\mathbf{r'}_y y_c \\ +\mathbf{r'}_z z_c \end{pmatrix} \mathbf{k} = \begin{pmatrix} \mathbf{r'}_x x_\mathbf{a} \\ +\mathbf{r'}_y y_\mathbf{a} \\ +\mathbf{r'}_z z_\mathbf{a} \end{pmatrix} = \mathbf{r'_r} \cdot \mathbf{r_a},$$

hence,

$$J_\mathbf{a}^{\mathbf{r'}} = J_\mathbf{r}^{\mathbf{r'}} J_\mathbf{a}^{\mathbf{r}} \quad \text{(the identity (1.3.1))},$$

identities (3.1.3), (3.1.4), (2.1.2), and (2.2.3) mean that

$$\begin{aligned} \mathbf{r'}_b \times \mathbf{r'}_c &= J_\mathbf{a}^{\mathbf{r'}} a_{\mathbf{r'}} = J_\mathbf{r}^{\mathbf{r'}} J_\mathbf{a}^{\mathbf{r}} a_{\mathbf{r'}} = J_\mathbf{r}^{\mathbf{r'}} J_\mathbf{a}^{\mathbf{r}} \left(a_x x_{\mathbf{r'}} + a_y y_{\mathbf{r'}} + a_z z_{\mathbf{r'}} \right) \\ &= J_\mathbf{r}^{\mathbf{r'}} J_\mathbf{a}^{\mathbf{r}} \left(a_x \mathbf{i}^- + a_y \mathbf{j}^- + a_z \mathbf{k}^- \right), \end{aligned}$$

and

$$\begin{aligned} \mathbf{r'}_b \times \mathbf{r'}_c &= J_\mathbf{r}^{\mathbf{r'}} J_\mathbf{a}^{\mathbf{r}} a_\mathbf{r} \cdot \left(\mathbf{i}\mathbf{i}^- + \mathbf{j}\mathbf{j}^- + \mathbf{k}\mathbf{k}^- \right) \\ &= J_\mathbf{r}^{\mathbf{r'}} \left(\mathbf{i}^-\mathbf{i} + \mathbf{j}^-\mathbf{j} + \mathbf{k}^-\mathbf{k} \right) \cdot J_\mathbf{a}^{\mathbf{r}} a_\mathbf{r} = J_\mathbf{r}^{\mathbf{r'}} \vec{\mathbf{e}}_*^- \cdot \mathbf{r}_b \times \mathbf{r}_c, \end{aligned}$$

respectively, so that

$$d\mathbf{S'} = \mathbf{r'}_b \times \mathbf{r'}_c dbdc = J\vec{\mathbf{e}}_*^- \cdot d\mathbf{S}, \quad J = J_\mathbf{r}^{\mathbf{r'}}.$$

Then, owing to (2.3.6) and (2.3.7), we have

$$\frac{d}{dt} d\mathbf{S} = J_t \, \vec{\mathbf{e}}_*^- \big|_{t'=t} \cdot d\mathbf{S} + J|_{t'=t} \, \vec{\mathbf{e}}_{*t}^- \cdot d\mathbf{S} = (\nabla \cdot \mathbf{u}) \, \vec{\mathbf{e}} \cdot d\mathbf{S} - \nabla \mathbf{u} \cdot d\mathbf{S} \tag{3.2.2}$$

and

$$\frac{d}{dt} dV = J_t dadbdc = (\nabla \cdot \mathbf{u}) \, dV. \tag{3.2.3}$$

This is the case with rates of *metric deformation* of *length*,

$$d\mathbf{r} = d\mathbf{r'} \cdot \vec{\mathbf{e}}^+ = \vec{\mathbf{e}}^+ \cdot d\mathbf{r} \cdot \vec{\mathbf{e}}^+ = \vec{\mathbf{e}}_*^+ \cdot \vec{\mathbf{e}}^+ \cdot d\mathbf{r} = \vec{\mathbf{g}} \cdot d\mathbf{r} = d\mathbf{r} \cdot \vec{\mathbf{g}},$$

and *area*,

$$d\mathbf{S} = \vec{\mathrm{e}}^{-} \cdot d\mathbf{S}' = J\vec{\mathrm{e}}^{-} \cdot d\mathbf{S} \cdot \vec{\mathrm{e}}^{-} = Jd\mathbf{S} \cdot \vec{\mathrm{e}}^{-} \cdot \vec{\mathrm{e}}_{*}^{-}$$
$$= Jd\mathbf{S} \cdot \vec{\mathrm{g}}^{-} = J\vec{\mathrm{g}}^{-} \cdot d\mathbf{S}, \tag{3.2.4}$$

whose volume

$$dV = d\mathbf{r} \cdot d\mathbf{S} = Jd\mathbf{r} \cdot \vec{\mathrm{g}} \cdot \vec{\mathrm{g}}^{-} \cdot d\mathbf{S} = Jd\mathbf{r} \cdot d\mathbf{S} = JdV = dV'.$$

In this case, we have, with (2.3.8) and (2.3.9), that

$$\frac{d}{dt}d\mathbf{r} = \vec{\mathrm{g}}_t \cdot d\mathbf{r} = \vec{\tau} \cdot d\mathbf{r}$$

and

$$\frac{d}{dt}d\mathbf{S} = \frac{d}{dt}J\vec{\mathrm{g}}^{-} \cdot d\mathbf{S} = J_t\vec{\mathrm{e}} \cdot d\mathbf{S} + \vec{\mathrm{g}}_t^{-} \cdot d\mathbf{S} = (\nabla \cdot \mathbf{u})\, d\mathbf{S} - \vec{\tau} \cdot d\mathbf{S}. \tag{3.2.5}$$

3.3. Contour, contact, and volume flows

As is usual, due to the differentiation by parts, the rate of physical element $\mathrm{b} \cdot da$, or the *element rate*, consists of the *translation rate* of its geometric element da with *full velocity* $d\mathrm{b}/dt$ and the above *deformation rate* of da with density b:

$$\frac{d}{dt}\mathrm{b} \cdot da = \frac{d\mathrm{b}}{dt} \cdot da + \mathrm{b} \cdot \frac{d}{dt}da \quad \text{for} \quad \frac{d\mathrm{b}}{dt} = \frac{\partial \mathrm{b}}{\partial t} + \mathbf{u} \cdot \nabla \mathrm{b}. \tag{3.3.1}$$

Besides, it is evident that the rate of the limiting sum of elements is made up of their rates (provided that the limiting sum is convergent), or *the integral rate is the integral of rates*:

$$\frac{d}{dt}\int_{\mathrm{c}} \mathrm{b} \cdot da = \int_{\mathrm{c}} \frac{d}{dt}\mathrm{b} \cdot da \quad \text{for} \quad \mathrm{c} = \mathrm{c}^t \quad (\text{or} \quad \mathbf{r} = \mathbf{r}^t). \tag{3.3.2}$$

So, as an example, we can consider the *linear flow of velocity*

$$\int_l \mathbf{u} \cdot d\mathbf{r}, \quad l = l^{\mathbf{u}}_{\mathbf{r}_-,\mathbf{r}_+} = \{\mathbf{r} : \mathbf{r}_t = \mathbf{u}\,(t,\mathbf{r})\}, \quad \partial l = \{\mathbf{r}_-, \mathbf{r}_+\},$$

taken over a piece-wise smooth *fluid contour* l carried with velocity $\mathbf{u}\,(t,\mathbf{r})$ by every point $\mathbf{r}$ including the initial $\mathbf{r}_-$ and the end $\mathbf{r}_+$ of l.

The rate of its element is

$$\frac{d}{dt}\mathbf{u}\cdot d\mathbf{r} = \frac{d\mathbf{u}}{dt}\cdot d\mathbf{r} + \mathbf{u}\cdot\frac{d}{dt}d\mathbf{r} = \mathbf{u}_t\cdot d\mathbf{r} + (\mathbf{u}\cdot\nabla\mathbf{u})\cdot d\mathbf{r} + \mathbf{u}\cdot(\mathbf{u}_{\mathbf{r}}\cdot d\mathbf{r}).$$

At that, the *vortex acceleration*

$$\omega\times\mathbf{u} = \mathbf{u}\cdot\nabla\mathbf{u} - \nabla\frac{|\mathbf{u}|^2}{2},\quad |\mathbf{u}|^2 = \mathbf{u}\cdot\mathbf{u},\quad \omega = \nabla\times\mathbf{u} \qquad (3.3.3)$$

(the *Gromeka–Lamb identity*), which is provided by the formula of the triple vector product (1.1.1), or

$$\mathbf{u}\times(\nabla\times\boldsymbol{v}) = \nabla\mathbf{u}\cdot\boldsymbol{v}|_{\mathbf{u}=\mathbf{const}} - \mathbf{u}\cdot\nabla\boldsymbol{v},$$

hence,

$$\mathbf{u}\times(\nabla\times\boldsymbol{v}) + \boldsymbol{v}\times(\nabla\times\mathbf{u}) = \nabla(\mathbf{u}\cdot\boldsymbol{v}) - \mathbf{u}\cdot\nabla\boldsymbol{v} - \boldsymbol{v}\cdot\nabla\mathbf{u},$$

implying (3.3.3) for $\boldsymbol{v} = \mathbf{u}$.

The last term of the above rate,

$$\mathbf{u}\cdot(\mathbf{u}_{\mathbf{r}}\cdot d\mathbf{r}) = (\mathbf{u}\cdot\mathbf{u}_{\mathbf{r}})\cdot d\mathbf{r},$$

reveals a vector density $\mathbf{u}\cdot\mathbf{u}_{\mathbf{r}}$ reduced to the *Bernoulli potential* $|\mathbf{u}|^2/2$ as

$$\mathbf{u}\cdot\mathbf{u}_{\mathbf{r}} = \mathbf{u}\cdot\mathbf{u}_x\mathbf{i} + \cdots = \left(\frac{|\mathbf{u}|^2}{2}\right)_x \mathbf{i} + \cdots = \nabla\frac{|\mathbf{u}|^2}{2}.$$

When taken on a *fluid curve* l free of the *vortex acceleration* $\omega\times\mathbf{u}$ as

$$\omega\times\mathbf{u}\cdot d\mathbf{r}|_l = \omega\times\mathbf{u}\cdot\mathbf{u}dt = 0,$$

the initial rates of element and flow reduce to

$$\frac{d}{dt}\mathbf{u}\cdot d\mathbf{r}\bigg|_l = \left(\mathbf{u}_t + \nabla|\mathbf{u}|^2\right)\cdot d\mathbf{r}$$

$$\text{and}\quad \frac{d}{dt}\int_l \mathbf{u}\cdot d\mathbf{r} = \int_l \mathbf{u}_t\cdot d\mathbf{r} + |\mathbf{u}|^2\Big|_{\mathbf{r}=\mathbf{r}_-}^{\mathbf{r}=\mathbf{r}_+},$$

respectively.

Then, every *closed fluid curve* l ($\partial l = \emptyset$, or $\mathbf{r}_- = \mathbf{r}_+$) is carried by flow $\mathbf{u}$ only with the *relative acceleration* $\mathbf{u}_t$ in the *circulation*

$$\frac{d}{dt}\int_l \mathbf{u}\cdot d\mathbf{r} = \int_l \mathbf{u}_t \cdot d\mathbf{r} \quad \text{for} \quad \partial l = \emptyset \ (\mathbf{r}_- = \mathbf{r}_+).$$

Finally, when taken on the *boundary* $l = \partial S$ of a surface region $S \subset \partial V$ where the *Stokes theorem*

$$\int_{\partial S} \mathbf{f}\cdot d\mathbf{r} = \int_S (\nabla \times \mathbf{f})\cdot d\mathbf{S}, \quad S \subset \partial V \ \ (\mathbf{f} = \mathbf{u}), \tag{3.3.4}$$

is valid, the flow rate reduces to the contact flow through S with the *relative acceleration of vorticity* ω_t:

$$\frac{d}{dt}\int_{\partial S} \mathbf{u}\cdot d\mathbf{r} = \int_S \omega_t \cdot d\mathbf{S}, \quad \omega_t = \nabla \times \mathbf{u}_t.$$

In particular, we have the following.

Theorem 3.3.1 (a variant of the Kelvin circulation theorem). *The velocity circulation* $\int_{\partial S} \mathbf{u}\cdot d\mathbf{r}$ *on every fluid boundary circle* ∂S *of a stationary vortex flow* $\mathbf{u}$ *is unchanged:*

$$\frac{d}{dt}\int_{\partial S} \mathbf{u}\cdot d\mathbf{r} = 0 \quad \text{for } \omega_t = \mathbf{0}.$$

Note that like the principle of least deformation (theorem 2.4.1), this statement does not depend on any physical law.

For any *contact closed flow*

$$\int_{\partial V} \vec{\mathbf{p}}\cdot d\mathbf{S}, \quad \vec{\mathbf{p}} = \mathbf{p}^x\mathbf{i} + \cdots = \mathbf{i}\mathbf{p}^x_* + \cdots = \vec{\mathbf{p}}\,(t, \mathbf{r}),$$

on a piece-wise smooth boundary ∂V of finite *volume* V (to be a domain and a body), the relevant *divector variant*

$$\int_{\partial V} \vec{\mathbf{p}}\cdot d\mathbf{S} = \int_V (\nabla \cdot \vec{\mathbf{p}}_*)\, dV \tag{3.3.5}$$

of the *divergence theorem*

$$\int_{\partial V} \mathbf{f}\cdot d\mathbf{S} = \int_V (\nabla \cdot \mathbf{f})\, dV, \quad \mathbf{f} = \mathbf{p}^{x,y,z}_*$$

of Gauss–Ostrogradsky (1826) is evidently valid for immediately verified identities

$$\int_{\partial V} \vec{\mathbf{p}} \cdot d\mathbf{S} = \mathbf{i} \int_{\partial V} \mathbf{p}_*^x \cdot d\mathbf{S} + \cdots$$
$$= \int_V (\nabla \cdot \mathbf{p}_*^x \mathbf{i} + \cdots)\, dV = \int_V (\nabla \cdot \vec{\mathbf{p}}_*)\, dV$$

to convert a *contact* (or surface) *density* $\vec{\mathbf{p}}$ into a *volume one* $\nabla \cdot \vec{\mathbf{p}}_*$, or to make a volume flow $\int_V (\nabla \cdot \vec{\mathbf{p}}_*)\, dV$ to a contact one $\int_{\partial V} \vec{\mathbf{p}} \cdot d\mathbf{S}$ (and vise versa), which after the principle of least measure and the Stokes theorem again falls into no physical law despite being directly related to physics.

Finally, as to the volume rates, when taken for a smooth scalar function of *body density* $\rho = \rho\,(t, \mathbf{r}) > 0$, the *mass conservation*, or *continuity equation*

$$\frac{d}{dt} \int_V \rho dV = 0 \quad \text{for any fluid volume } V = V^t \tag{3.3.6}$$

with element and flow rates (3.2.3), (3.3.1), and (3.3.2), or

$$\frac{d}{dt} \int_V \rho dV = \int_V \frac{d}{dt} \rho dV = \int_V \frac{d\rho}{dt} dV + \int_V \rho \frac{d}{dt} dV$$
$$= \int_V \left(\frac{d\rho}{dt} + \rho \nabla \cdot \mathbf{u} \right) dV,$$

and identities

$$\frac{d\rho}{dt} + \rho \nabla \cdot \mathbf{u} = \rho_t + \mathbf{u} \cdot \nabla \rho + \rho \nabla \cdot \mathbf{u} = \rho_t + \nabla \cdot \rho \mathbf{u},$$

for any small V, assumes its differential *divergence form:*

$$\rho_t + \nabla \cdot \rho \mathbf{u} = 0. \tag{3.3.7}$$

PART 2

Rigid Body Rotations

Chapter 4

Angular Momentum Equilibrium

4.1. A point pendulum

Let us now turn to the *torque* of Archimedes in its simple realization.

Given the reference frames $\mathbf{i}^0 = \mathbf{i}_0$, $\mathbf{j}^0 = \mathbf{j}_0$, $\mathbf{k}^0 = \mathbf{k}_0$ with fixed and moving $\mathbf{i}, \mathbf{j}, \mathbf{k}$, we shall relate the latter firmly,

$$\mathbf{i}\cos\varphi - \mathbf{j}\sin\varphi = \mathbf{i}_0, \quad \mathbf{j}\cos\varphi + \mathbf{i}\sin\varphi = \mathbf{j}_0 \quad \text{and}$$
$$\mathbf{k} = \mathbf{k}_0 \quad \text{for } \varphi = \varphi(t),$$

to a pendulum of length l and mass m in a point

$$\mathbf{r} = l\mathbf{i},$$

depicted in Fig. 4.1.1.

Then, at every instant of time $t > 0$, while rotating with an angular velocity, linear velocity, and acceleration around a fixed point $\mathbf{0}$,

$$\boldsymbol{\Omega} = \Omega\mathbf{k}, \quad \Omega = \varphi_t, \quad \mathbf{u} = \boldsymbol{\Omega} \times \mathbf{r} = \Omega l\mathbf{j} \quad \text{and} \quad \mathbf{u}_t = \boldsymbol{\Omega}_t \times \mathbf{r}$$

under Archimedean torque $\mathbf{r} \times m\mathbf{g}$ applied with force $m\mathbf{g}$ of constant gravity acceleration $\mathbf{g} = g\mathbf{i}_0$, $g = 9.8\,m/s^2$, to the *fulcrum* $\mathbf{0}$ through the *arm* $\mathbf{r}$, the pendulum realizes the following unexpected *angular generalization* [Euler, 1765]:

$$\mathbf{r} \times m\frac{d\mathbf{u}}{dt} = \mathbf{r} \times m\mathbf{g} \quad \text{for } \frac{d\mathbf{u}}{dt} = \mathbf{u}_t + \mathbf{u}\cdot\nabla\mathbf{u} \quad \text{and} \quad \mathbf{u} = \boldsymbol{\Omega} \times \mathbf{r}$$

with the *total acceleration* $d\mathbf{u}/dt$ as the sum of the *local* acceleration in a void, $\mathbf{u}_t$, and the *convective* one in a smooth medium, $(\mathbf{u}\cdot\nabla)\,\mathbf{u}$,

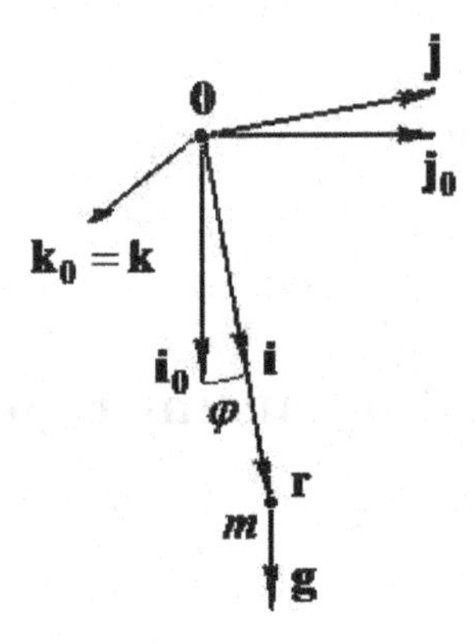

Figure 4.1.1. Point pendulum of mass m.

provided that the medium in hand is a rigid body of a pendulum rotating with velocity $\mathbf{u} = \boldsymbol{\Omega} \times \mathbf{r}$, so

$$(\mathbf{u} \cdot \nabla)\, \mathbf{u} = u\mathbf{u}_x + \cdots = u\boldsymbol{\Omega} \times \mathbf{r}_x + \cdots = \boldsymbol{\Omega} \times u\mathbf{i} + \cdots = \boldsymbol{\Omega} \times \mathbf{u},$$

and the *specific inertia moment*

$$\vec{\mathbf{a}} = \mathbf{r} \cdot \mathbf{r}\vec{\mathbf{e}} - \mathbf{r}\mathbf{r}, \quad \text{or} \quad \vec{\mathbf{a}} = l^2\,(\mathbf{j}\mathbf{j} + \mathbf{k}\mathbf{k}) \quad \text{for } \mathbf{r} = l\mathbf{i}$$

had come into existence to produce the *specific angular momentum*

$$\mathbf{l} = \mathbf{r} \times \mathbf{u} = \mathbf{r} \times (\boldsymbol{\Omega} \times \mathbf{r}) = (\mathbf{r} \cdot \mathbf{r})\, \boldsymbol{\Omega} - \mathbf{r}\mathbf{r} \cdot \boldsymbol{\Omega} = (\mathbf{r} \cdot \mathbf{r})\, \vec{\mathbf{e}} \cdot \boldsymbol{\Omega} - \mathbf{r}\mathbf{r} \cdot \boldsymbol{\Omega} = \vec{\mathbf{a}} \cdot \boldsymbol{\Omega}$$

and to give rise to

$$\mathbf{r} \times (\mathbf{u} \cdot \nabla)\, \mathbf{u} = \mathbf{r} \times (\boldsymbol{\Omega} \times \mathbf{u}) = \boldsymbol{\Omega}\,(\mathbf{r} \cdot \mathbf{u}) - \mathbf{u}\,(\mathbf{r} \cdot \boldsymbol{\Omega}) = -\boldsymbol{\Omega} \times \mathbf{r}\mathbf{r} \cdot \boldsymbol{\Omega},$$

or

$$\begin{aligned} \mathbf{r} \times (\mathbf{u} \cdot \nabla)\mathbf{u} &= -\boldsymbol{\Omega} \times \mathbf{r}\mathbf{r} \cdot \boldsymbol{\Omega} = \boldsymbol{\Omega} \times ((\mathbf{r} \cdot \mathbf{r})\boldsymbol{\Omega} - \mathbf{r}\mathbf{r} \cdot \boldsymbol{\Omega}) \\ &= \boldsymbol{\Omega} \times \vec{\mathbf{a}} \cdot \boldsymbol{\Omega} = \boldsymbol{\Omega} \times \mathbf{l}, \end{aligned}$$

which results in

$$\mathbf{r} \times \frac{d\mathbf{u}}{dt} = \mathbf{r} \times \mathbf{u}_t + \mathbf{r} \times (\mathbf{u} \cdot \nabla)\, \mathbf{u} = \mathbf{l}_t + \boldsymbol{\Omega} \times \mathbf{l} = \mathbf{r} \times \mathbf{g}$$

and for

$$\mathbf{l}_t = \vec{\mathbf{a}} \cdot \mathbf{k}\varphi_{tt} = l^2\varphi_{tt}, \quad \boldsymbol{\Omega} \times \mathbf{l} = \varphi_t^2\mathbf{k} \times \vec{\mathbf{a}} \cdot \mathbf{k} = 0 \quad \text{and}$$

$$\mathbf{r} \times \mathbf{g} = lg\mathbf{i} \times \mathbf{i}_0 = lg\mathbf{i} \times (\mathbf{i}\cos\varphi - \mathbf{j}\sin\varphi) = -\mathbf{k}lg\sin\varphi$$

leads to the required *pendulum equation*

$$\varphi_{tt} = -\frac{g}{l}\sin\varphi.$$

In the next section, following Euler [1765], we shall do the same with a rigid body rotating around a fulcrum.

4.2. The ideal top, or Euler's dynamic equations

For this purpose, first let us return to Theorem 2.4.1 that reveals the *rotations* $\mathbf{u} = \mathbf{\Omega} \times \mathbf{r}$ themselves for a homogeneous strain

$$\mathbf{u_r} = \mathbf{\Omega} \times \mathbf{r_r} = \mathbf{\Omega} \times \vec{\mathbf{e}} = \vec{\mathbf{\Omega}}(t).$$

When materialized into a rigid body $V = \{\mathbf{r}\} = V^t$ of constant density $\rho = \text{const} > 0$ (holding the mass conservation $\rho_t + \nabla\rho \cdot \mathbf{u} = 0$) and fixed point $\mathbf{r} = \mathbf{0}$, the incompressible smooth medium makes up what is usually referred to as a *top* V rotated with its *arms* $\mathbf{r} \in V$ around a *fulcrum* $\mathbf{0}$.

As mentioned above, such a materialization with become possible only when the *translational dynamic equilibrium* between

$$\text{rate } \frac{d}{dt}\overline{\mathbf{u}} \quad \text{of momentum flow } \overline{\mathbf{u}} = \int_V \mathbf{u}\rho dV$$

for a *homogeneous* velocity $\mathbf{u} = \mathbf{u}(t)$ and

$$\text{force } \overline{\mathbf{g}} = \int_V \mathbf{g}(t, \mathbf{r})\,\rho dV \quad \text{applied to points } \mathbf{r} \in V$$

with prescribed body *accelerations* $\mathbf{g} = \mathbf{g}(t, \mathbf{r})$ is transformed into the *angular dynamic equilibrium* between

$$\text{rate } \frac{d}{dt}\overline{\mathbf{r} \times \mathbf{u}} \quad \text{of angular momentum flow } \overline{\mathbf{r} \times \mathbf{u}} = \int_V \mathbf{r} \times \mathbf{u}\rho dV$$

for *linear* and *angular* velocities

$$\mathbf{u} = \mathbf{\Omega} \times \mathbf{r} \quad \text{and} \quad \mathbf{\Omega} = \mathbf{\Omega}(t) = \mathbf{i}\Omega^x + \mathbf{j}\Omega^y + \mathbf{k}\Omega^z, \tag{4.2.1}$$

and

torque $\overline{\mathbf{r}\times\mathbf{g}}$ applied to fulcrum $\mathbf{0}$ with forces $\mathbf{g}\rho dV$

through the arms $\mathbf{r}\in V$.

In other words, both the initial Newton's*translational* [Newton, 1687] and further Euler's *rotational* [Euler, 1765] dynamic laws,

$$\frac{d}{dt}\int_V \mathbf{u}\rho dV = \int_V \mathbf{g}\rho dV, \quad \text{or} \quad \frac{d}{dt}\overline{\mathbf{u}} = \overline{\mathbf{g}}, \quad \text{for } \mathbf{u} = \mathbf{u}(t)$$
$$\text{and} \quad \mathbf{u}_{\mathbf{r}} = \vec{\mathbf{0}}, \tag{4.2.2}$$

or

$$\frac{d}{dt}\int_V \mathbf{r}\times\mathbf{u}\rho dV = \int_V \mathbf{r}\times\mathbf{g}\rho dV, \quad \text{or} \quad \frac{d}{dt}\overline{\mathbf{r}\times\mathbf{u}} = \overline{\mathbf{r}\times\mathbf{g}},$$
$$\text{for } \mathbf{u} = \boldsymbol{\Omega}\times\mathbf{r} \quad \text{and} \quad \mathbf{u}_{\mathbf{r}} = \vec{\boldsymbol{\Omega}} = \boldsymbol{\Omega}\times\vec{\mathbf{e}} = \vec{\mathbf{e}}\times\boldsymbol{\Omega}, \quad \boldsymbol{\Omega} = \boldsymbol{\Omega}(t), \tag{4.2.3}$$

have to be valued for any rigid body V.

As this takes place, Equation (4.2.3) can be rewritten as follows.

Theorem 4.2.1 (Euler's form of the top equation). *Given the torque*

$$\mathbf{M} = \mathbf{i}M^x + \mathbf{j}M^y + \mathbf{k}M^z = \overline{\mathbf{r}\times\mathbf{g}} = \mathbf{i}\overline{yg^z - zg^y}$$
$$+\mathbf{j}\overline{zg^x - xg^z} + \mathbf{k}\overline{xg^y - yg^x},$$

moment of inertia, or inertia

$$\vec{\mathbf{A}} = \overline{\vec{\mathbf{a}}} = \int_V \vec{\mathbf{a}}\rho dV = \begin{pmatrix} \overline{y^2}+\overline{z^2} & -\overline{xy} & -\overline{xz} \\ -\overline{xy} & \overline{z^2}+\overline{x^2} & -\overline{yz} \\ -\overline{xz} & -\overline{yz} & \overline{x^2}+\overline{y^2} \end{pmatrix},$$
$$\vec{\mathbf{a}} = \mathbf{r}\cdot r\vec{\mathbf{e}} - \mathbf{r}\mathbf{r}, \quad \vec{\mathbf{A}}_t = \vec{\mathbf{0}},$$

and angular momentum

$$\mathbf{J} = \vec{\mathbf{A}}\cdot\boldsymbol{\Omega}, \quad \mathbf{J}_t = \vec{\mathbf{A}}\cdot\boldsymbol{\Omega}_t, \quad \boldsymbol{\Omega} = \boldsymbol{\Omega}(t),$$

the ideal top equation (4.2.3) *takes Euler's form:*

$$\mathbf{J}_t + \boldsymbol{\Omega} \times \mathbf{J} = \mathbf{M}, \quad or \quad \vec{\boldsymbol{A}} \cdot \boldsymbol{\Omega}_t + \boldsymbol{\Omega} \times \vec{\boldsymbol{A}} \cdot \boldsymbol{\Omega} = \mathbf{M}. \qquad (4.2.4)$$

Proof. Really, taking the rotating arm $\mathbf{r}$ of the top $V = V^t$ to be evidently at rest respectively to V, or independent of time t,

$$\frac{d\mathbf{r}}{dt} = \mathbf{r}_t + (\mathbf{u} \cdot \nabla)\mathbf{r} = (\mathbf{u} \cdot \nabla)\mathbf{r} = u\mathbf{r}_x + v\mathbf{r}_y + w\mathbf{r}_z$$
$$= u\mathbf{i} + v\mathbf{j} + w\mathbf{k} = \mathbf{u} = \boldsymbol{\Omega} \times \mathbf{r},$$

hence,

$$\frac{d\mathbf{r}}{dt} \times \mathbf{u} = \mathbf{u} \times \mathbf{u} = \mathbf{0} \quad \text{in } V = V^t,$$

we have the rate of angular momentum flow,

$$\frac{d}{dt}\int_V \mathbf{r} \times \mathbf{u}\rho dV = \int_V \left(\frac{d}{dt}\mathbf{r} \times \mathbf{u}\right) \rho dV,$$

equal to the volume flow of angular momentum rates,

$$\int_V \left(\frac{d}{dt}\mathbf{r} \times \mathbf{u}\right) \rho dV = \int_V \mathbf{r} \times \frac{d\mathbf{u}}{dt}\rho dV$$
$$= \int_V \mathbf{r} \times \mathbf{u}_t \rho dV + \int_V \mathbf{r} \times (\mathbf{u} \cdot \nabla)\mathbf{u}\rho dV,$$

in which torques of relative and fluid accelerations,

$$\mathbf{u}_t = \boldsymbol{\Omega}_t \times \mathbf{r} \quad \text{for } \mathbf{u} = \boldsymbol{\Omega} \times \mathbf{r} \quad \text{and} \quad \boldsymbol{\Omega} = \boldsymbol{\Omega}(t),$$

and

$$(\mathbf{u} \cdot \nabla)\mathbf{u} = u\mathbf{u}_x + \cdots = u\boldsymbol{\Omega} \times \mathbf{r}_x + \cdots = \boldsymbol{\Omega} \times u\mathbf{i} + \cdots = \boldsymbol{\Omega} \times \mathbf{u},$$

prove to be

$$\mathbf{r} \times \mathbf{u}_t = \mathbf{r} \times (\boldsymbol{\Omega}_t \times \mathbf{r}) = |\mathbf{r}|^2\boldsymbol{\Omega}_t - \mathbf{r}\mathbf{r} \cdot \boldsymbol{\Omega}_t = \vec{\mathbf{a}} \cdot \boldsymbol{\Omega}_t, \quad \vec{\mathbf{a}} = |\mathbf{r}|^2\vec{\mathbf{e}} - \mathbf{r}\mathbf{r},$$

and

$$\mathbf{r} \times (\mathbf{u} \cdot \nabla)\mathbf{u} = \mathbf{r} \times (\boldsymbol{\Omega} \times \mathbf{u}) = \boldsymbol{\Omega}(\mathbf{r} \cdot \mathbf{u}) - \mathbf{u}(\mathbf{r} \cdot \boldsymbol{\Omega}) = -\boldsymbol{\Omega} \times \mathbf{r}\mathbf{r} \cdot \boldsymbol{\Omega},$$

or

$$\mathbf{r} \times (\mathbf{u} \cdot \nabla)\,\mathbf{u} = -\boldsymbol{\Omega} \times \mathbf{rr} \cdot \boldsymbol{\Omega} = \boldsymbol{\Omega} \times ((\mathbf{r} \cdot \mathbf{r})\,\boldsymbol{\Omega} - \mathbf{rr} \cdot \boldsymbol{\Omega}) = \boldsymbol{\Omega} \times \vec{\mathbf{a}} \cdot \boldsymbol{\Omega},$$

so

$$\begin{aligned}\frac{d}{dt}\int_V \mathbf{r} \times \mathbf{u}\rho dV &= \int_V \vec{\mathbf{a}} \cdot \boldsymbol{\Omega}_t \rho dV + \int_V \boldsymbol{\Omega} \times \vec{\mathbf{a}} \cdot \boldsymbol{\Omega}\rho dV \\ &= \vec{\mathbf{A}} \cdot \boldsymbol{\Omega}_t + \boldsymbol{\Omega} \times \vec{\mathbf{A}} \cdot \boldsymbol{\Omega}, \quad \vec{\mathbf{A}} = \int_V \vec{\mathbf{a}}\rho dV,\end{aligned}$$

which concludes the proof. □

4.3. Inertia forms

As is usual, from the general symmetric inertia $\vec{\mathbf{A}} = \vec{\mathbf{A}}^*$ consisting of components, or *moments* $\overline{x^2 + y^2} = \overline{x^2} + \overline{y^2}, -\overline{xy}, \ldots$ as satisfying the evident conditions $\overline{x^2}, \ldots, \geq 0$ and the Cushy–Bunyakovsky inequalities $\overline{x^2y^2} \geq \overline{xy}^2, \ldots$, it is natural to turn to the *generic* $\vec{\mathbf{A}}$, or to the *typical rigid body* V to be a top (4.2.4), which moments are *strictly positive* as

$$\overline{x^2}, \overline{y^2}, \overline{z^2} > 0 \quad \text{and} \quad \overline{x^2y^2} > \overline{xy}^2, \ldots. \tag{4.3.1}$$

On the other hand, we have that

Theorem 4.3.1 (on principal moments). *As a rigid body* V*, a top* (4.2.4) *is typical to be of strictly positive moments* (4.3.1) *if and only if its inertia* $\vec{\boldsymbol{A}}$ *is definite* (*positively*) *to satisfy conditions*

$$\mathbf{a} \cdot \vec{\mathbf{A}} \cdot \mathbf{a} > 0 \quad \textit{and} \quad \mathbf{a} \cdot \vec{\mathbf{A}} \cdot \mathbf{b} = \mathbf{b} \cdot \vec{\mathbf{A}} \cdot \mathbf{a} \ \textit{for any vectors}\ \mathbf{a}, \mathbf{b} \neq \mathbf{0}. \tag{4.3.2}$$

Proof. Really, in this case, all the principal minors $\Delta_{1,2,3}$ of $\vec{\mathbf{A}}$ are positive:

$$\begin{aligned}\Delta_1 = \overline{y^2} + \overline{z^2} > 0, \quad \Delta_2 &= (\overline{y^2} + \overline{z^2})(\overline{z^2} + \overline{x^2}) \\ &\quad - \overline{xy}^2 > \overline{x^2y^2} - \overline{xy}^2 > 0\end{aligned}$$

and

$$\Delta_3 = \begin{pmatrix} 2(\overline{y^2}\,\overline{z^2}\,\overline{x^2} - \overline{xy}\,\overline{yz}\,\overline{zx}) + (\overline{x^2}\,\overline{y^2} - \overline{xy}^2)(\overline{x^2} + \overline{y^2}) \\ +(\overline{y^2}\,\overline{y^2} - \overline{yz}^2)(\overline{y^2} + \overline{z^2}) \\ +(\overline{z^2}\,\overline{x^2} - \overline{zx}^2)\,(\overline{z^2} + \overline{x^2}) \end{pmatrix}$$

$$> 2\begin{pmatrix} \overline{x^2 y^2 z^2} \\ -|\overline{xy}|\,|\overline{yz}|\,|\overline{zx}| \end{pmatrix} > 0$$

since

$$3\begin{pmatrix} \overline{x^2}\,\overline{y^2}\,\overline{z^2} \\ -\,|\overline{xy}|\,|\overline{yz}|\,|\overline{zx}| \end{pmatrix} = \overline{x^2}\,\overline{y^2}\,\overline{z^2} - |\overline{xy}|\,|\overline{yz}|\,|\overline{zx}| + \cdots \geq \overline{x^2}\,\overline{y^2}\,\overline{z^2}$$

$$- \sqrt{\overline{x^2}}\sqrt{\overline{y^2}}\,|\overline{yz}|\,\sqrt{\overline{z^2}}\sqrt{\overline{x^2}} + \cdots,$$

and

$$\overline{x^2}\,\overline{y^2}\,\overline{z^2} - \sqrt{\overline{x^2}}\sqrt{\overline{y^2}}\,|\overline{yz}|\,\sqrt{\overline{z^2}}\sqrt{\overline{x^2}} + \cdots$$

$$= \overline{x^2}\sqrt{\overline{y^2}}\sqrt{\overline{z^2}}\left(\sqrt{\overline{y^2}}\sqrt{\overline{z^2}} - |\overline{yz}|\right) + \cdots > 0,$$

which proves the required equivalence of (4.3.1) and (4.3.2) due to the Sylvester criterion. The proof is complete. □

As a consequence, we have that

Theorem 4.3.2 (on inertia orts). *Every typical rigid body* V *rotating around a fixed point* $\mathbf{0}$ *with a generic inertia* $\vec{\mathbf{A}}$ *and at angular and linear velocities* $\boldsymbol{\Omega} = \boldsymbol{\Omega}(t)$ *and* $\mathbf{u} = \boldsymbol{\Omega} \times \mathbf{r}$ *has got its own small, middle, and big inertia orts*

$$\mathbf{i} = \mathbf{i}_0 + \varphi \times \mathbf{i}_0, \quad \mathbf{j} = \mathbf{j}_0 + \varphi \times \mathbf{j}_0 \quad \textit{and}$$

$$\mathbf{k} = \mathbf{k}_0 + \varphi \times \mathbf{k}_0 \;\; \textit{for}\; \varphi(t) = \int_0^t \boldsymbol{\Omega}(s)\, ds, \quad t \geq 0,$$

or eigenvectors of $\vec{A}$ *that belong to the small, middle, and big principal moments* $\lambda_{1,2,3}$*, respectively, as*

$$\vec{A} \cdot \mathbf{i} = \lambda_1 \mathbf{i}, \quad \vec{A} \cdot \mathbf{j} = \lambda_2 \mathbf{j} \quad \textit{and}$$
$$\vec{A} \cdot \mathbf{k} = \lambda_3 \mathbf{k}, \quad 0 < \lambda_1 \leq \lambda_2 \leq \lambda_3, \tag{4.3.3}$$

when rigidly attached to the moving body $V = V^t$ *with rates*

$$\mathbf{i}_t = \mathbf{u_r} \cdot \mathbf{i} = \vec{\Omega} \cdot \mathbf{i} = \mathbf{\Omega} \times \mathbf{i}, \ldots, \quad t > 0, \quad \textit{for } \mathbf{i}|_{t=0} = \mathbf{i}_0, \ldots \tag{4.3.4}$$

Proof. Really, the existence of $\mathbf{i}_0$, $\mathbf{j}_0$, $\mathbf{k}_0$ for a definite symmetric matrix $\vec{\mathbf{A}}$ follows immediately from Theorem 4.3.1. Then

$$\vec{\mathbf{A}} \cdot \mathbf{i} = \vec{\mathbf{A}} \cdot \mathbf{i}_0 + \vec{\mathbf{A}} \cdot \varphi \times \mathbf{i}_0 = \lambda_1 \mathbf{i}_0 - \vec{\mathbf{A}} \cdot \mathbf{i}_0 \times \varphi$$
$$= \lambda_1 \mathbf{i}_0 - \lambda_1 \mathbf{i}_0 \times \varphi = \lambda_1 \mathbf{i}_0 + \lambda_1 \varphi \times \mathbf{i}_0 = \lambda_1 \mathbf{i}, \ldots,$$

which concludes the proof. □

Due to theorems 4.3.1 and 4.3.2, it is evident that

Theorem 4.3.3 (on inertia forms). *Together with principal moments* $\lambda_{1,2,3}$*, inertia orts* $\mathbf{i}, \mathbf{j}, \mathbf{k}$ *and internal moment*

$$\mathbf{\Omega} \times \mathbf{J} = \begin{pmatrix} \mathbf{i}\,(\lambda_3 - \lambda_2)\,\Omega^y \Omega^z \\ +\mathbf{j}\,(\lambda_1 - \lambda_3)\,\Omega^z \Omega^x \\ +\mathbf{k}\,(\lambda_2 - \lambda_1)\,\Omega^x \Omega^y \end{pmatrix}$$
$$\times (\mathbf{J} = \vec{\boldsymbol{A}} \cdot \mathbf{\Omega} = \mathbf{i}\lambda_1 \Omega^x + \mathbf{j}\lambda_2 \Omega^y + \mathbf{k}\lambda_3 \Omega^z), \tag{4.3.5}$$

in Theorems 4.3.1 and 4.3.2, as a rigid body V*, every top acquires a form* V *to be*

(**f1**) *a spheroid* V *for* $\lambda_1 = \lambda_2 = \lambda_3$*,*
(**f2**) *a disk* V *for* $\lambda_1 = \lambda_2 < \lambda_3$*,*
(**f3**) *a spindle* V *for* $\lambda_1 < \lambda_2 = \lambda_3$*,*
(**f4**) *a stone* V *for* $\lambda_1 < \lambda_2 < \lambda_3$*.*

For example, the *coin* as a solid circular cylinder V of height h and diameter d when fixed at its centre $\mathbf{0}$ ($x = y = z = 0$) of mass

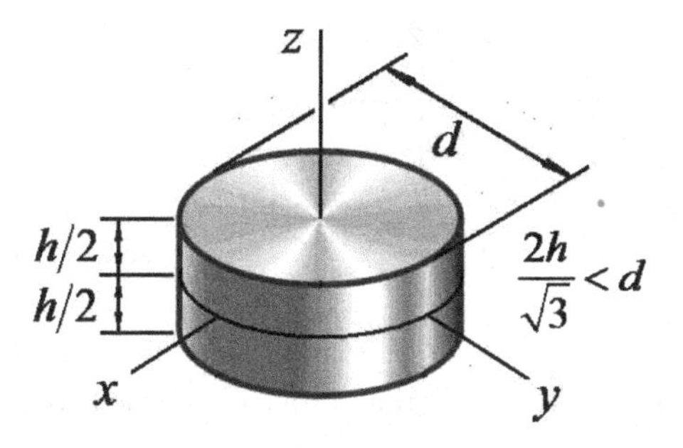

Figure 4.3.1. A coin as a disc **f2**.

$M = \rho\pi d^2 h/4$ in Fig. 4.3.1 is a disk **f2** only for a sufficiently small, or a *subcritical* h:

$$\vec{\mathbf{A}} = \begin{pmatrix} \lambda_1 = M\left(d^2/16 + h^2 12\right) & 0 & 0 \\ 0 & \lambda_2 = \lambda_1 < \lambda_3 & 0 \\ 0 & 0 & \lambda_3 = Md^2/8 \end{pmatrix}$$

for $h < d\sqrt{3}/2$.

The same coin becomes a spindle **f3** when *supercritical,*

$$\vec{\mathbf{A}} = \begin{pmatrix} \lambda_1 = Md^2/8 & 0 & 0 \\ 0 & \lambda_1 < \lambda_2 = \lambda_3 & 0 \\ 0 & 0 & \lambda_3 = M\left(Md^2/16 + Mh^2/12\right) \end{pmatrix}$$

for $h > d\sqrt{3}/2$,

and a spheroid **f1** when *critical,*

$$\vec{\mathbf{A}} = \begin{pmatrix} \lambda_1 = \lambda_2 & 0 & 0 \\ 0 & \lambda_2 = \lambda_3 & 0 \\ 0 & 0 & \lambda_3 = Md^2/8 \end{pmatrix} \quad \text{for } h = d\sqrt{3}/2,$$

but not a stone **f4**.

To become a stone **f4**, a coin has to be slightly deformed into a solid elliptic cylinder V of big and small semi-axes, $d/2$ and $\delta/2 < d/2$, and subcritical height $h < \delta\sqrt{3}/2$, fixed at the same centre of mass $M = \rho\pi d\delta h/4$ with inertia

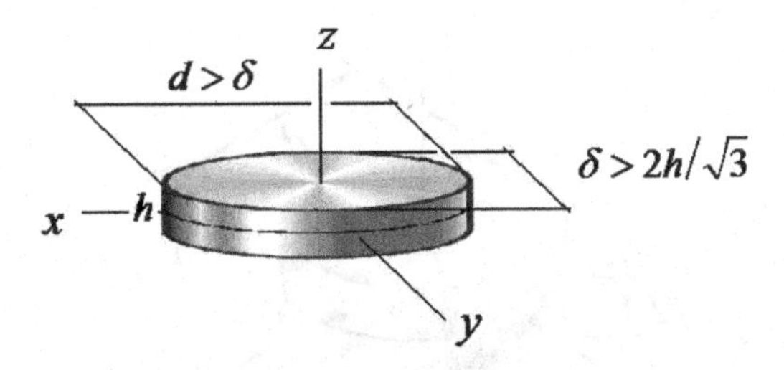

Figure 4.3.2. A stone **f4** as a deformed coin.

$$
\vec{\mathbf{A}} = \begin{pmatrix} \lambda_1 = M\left(\frac{\delta^2}{16} + \frac{h^2}{12}\right) < \lambda_2 & 0 & 0 \\ 0 & \lambda_2 = M\left(\frac{d^2}{16} + \frac{h^2}{12}\right) < \lambda_3 & 0 \\ 0 & 0 & \lambda_3 = M\frac{\delta^2 + d^2}{16} \end{pmatrix}
$$

for $\delta < d$ and $h < \delta\sqrt{3}/2$.

as in Fig. 4.3.2.

Another stone is made up of a rectangular bar V of big height $l > h$, middle width $h > d$, small thickness d, and centered mass M with inertia

$$
\vec{\mathbf{A}} = \begin{pmatrix} \lambda_1 = \frac{M}{12}\left(h^2 + d^2\right) < \lambda_2 & 0 & 0 \\ 0 & \lambda_2 = \frac{M}{12}\left(d^2 + l^2\right) < \lambda_3 & 0 \\ 0 & 0 & \lambda_3 = \frac{M}{12}\left(l^2 + h^2\right) \end{pmatrix},
$$

as in Fig. 4.3.3.

4.4. Polhodes

Following Euler [1765] and [Arnold, 1989], we are interested mainly in the *ideal free top*

$$
\mathbf{J}_t + \mathbf{\Omega} \times \mathbf{J} = \mathbf{0}, \quad \mathbf{J} = \vec{\mathbf{A}} \cdot \mathbf{\Omega}, \quad t > 0, \quad \mathbf{\Omega}|_{t=0} = \mathbf{\Omega}_0, \qquad (4.4.1)
$$

to be the equilibrium (4.2.3), or (4.2.4), free of torque $\mathbf{M} = \mathbf{0}$:

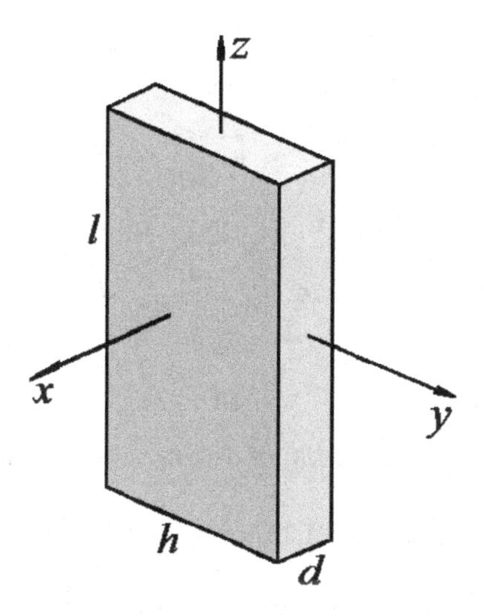

Figure 4.3.3. A stone **f4** as a bar.

Theorem 4.4.1 (on inertia orts as principal rotations). *Each of the inertia orts* $\mathbf{i}, \mathbf{j}, \mathbf{k}$ *is a stationary (as independent of time) solution of Euler's equation* (4.4.1) *to be a principal rotation of the corresponding ideal free top rotating with angular velocity* $\boldsymbol{\Omega} = \mathbf{i}, \mathbf{j}, \mathbf{k}$*, respectively.*

Proof. Really, we have, from (4.3.4) and (4.3.5), that

$$\mathbf{i}_t = \boldsymbol{\Omega} \times \mathbf{i} = \mathbf{I} \times \mathbf{i} = \mathbf{0}, \ldots \quad \text{and} \quad \boldsymbol{\Omega} \times \vec{\mathbf{A}} \cdot \boldsymbol{\Omega} = \lambda_1 \mathbf{i} \times \mathbf{i} = \mathbf{0}, \ldots$$
$$\text{for } \boldsymbol{\Omega} = \mathbf{i}, \ldots, \; t > 0,$$

so (4.4.1) is fulfilled, which concludes the proof. □

In turn, the top (4.4.1) is fully integrated with *kinetic energy* and *energy moment*,

$$K = \frac{1}{2}\boldsymbol{\Omega} \cdot \mathbf{J} \quad \text{and} \quad L = \frac{1}{2}\mathbf{J} \cdot \mathbf{J},$$

respectively, as its *first integrals*, remaining unchanged in time:

$$K_t = \boldsymbol{\Omega} \cdot (\mathbf{J}_t + \boldsymbol{\Omega} \times \mathbf{J}) = 0 \quad \text{and}$$
$$L_t = \mathbf{J} \cdot (\mathbf{J}_t + \boldsymbol{\Omega} \times \mathbf{J}) = 0 \text{ for } t > 0.$$

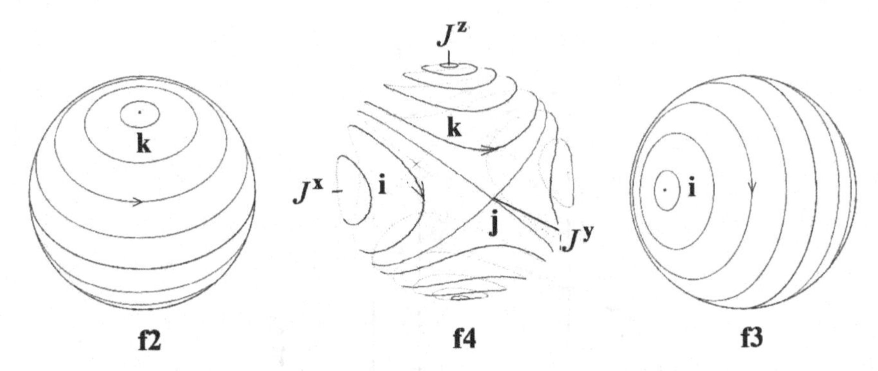

Figure 4.4.1. Polhodes (4.4.3) of disk **f2**, stone **f4**, and spindle **f3**.

The integral curves of (4.4.1), or the so-called *polhodes*

$$\mathbf{J}\,(t) = J^x\mathbf{i} + J^y\mathbf{j} + J^z\mathbf{k}, \quad t \geq 0,$$

are delivered by the common lines (of intersection) of corresponding surfaces

$$\boldsymbol{\Omega} \cdot \mathbf{J} = 2K = \text{const} \quad \text{and} \quad \mathbf{J} \cdot \mathbf{J} = 2L = \text{const}, \tag{4.4.2}$$

or

$$\frac{J^{x2}}{\lambda_1} + \frac{J^{y2}}{\lambda_2} + \frac{J^{z2}}{\lambda_3} = \text{const} \quad \text{and} \quad J^{x2} + J^{y2} + J^{z2} = \text{const},$$

including their *poles*, or *stagnation points*

$$J^x = \lambda_1\mathbf{i}, \quad J^y = \lambda_2\mathbf{j}, \quad \text{and} \quad J^z = \lambda_3\mathbf{k}.$$

As evidenced by integrals (4.4.2) in Fig. 4.4.1, all the polhodes prove to be closed curves. Therefore, the free rotation of the bar around its intermediate, or middle inertia axes y in Fig. 4.3.3 will be surprisingly unstable in that when slightly deprived of the middle principal rotation $\mathbf{j}_0$, any angular velocity vector $\boldsymbol{\Omega} = \boldsymbol{\Omega}\,(t)$ of top (4.4.1) gets, at once, into the nearest polhode that, then, transports $\boldsymbol{\Omega}$ to the opposite middle $-\mathbf{j}_0$, and so on.

This can be illustrated by the well-known bizarre turns of the popular *wing nut of Dzhanibekov* and *tippe top* caused by the same reason (the rotation $\boldsymbol{\Omega}\,(t)$ starts near the middle $\mathbf{j}_0$, or around the intermediate principal axis y, and proceeds along the nearest polhode to $\mathbf{j}_0$) and both result in the same effect (the rotation $\boldsymbol{\Omega}(t)$ turns in time from the initial $\mathbf{j}_0$ to the opposite $-\mathbf{j}_0$).

Such a "freaky" *middle instability* of $\mathbf{j}_0$ is kindly complemented by the alternative "calm" *extremal stabilities* of $\mathbf{k}_0$ and $\mathbf{i}_0$ as rotations around the maximal and minimal principal axes z and x, respectively, as is evident from the *pattern of trajectories* (here, polhodes) in Fig. 4.4.1 and delivered by *Euler's theorem on stability* [Arnold, 1989].

4.5. Principal rotations

So, in line with the Lyapunov stability [Lyapunov, 1892; Pontryagin, 1962], we have that

Theorem 4.5.1 (on principal rotations). *In the case of the ideal free top* (4.4.1), *every linear combination*

$$\begin{aligned}
&\textbf{(f1a)}\ \boldsymbol{\Omega}_* = \Omega^x\mathbf{i} + \Omega^y\mathbf{j} + \Omega^z\mathbf{k}, && \lambda_* = \lambda_1 = \lambda_2 = \lambda_3,\\
&\textbf{(f2a)}\ \boldsymbol{\Omega}_* = \Omega^z\mathbf{k}, && \lambda_* = \lambda_3 > \lambda_1 = \lambda_2,\\
&\textbf{(f3b)}\ \boldsymbol{\Omega}_* = \Omega^x\mathbf{i}, && \lambda_* = \lambda_1 < \lambda_2 = \lambda_3,\\
&\textbf{(f2c)}\ \boldsymbol{\Omega}_* = \Omega^x\mathbf{i} + \Omega^y\mathbf{j}, && \lambda_* = \lambda_1 = \lambda_2 < \lambda_3,\\
&\textbf{(f3d)}\ \boldsymbol{\Omega}_* = \Omega^y\mathbf{j} + \Omega^z\mathbf{k}, && \lambda_* = \lambda_3 = \lambda_2 > \lambda_1,\\
&\textbf{(f4a)}\ \boldsymbol{\Omega}_* = \Omega^z\mathbf{k}, && \lambda_* = \lambda_3 > \lambda_2 > \lambda_1,\\
&\textbf{(f4b)}\ \boldsymbol{\Omega}_* = \Omega^x\mathbf{i}, && \lambda_* = \lambda_1 < \lambda_2 < \lambda_3,
\end{aligned} \qquad (4.5.1)$$

of inertia orts $\mathbf{i}$, $\mathbf{j}$ *or* $\mathbf{k}$, *with arbitrary coefficients* $\Omega^{x,y,z} = \text{const}$ *and either the least or largest principal moment,*

$$\lambda_* = \lambda_1 \quad \textit{or} \quad \lambda_* = \lambda_3,$$

or an ideal extreme rotation $\boldsymbol{\Omega}_*$ *proves to be that of top* (4.4.1), *and what is more, stable by Lyapunov* (*i.e. for every* $\varepsilon > 0$ *there exists* $\delta > 0$ *such that* $|\boldsymbol{\Omega}_0 - \boldsymbol{\Omega}_*| < \delta$ *implies* $|\boldsymbol{\Omega} - \boldsymbol{\Omega}_*| < \varepsilon$ *for all* $t > 0$, *whatever* $\boldsymbol{\Omega}$ *from* (4.4.1)), *whereas the remaining ideal middle rotation*

$$\textbf{(f4c)}\quad \boldsymbol{\Omega}_* = \Omega^y\mathbf{j}, \quad \lambda_1 < \lambda_* = \lambda_2 < \lambda_3, \qquad (4.5.2)$$

of (4.4.1) *is unstable.*

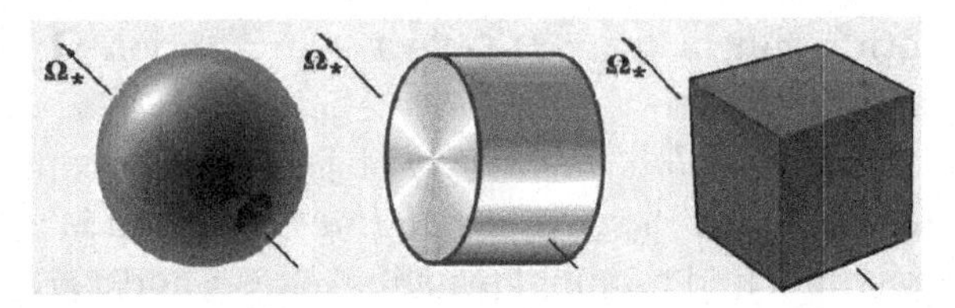

Figure 4.5.1. The spheroid **f1** to be either a rigid ball of diameter d, a solid circular cylinder of height $d\sqrt{3}/2$ and diameter d, or a cube of side d (fixed at their centers of mass).

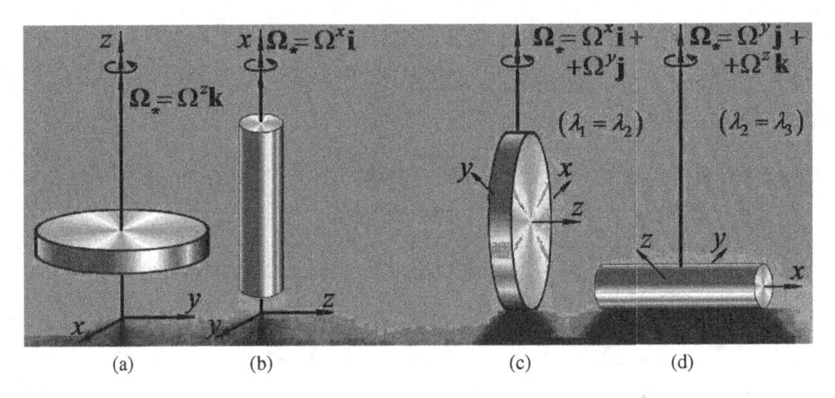

Figure 4.5.2. Basic rotations **f2a**, **f2c** and **f3b**, **f3d** of disk **f2** and spindle **f3**.

Proof. Really, we have in (4.5.1) and (4.5.2) for $\boldsymbol{\Omega}_*$ and $\mathbf{u} = \boldsymbol{\Omega}_* \times \mathbf{r}$ that the internal moment

$$\boldsymbol{\Omega}_* \times \vec{\mathbf{A}} \cdot \boldsymbol{\Omega}_* = \lambda_* \boldsymbol{\Omega}_* \times \boldsymbol{\Omega}_* = \mathbf{0}$$

and the velocity flow deformation

$$\mathbf{u_r} = \vec{\boldsymbol{\Omega}}_* = \boldsymbol{\Omega}_* \times \vec{\mathbf{e}} = \vec{\mathbf{e}} \times \boldsymbol{\Omega}_*, \text{ hence,}$$

$$\boldsymbol{\Omega}_{*t} = \mathbf{u_r} \cdot \boldsymbol{\Omega}_* = \vec{\boldsymbol{\Omega}}_* \cdot \boldsymbol{\Omega}_* = \boldsymbol{\Omega}_* \times \boldsymbol{\Omega}_* = \mathbf{0},$$

so the equilibrium (4.4.1) is fulfilled. The polhodes of Fig. 4.4.1 conclude the proof. □

In detail, the stabilities of basic rotations $\boldsymbol{\Omega}_*$ from (4.5.1) will be discussed later. As to their examples, *every* free stationary rotation

$\boldsymbol{\Omega}_*$ of the spheroid **f1** around its mass centre, as in Fig. 4.5.1, proves to be basic as **f1a** in (4.5.1), and hence, stable.

This is the case with all the basic rotations (around the mass center) **f2a**, **f2c** and **f3b**, **f3d** (of disk **f2** and spindle **f3**, respectively) shown in Fig. 4.5.2.

Chapter 5

Angular Friction

5.1. A dissipative top

Despite being similar in inertia forms **f1**–**f4**, all real tops yet differ from free ideal ones by friction. Indeed, to be in use as *freely working* (routinely) instead of *freely rotating* (ideally), such industrial "nuts and bolts" as lathe or turbine shafts, propeller or extruder screws, drills, rotors, flywheels, bearings, crankshafts, centrifugal governors, and gyroscopes have to account for the unavoidable *friction angular momentum* $\nu\mathbf{F}$ and the necessary *antifriction torque* $\mathbf{M} = \nu\mathbf{T}$, the latter being applied *stationarily* ($\mathbf{T}_t = \mathbf{0}$) and regularly, with the same *friction factor* $\nu = \text{const} > 0$, to balance the former in the *angular equilibrium with friction*

$$\mathbf{J}_t + \nu\mathbf{F} + \boldsymbol{\Omega} \times \mathbf{J} = \nu\mathbf{T}, \quad \mathbf{J} = \vec{\mathbf{A}} \cdot \boldsymbol{\Omega}, \quad \mathbf{T}_t = \mathbf{0},$$

$$t > 0, \quad \boldsymbol{\Omega}|_{t=0} = \boldsymbol{\Omega}_0, \quad \nu = \text{const} > 0,$$

in which the limiting case $\nu \to +0$ is the ideal one (4.4.1).

Generally, *friction* resembles a tangential contact *force* that is acting on a sliding but rubbing surface and that rises in proportion to the applied normal load and in the opposite manner to the surface displacement. So, the *antifriction torque* $-\nu\mathbf{F}$ can be determined with the relation

$$\mathbf{F} = \vec{\mathbf{B}} \cdot \boldsymbol{\Omega}$$

due to a definite symmetric *dissipation* $\vec{\mathbf{B}}$,

$$\mathbf{a} \cdot \vec{\mathbf{B}} \cdot \mathbf{a} > 0 \quad \text{and} \quad \mathbf{a} \cdot \vec{\mathbf{B}} \cdot \mathbf{b} = \mathbf{b} \cdot \vec{\mathbf{B}} \cdot \mathbf{a} \text{ for any } \mathbf{a}, \mathbf{b} \neq \mathbf{0},$$

assumed additionally to be *permutable* with inertia $\vec{\mathbf{A}}$,

$$\vec{\mathbf{A}} \cdot \vec{\mathbf{B}} = \vec{\mathbf{B}} \cdot \vec{\mathbf{A}},$$

as in the case when the friction moment $\nu\mathbf{F}$ proves to be a constant proportion of the angular velocity, the angular momentum, or the moment of the latter,

$$\mathbf{F} = \boldsymbol{\Omega}(\vec{\mathbf{B}} = \vec{\mathbf{e}}), \mathbf{F} = \mathbf{J}(\vec{\mathbf{B}} = \vec{\mathbf{A}}) \quad \text{or} \quad \mathbf{F} = \vec{\mathbf{A}} \cdot \mathbf{J}(\vec{\mathbf{B}} = \vec{\mathbf{A}} \cdot \vec{\mathbf{A}}),$$

respectively (in which the second case seems most acceptable for direct applications in connection with the local property of fiction mentioned initially).

As a result, we have got the *dissipative top*

$$\vec{\mathbf{A}} \cdot \boldsymbol{\Omega}_t + \nu\vec{\mathbf{B}} \cdot \boldsymbol{\Omega} + \boldsymbol{\Omega} \times \vec{\mathbf{A}} \cdot \boldsymbol{\Omega} = \nu\mathbf{T},$$
$$\mathbf{T}_t = \mathbf{0}, \quad t > 0, \quad \boldsymbol{\Omega}|_{t=0} = \boldsymbol{\Omega}_0, \tag{5.1.1}$$

commutative in the sense that its dissipation $\vec{\mathbf{B}}$ is commutable with inertia $\vec{\mathbf{A}}$, and hence, acquires the same inertia orts,

$$\vec{\mathbf{B}} \cdot \mathbf{i} = \eta_1 \mathbf{i}, \ldots, \quad \text{or} \quad \vec{\mathbf{B}} \cdot \mathbf{i} = \mu_1 \vec{\mathbf{A}} \cdot \mathbf{i}, \ldots, \quad \mu_{1,2,3} = \eta_{1,2,3}/\lambda_{1,2,3}, \tag{5.1.2}$$

and the corresponding *affiliated moments* $\mu_{1,2,3} > 0$.

According to the Vyshnegradsky theory of Watt's centrifugal governor [Pontryagin, 1962], friction itself does not destroy but improve its work, making rotations more stable in the sense that circular *polhodes* in Fig. 4.4.1 become *spirals* convergent to the minimal and maximal principal rotations $\mathbf{i}$ and $\mathbf{k}$ *asymptotically* approaching them, as in Fig. 5.1.1.

We shall aim at verifying such an improvement for seven stable basic rotations (4.5.1) taking in to account the following.

Theorem 5.1.1 (on basic rotations with friction). *In addition to Theorem* 4.5.1, *every stationary basic rotation* $\boldsymbol{\Omega}_*$ *from* (4.5.1) *remains the same for the dissipative top* (5.1.1) *whenever the applied*

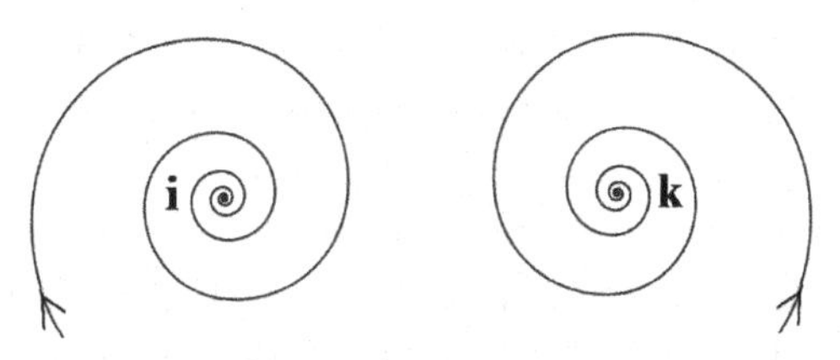

Figure 5.1.1. A local transformation of every circular polhode around a minimum or maximum principal rotation into a spiral asymptotically approaching $\mathbf{i}$ or $\mathbf{k}$, respectively, when the ideal free top, with the friction factor $\nu = 0$, becomes a dissipative one, with $\nu > 0$.

torque $\nu\mathbf{T}$ *balances the friction angular momentum* $\nu\vec{\boldsymbol{B}}\cdot\boldsymbol{\Omega}_*$, *namely,*

$$\boldsymbol{\Omega}_{*t} = \mathbf{0}, \quad \boldsymbol{\Omega}_* \times \vec{\boldsymbol{A}} \cdot \boldsymbol{\Omega}_* = \mathbf{0}$$

and, if

$$\boldsymbol{\Omega}_0 = \boldsymbol{\Omega}_* \quad \text{and} \quad \mathbf{T} = \vec{\boldsymbol{B}} \cdot \boldsymbol{\Omega}_* \quad \text{for } \nu > 0,$$

then

$$\nu\vec{\boldsymbol{B}} \cdot \boldsymbol{\Omega}_* + \boldsymbol{\Omega}_* \times \vec{\boldsymbol{A}} \cdot \boldsymbol{\Omega}_* = \nu\mathbf{T} \text{ for } t > 0 \quad \text{and} \quad \boldsymbol{\Omega}_* = \boldsymbol{\Omega}_0 \text{ for } t \geq 0. \tag{5.1.3}$$

Proof. Really, here, the first and second identities have already been verified earlier, which implies (5.1.3) and, hence, (5.1.1), which concludes the proof. □

Thus, we have the stationary *equilibrium rotation* $\boldsymbol{\Omega}_*$ and its nonstationary *disturbance* $\boldsymbol{\Omega}$ as delivered by the dynamical system (5.1.1). The *deviation*

$$\xi = \boldsymbol{\Omega} - \boldsymbol{\Omega}_*$$

of $\boldsymbol{\Omega}$ from $\boldsymbol{\Omega}_*$ satisfies the *equation in variations*

$$\vec{\mathbf{A}} \cdot \xi_t + \nu\vec{\mathbf{B}} \cdot \xi + \boldsymbol{\Omega}_* \times \vec{\mathbf{A}} \cdot \xi + \xi \times \vec{\mathbf{A}} \cdot \boldsymbol{\Omega}_* + \xi \times \vec{\mathbf{A}} \cdot \xi = \mathbf{0}, \quad t > 0,$$
$$\xi|_{t=0} = \xi_0 = \boldsymbol{\Omega}_0 - \boldsymbol{\Omega}_*, \tag{5.1.4}$$

equivalent to (5.1.1) as the difference between (5.1.1) and (5.1.3).

As we have already seen with patterns of polhodes in Fig. 4.4.1, when free of friction ($\nu = 0$), each equilibrium $\boldsymbol{\Omega}_*$ from (4.5.1) is *stable* by Lyapunov [Lyapunov, 1892] (as in Theorem 4.5.1).

The above-mentioned improvement of friction $\nu > 0$ reduces to the *asymptotic stability* (by Lyapunov [1892]) of $\boldsymbol{\Omega}_*$, or to the additional property of $|\xi| \to 0$ as $t \to \infty$ (depicted in Fig. 5.1.1), left to be verified in the remaining three sections for every *basic rotation* from (4.5.1), namely, for

$$\begin{gathered}\textit{\textbf{even}} \text{ rotation such as that of a spheroid } \mathbf{f1a}, \text{ disk } \mathbf{f2a}, \\ \text{and spindle } \mathbf{f3b}, \\ \textit{\textbf{odd}} \text{ one such as that of a disk } \mathbf{f2c} \text{ and spindle } \mathbf{f3d}, \\ \textit{\textbf{extreme}} \text{ one such as that of a stone } \mathbf{f4a} \text{ and } \mathbf{f4b}.\end{gathered} \tag{5.1.5}$$

We note that they all turn out to be *minimal* or *maximal* rotations in the standard classification [Arnold, 1989] that remain stable for the ideal free top (4.4.1), as was proved first by Euler and evidenced with polhodes in Fig. 4.4.1.

Now, let us return to the dissipative top (5.1.1) and prove the respective stability of its even, odd and extreme basic rotations.

5.2. Even stability

Every *even rotation* from (5.1.5) in (4.5.1) satisfies the *parity identity*

$$\boldsymbol{\Omega}_* \cdot \xi \times \vec{\mathbf{A}} \cdot \xi = 0 \quad \text{for any } \xi. \tag{5.2.1}$$

Really, we have that

$$\xi \times \vec{\mathbf{A}} \cdot \xi = \lambda \xi \times \xi = \mathbf{0} \quad \text{for } \mathbf{f1a},$$

further,

$$\boldsymbol{\Omega}_* \cdot \xi \times \vec{\mathbf{A}} \cdot \xi = \Omega^z \mathbf{k} \cdot \begin{pmatrix} \mathbf{i}\,(\lambda_3 - \lambda_1)\,\xi^y \xi^z \\ +\mathbf{j}\,(\lambda_1 - \lambda_3)\,\xi^z \xi^x \end{pmatrix} = 0 \quad \text{for } \mathbf{f2a}$$

and

$$\boldsymbol{\Omega}_* \cdot \xi \times \vec{\mathbf{A}} \cdot \xi = \Omega^x \mathbf{i} \cdot \begin{pmatrix} \mathbf{j}\,(\lambda_1 - \lambda_3)\,\xi^z \xi^x \\ +\mathbf{k}\,(\lambda_3 - \lambda_1)\,\xi^x \xi^y \end{pmatrix} = 0 \quad \text{for } \mathbf{f3b}.$$

Besides, multiplying (5.1.4) scalar by ξ and using the symmetry of $\vec{\mathbf{A}}$,

$$\xi \cdot \vec{\mathbf{A}} \cdot \xi_t = \xi_t \cdot \vec{\mathbf{A}} \cdot \xi = \frac{1}{2}\left(\xi \cdot \vec{\mathbf{A}} \cdot \xi\right)_t,$$

with the evident volume invariance

$$\xi \cdot \boldsymbol{\Omega}_* \times \vec{\mathbf{A}} \cdot \xi = \boldsymbol{\Omega}_* \cdot \vec{\mathbf{A}} \cdot \xi \times \xi = -\boldsymbol{\Omega}_* \cdot \xi \times \vec{\mathbf{A}} \cdot \xi,$$

and volume degenerations

$$\xi \cdot \xi \times \vec{\mathbf{A}} \cdot \boldsymbol{\Omega}_* = \xi \cdot \xi \times \vec{\mathbf{A}} \cdot \xi = 0,$$

we come to the *energy equation*

$$\frac{1}{2}\left(\xi \cdot \vec{\mathbf{A}} \cdot \xi\right)_t + \nu \xi \cdot \vec{\mathbf{B}} \cdot \xi = \boldsymbol{\Omega}_* \cdot \xi \times \vec{\mathbf{A}} \cdot \xi. \tag{5.2.2}$$

As a consequence, we come to the following.

Theorem 5.2.1 (on the stability of even basic rotations). *Every even basic rotation from* (5.1.5) *of the dissipative top* (4.5.1) *is non-linearly and asymptotically stable in the sense that, whatever the initial disturbance* $\boldsymbol{\Omega}_0$, *the deviation* $\xi = \boldsymbol{\Omega} - \boldsymbol{\Omega}_*$ *exponentially decreases in time as*

$$|\xi| \le |\xi_0| \sqrt{\frac{\lambda_3}{\lambda_1}} e^{-\nu\mu t}, \quad t \ge 0, \quad \text{where } \mu = \min \mu_{1,2,3} \text{ from (5.1.2).} \tag{5.2.3}$$

Proof. Really, with the above parity identity (5.2.1) and positive definiteness

$$\xi \cdot \vec{\mathbf{B}} \cdot \xi \geq \mu \xi \cdot \vec{\mathbf{A}} \cdot \xi,$$

the energy equation (5.2.2) results in the relation

$$\frac{1}{2} e^{-2\nu\mu t} \left(e^{2\nu\mu t} \xi \cdot \vec{\mathbf{A}} \cdot \xi \right)_t = \frac{1}{2} \left(\xi \cdot \vec{\mathbf{A}} \cdot \xi \right)_t + \nu\mu \xi \cdot \vec{\mathbf{A}} \cdot \xi \leq \xi \cdot \vec{\mathbf{A}} \cdot \xi_t + \nu \xi \cdot \vec{\mathbf{B}} \cdot \xi = 0$$

that leads to the required estimate (5.2.3):

$$e^{2\nu\mu t} \lambda_1 |\xi|^2 \leq e^{2\nu\mu t} \xi \cdot \vec{\mathbf{A}} \cdot \xi \leq e^{2\nu\mu t} \xi \cdot \vec{\mathbf{A}} \cdot \xi \Big|_{t=0} = \xi_0 \cdot \vec{\mathbf{A}} \cdot \xi_0 \leq \lambda_3 |\xi_0|^2,$$

which concludes the proof. □

5.3. Odd stability

The remaining odd basic rotations of the *axis-symmetric* top **f2** or **f3** in (5.1.5) are *abnormal*:

$$\mathbf{\Omega}_* \cdot \xi \times \vec{A} \cdot \xi = \begin{pmatrix} \Omega^x \mathbf{i} \\ +\Omega^y \mathbf{j} \end{pmatrix} \cdot \begin{pmatrix} \mathbf{i} (\lambda_3 - \lambda_1) \xi^y \xi^z \\ +\mathbf{j} (\lambda_1 - \lambda_3) \xi^z \xi^x \end{pmatrix} = (\lambda_3 - \lambda_1) \begin{vmatrix} \Omega^x & \xi^x \\ \Omega^y & \xi^y \end{vmatrix} \xi^z \neq 0 \quad \text{for } \mathbf{f2c}$$

when $\xi^x = -\Omega^y$, $\xi^y = \Omega^x$, $\Omega^{x2} + \Omega^{y2}, \xi^z \neq 0$, and

$$\mathbf{\Omega}_* \cdot \xi \times \vec{\mathbf{A}} \cdot \xi = \begin{pmatrix} \Omega^y \mathbf{j} \\ +\Omega^z \mathbf{k} \end{pmatrix} \cdot \begin{pmatrix} \mathbf{j} (\lambda_1 - \lambda_3) \xi^z \xi^x \\ +\mathbf{k} (\lambda_3 - \lambda_1) \xi^x \xi^y \end{pmatrix} = \xi^x (\lambda_3 - \lambda_1) \begin{vmatrix} \Omega^y & \xi^y \\ \Omega^z & \xi^z \end{vmatrix} \neq 0 \quad \text{for } \mathbf{f3d}$$

when $\xi^y = \Omega^z$, $\xi^z = -\Omega^y$, $\Omega^{y2} + \Omega^{z2}, \xi^x \neq 0$.

In this case, while decomposing ξ into the basic and principal rotations as

$$\xi = \xi_* + \xi', \quad \xi_* \cdot \xi' = 0,$$

and

$$\xi_* = \xi^x \mathbf{i} + \xi^y \mathbf{j} \quad \text{and} \quad \xi' = \xi^z \mathbf{k} \quad \text{for } \mathbf{f2c},$$
$$\text{or for } \lambda_1 = \lambda_2 = \lambda_* \quad \text{and} \quad \lambda_3 = \lambda' > \lambda_*,$$

or

$$\xi_* = \xi^y \mathbf{j} + \xi^z \mathbf{k}, \quad \xi' = \xi^x \mathbf{i} \text{ for } \mathbf{f3d},$$
$$\text{or for } \lambda_2 = \lambda_3 = \lambda_* \quad \text{and} \quad \lambda_1 = \lambda' < \lambda_*,$$

respectively, so that

$$\vec{\mathbf{A}} \cdot \xi_* = \lambda_* \xi_*, \quad \vec{\mathbf{A}} \cdot \boldsymbol{\Omega}_* = \lambda_* \boldsymbol{\Omega}_*,$$
$$\vec{\mathbf{A}} \cdot \xi' = \lambda' \xi^z \quad \text{and} \quad \vec{\mathbf{B}} \cdot \xi' = \mu' \vec{\mathbf{A}} \cdot \xi'$$

where

$$\mu' = \mu_3 \text{ for } \mathbf{f2c} \quad \text{or} \quad \mu' = \mu_1 \quad \text{for} \quad \mathbf{f3d},$$

then, multiplying (5.1.4) scalar by ξ' as

$$\xi' \cdot \vec{\mathbf{A}} \cdot \xi_t + \nu \xi' \cdot \vec{\mathbf{B}} \cdot \xi + \xi' \cdot \boldsymbol{\Omega}_* \times \vec{\mathbf{A}} \cdot \xi$$
$$+ \xi' \cdot \xi \times \vec{\mathbf{A}} \cdot \boldsymbol{\Omega}_* + \xi' \cdot \xi \times \vec{\mathbf{A}} \cdot \xi = 0, \quad \text{for } \mathbf{f2c}, \ \mathbf{f3d}$$

and

$$\xi' \cdot \boldsymbol{\Omega}_* \times \vec{\mathbf{A}} \cdot \xi = \xi' \cdot \boldsymbol{\Omega}_* \times \left(\lambda_* \xi_* + \lambda' \xi' \right)$$
$$= \lambda_* \xi' \cdot \boldsymbol{\Omega}_* \times \xi_* = \lambda_* \boldsymbol{\Omega}_* \cdot \xi_* \times \xi',$$
$$\xi' \cdot \xi \times \vec{\mathbf{A}} \cdot \boldsymbol{\Omega}_* = \xi' \cdot \left(\xi_* + \xi' \right) \times \lambda_* \boldsymbol{\Omega}_*$$
$$= \lambda_* \xi' \cdot \xi_* \times \boldsymbol{\Omega}_* = -\lambda_* \boldsymbol{\Omega}_* \cdot \xi_* \times \xi',$$
$$\xi' \cdot \xi \times \vec{\mathbf{A}} \cdot \xi = \lambda_* \xi' \cdot \xi_* \times \xi_* + \lambda' \xi' \cdot \xi_* \times \xi' = \lambda' \xi_* \cdot \xi' \times \xi' = 0,$$

hence,

$$\xi' \cdot \boldsymbol{\Omega}_* \times \vec{\mathbf{A}} \cdot \xi + \xi' \cdot \xi \times \vec{\mathbf{A}} \cdot \boldsymbol{\Omega}_* + \xi\prime \cdot \xi \times \vec{\mathbf{A}} \cdot \xi = 0,$$

and using the symmetry of $\vec{\mathbf{A}}$,

$$\xi' \cdot \vec{\mathbf{A}} \cdot \xi_t = \xi' \cdot \vec{\mathbf{A}} \cdot \xi'_t = \xi'_t \cdot \vec{\mathbf{A}} \cdot \xi' = \frac{1}{2}\left(\xi' \cdot \vec{\mathbf{A}} \cdot \xi'\right)_t,$$

we come to the *residue energy equation*

$$\frac{1}{2}\left(\xi' \cdot \vec{\mathbf{A}} \cdot \xi'\right)_t + \nu\xi' \cdot \vec{\mathbf{B}} \cdot \xi' = 0, \quad \text{or}$$
$$\frac{1}{2}\left(\xi' \cdot \vec{\mathbf{A}} \cdot \xi'\right)_t + \nu\mu'\xi' \cdot \vec{\mathbf{A}} \cdot \xi' = 0. \tag{5.3.1}$$

At that, with identities

$$\xi \cdot \xi \times \vec{\mathbf{A}} \cdot \boldsymbol{\Omega}_* = \xi \cdot \xi \times \vec{\mathbf{A}} \cdot \xi = 0$$

and

$$\boldsymbol{\Omega}_* \cdot \xi \times \vec{\mathbf{A}} \cdot \xi = \boldsymbol{\Omega}_* \cdot (\xi_* + \xi') \times (\lambda_* \xi_* + \lambda' \xi') = (\lambda' - \lambda_*)\, \boldsymbol{\Omega}_* \cdot \xi_* \times \xi',$$

the energy equation (5.2.2) takes the form

$$\frac{1}{2}(\xi \cdot \vec{A} \cdot \xi)_t + \nu\xi \cdot \vec{\mathbf{B}} \cdot \xi = (\lambda_* - \lambda')\Omega_* \cdot \xi_* \times \xi'. \tag{5.3.2}$$

As a result, we have that

Theorem 5.3.1 (on the stability of odd basic rotations). *As either the least basic rotation* **f2c** *of a disk or the largest rotation* **f3d** *of a spindle from* (4.5.1), *every added basic rotation from* (5.1.5) *is a non-linearly and asymptotically stable rotation* Ω_* *of the dissipative top* (4.5.1) *in the sense that the deviation* $\xi = \Omega - \Omega_*$ *exponentially decreases in time for any initial disturbance* Ω_0 *in accordance with the estimate:*

$$|\xi| \leq \left(|\xi_0| + |\xi'_0| \left(\frac{\lambda_3}{\lambda_1} - 1\right) |\Omega_*| \int_0^t e^{-\nu(\mu'-\mu)s} ds\right)$$
$$\times \sqrt{\frac{\lambda_3}{\lambda_1}} e^{-\nu\mu t}, \; t > 0, \; \xi_0 = \Omega_0 - \Omega_*, \tag{5.3.3}$$

where $\xi'_0 = \xi_0^z \mathbf{k}$, $\mu' = \mu_3$ for **f2c** or $\xi'_0 = \xi_0^x \mathbf{i}$, $\mu' = \mu_1$ for **f3d**.

Proof. Really, we have, from (5.3.1) and (5.3.2), that

$$\lambda_1|\xi'|^2 \le \xi' \cdot \vec{\mathbf{A}} \cdot \xi' = \xi'_0 \cdot \vec{\mathbf{A}} \cdot \xi'_0 e^{-2\nu\mu' t}$$
$$\le \lambda_3 |\xi'_0|^2 e^{-2\nu\mu' t}, \quad \text{or} \quad |\xi'| \le |\xi'_0| \sqrt{\frac{\lambda_3}{\lambda_1}} e^{-\nu\mu' t},$$

and

$$\frac{1}{2} e^{-2\nu\mu t} \left(e^{2\nu\mu t} \xi \cdot \vec{\mathbf{A}} \cdot \xi \right)_t$$
$$= \frac{1}{2} \left(\xi \cdot \vec{\mathbf{A}} \cdot \xi \right)_t + \nu\mu \xi \cdot \vec{A} \cdot \xi \le \frac{1}{2} \left(\xi \cdot \vec{\mathbf{A}} \cdot \xi \right)_t + \nu \xi \cdot \vec{\mathbf{B}} \cdot \xi$$
$$= \left(\lambda_* - \lambda' \right) \boldsymbol{\Omega}_* \cdot \xi_* \times \xi' \le \left(\lambda_3 - \lambda_1 \right) |\boldsymbol{\Omega}_*| \, |\xi| \, |\xi'|$$
$$\le \left(\frac{\lambda_3}{\lambda_1} - 1 \right) \sqrt{\lambda_3} \, |\boldsymbol{\Omega}_*| \sqrt{\xi \cdot \vec{\mathbf{A}} \cdot \xi} \, |\xi'_0| \, e^{-\nu\mu' t},$$

respectively; so,

$$\frac{1}{2} \left(\left(e^{\nu\mu t} \sqrt{\xi \cdot \vec{\mathbf{A}} \cdot \xi} \right)^2 \right)_t \le e^{\nu\mu t} \sqrt{\xi \cdot \vec{\mathbf{A}} \cdot \xi} \left(\frac{\lambda_3}{\lambda_1} - 1 \right)$$
$$\times \sqrt{\lambda_3} \, |\xi'_0| \, |\boldsymbol{\Omega}_*| \, e^{-\nu(\mu' - \mu) t},$$

or

$$\left(e^{\nu\mu t} \sqrt{\xi \cdot \vec{\mathbf{A}} \cdot \xi} \right)_t \le \left(\frac{\lambda_3}{\lambda_1} - 1 \right) \sqrt{\lambda_3} \, |\xi'_0| \, |\boldsymbol{\Omega}_*| \, e^{-\nu(\mu' - \mu) t},$$

which results either in

$$e^{\nu\mu t} |\xi| \sqrt{\lambda_1} - |\xi_0| \sqrt{\lambda_3} \le e^{\nu\mu t} \sqrt{\xi \cdot \vec{\mathbf{A}} \cdot \xi} - \sqrt{\xi_0 \cdot \vec{\mathbf{A}} \cdot \xi_0}$$
$$\le \left(\frac{\lambda_3}{\lambda_1} - 1 \right) \sqrt{\lambda_3} \, |\xi'_0| \, |\boldsymbol{\Omega}_*| \int_0^t e^{-\nu(\mu' - \mu) s} ds,$$

or in (5.3.3).

The proof is complete. □

5.4. Extreme stability

In the remaining cases **f4a** and **f4b** in (4.5.1), while multiplying the equation in variations (5.1.4) scalar by the moment $\vec{\mathbf{A}} \cdot \xi$ and accounting for degenerations of volumes

$$(\vec{\mathbf{A}} \cdot \xi) \cdot \boldsymbol{\Omega}_* \times \vec{\mathbf{A}} \cdot \xi = (\vec{\mathbf{A}} \cdot \xi) \cdot \xi \times \vec{\mathbf{A}} \cdot \xi = 0,$$

and volume invariance

$$\left(\vec{\mathbf{A}} \cdot \xi\right) \cdot \xi \times \vec{\mathbf{A}} \cdot \boldsymbol{\Omega}_* = \vec{\mathbf{A}} \cdot \boldsymbol{\Omega}_* \cdot \left(\vec{\mathbf{A}} \cdot \xi\right) \times \xi = -\vec{\mathbf{A}} \cdot \boldsymbol{\Omega}_* \cdot \xi \times \vec{\mathbf{A}} \cdot \xi,$$

we come to the *energy moment equation*

$$\frac{1}{2}\left(\left(\vec{\mathbf{A}} \cdot \xi\right) \cdot \vec{\mathbf{A}} \cdot \xi\right)_t + \nu \left(\vec{\mathbf{A}} \cdot \xi\right) \cdot \vec{\mathbf{B}} \cdot \xi = \vec{\mathbf{A}} \cdot \boldsymbol{\Omega}_* \cdot \xi \times \vec{\mathbf{A}} \cdot \xi. \tag{5.4.1}$$

In turn, when multiplied by

$$\lambda_* = \lambda_3 \quad \text{in } \mathbf{f4a} \quad \text{or} \quad \lambda_* = \lambda_1 \quad \text{in } \mathbf{f4b},$$

energy equation (5.4.1) can be reduced by (5.2.2) to give rise, with

$$\vec{\mathbf{A}} \cdot \boldsymbol{\Omega}_* = \lambda_* \boldsymbol{\Omega}_* \quad \text{for } \boldsymbol{\Omega}_* = \Omega^z \mathbf{k} \quad \text{in } \mathbf{f4a} \quad \text{or} \quad \boldsymbol{\Omega}_* = \Omega^x \mathbf{i} \quad \text{in } \mathbf{f4b},$$

to the following *energy balance*:

$$\frac{1}{2}\left(\left(\vec{\mathbf{A}}^{\mp} \cdot \xi\right) \cdot \vec{\mathbf{A}} \cdot \xi\right)_t + \nu \left(\vec{\mathbf{A}}^{\mp} \cdot \xi\right) \cdot \vec{\mathbf{B}} \cdot \xi = 0 \quad \text{for } \vec{\mathbf{A}}^{\mp} = \mp\left(\vec{\mathbf{A}} - \lambda_* \vec{\mathbf{e}}\right),$$

reduced to that for the orthogonal complement

$$\xi' = \xi^x \mathbf{i} + \xi^y \mathbf{j} \quad \text{in } \mathbf{f4a} \quad \text{or} \quad \xi' = \xi^y \mathbf{j} + \xi^z \mathbf{k} \quad \text{in } \mathbf{f4b},$$

as

$$\begin{aligned} &\frac{1}{2}\left(\left(\vec{\mathbf{A}}^{\mp} \cdot \xi'\right) \cdot \vec{\mathbf{A}} \cdot \xi'\right)_t + \nu \left(\vec{\mathbf{A}}^{\mp} \cdot \xi'\right) \cdot \vec{\mathbf{B}} \cdot \xi' = 0, \\ &\vec{\mathbf{A}}^{\mp} = \mp\left(\vec{\mathbf{A}} - \lambda_* \vec{\mathbf{e}}\right), \end{aligned} \tag{5.4.2}$$

respectively, since

$$\xi = \xi_* + \xi', \quad \vec{\mathbf{A}} \cdot \xi_* = \lambda_* \xi_* \quad \text{and} \quad \xi_* = \xi^z \mathbf{k} \text{ in } \mathbf{f4a} \text{ or } \xi_* = \xi^x \mathbf{i} \text{ in } \mathbf{f4b}.$$

In doing so, vectors $\boldsymbol{\Omega}_*$ and ξ_* remain collinear,

$$\boldsymbol{\Omega}_* \times \xi_* = \mathbf{0},$$

which reduces the right-hand side of (5.2.2) to that depending only on the complement ξ':

$$\boldsymbol{\Omega}_* \cdot \xi \times \vec{\mathbf{A}} \cdot \xi = \boldsymbol{\Omega}_* \cdot (\xi_* + \xi') \times \left(\lambda_* \xi_* + \vec{\mathbf{A}} \cdot \xi'\right) = \boldsymbol{\Omega}_* \cdot \xi' \times \vec{\mathbf{A}} \cdot \xi'.$$

As a result, we come to the following.

Theorem 5.4.1 (on stability of extremal rotation). *As either maximum or minimum rotations,* **f4a** *and* **f4b**, *of a stone from* (4.5.1), *every extreme basic rotation from* (5.1.5) *is a non-linear and asymptotically stable basic rotation* $\boldsymbol{\Omega}_*$ *of a dissipative top* (4.5.1) *in the sense that the deviation* $\xi = \boldsymbol{\Omega} - \boldsymbol{\Omega}_*$ *exponentially decreases in time for any initial disturbance* $\boldsymbol{\Omega}_0$ *in accordance with the estimate:*

$$|\xi|^2 \leq \left(|\xi_0|^2 + 2|\xi'_0|^2 \kappa |\boldsymbol{\Omega}_*| \int_0^t e^{-2\nu(\mu'-\mu)s} ds\right) \frac{\lambda_3}{\lambda_1} e^{-2\nu\mu t},$$

$$t > 0, \quad \xi_0 = \boldsymbol{\Omega}_0 - \boldsymbol{\Omega}_*,$$

$$\text{where} \quad \boldsymbol{\Omega}_* = \Omega_*^z \mathbf{k}, \xi'_0 = \xi_0^x \mathbf{i} + \xi_0^y \mathbf{j},$$

$$\kappa = \frac{(\lambda_3 - \lambda_1)\lambda_2}{(\lambda_3 - \lambda_2)\lambda_1}, \mu' = \min \mu_{1,2}$$

$$\text{for} \quad \mathbf{f4a} \quad \text{or} \quad \boldsymbol{\Omega}_* = \Omega_*^x \mathbf{i}, \xi'_0 = \xi_0^y \mathbf{j} + \xi_0^z \mathbf{k},$$

$$\kappa = \frac{(\lambda_3 - \lambda_1)\lambda_3}{(\lambda_2 - \lambda_1)\lambda_2}, \quad \mu' = \min \mu_{2,3} \quad \text{for} \quad \mathbf{f4b}. \tag{5.4.3}$$

Proof. Really, we have, in the case of **f4a**, that

$$(\lambda_3 - \lambda_2)\lambda_1 |\xi'|^2 \leq (\vec{\mathbf{A}}^- \cdot \xi') \cdot \vec{\mathbf{A}} \cdot \xi' = (\lambda_3 - \lambda_1)\lambda_1 \xi^{x2} + (\lambda_3 - \lambda_2)\lambda_2 \xi^{y2} \leq (\lambda_3 - \lambda_1)\lambda_2 |\xi'|^2$$

or in the case of **f4b** that

$$(\lambda_2 - \lambda_1)\lambda_2 |\xi'|^2 \leq (\vec{A}^+ \cdot \xi') \cdot \vec{\mathbf{A}} \cdot \xi' = (\lambda_2 - \lambda_1)\lambda_2 \xi^{y2} + (\lambda_3 - \lambda_1)\lambda_3 \xi^{z2} \leq (\lambda_3 - \lambda_1)\lambda_3 |\xi'|^2.$$

Then

$$(\vec{\mathbf{A}}^- \cdot \xi') \cdot \vec{\mathbf{B}} \cdot \xi' = \mu_1(\lambda_3 - \lambda_1)\lambda_1 \xi^{x2} + \mu_2(\lambda_3 - \lambda_2)\lambda_2 \xi^{y2}$$
$$\geq \mu'(\vec{\mathbf{A}}^- \cdot \xi') \cdot \vec{\mathbf{A}} \cdot \xi' \quad in \quad \mathbf{f4a}$$

and

$$(\vec{\mathbf{A}}^+ \cdot \xi') \cdot \vec{\mathbf{B}} \cdot \xi' = \mu_2(\lambda_2 - \lambda_1)\lambda_2 \xi^{y2} + \mu_3(\lambda_3 - \lambda_1)\lambda_3 \xi^{z2}$$
$$\geq \mu'(\vec{\mathbf{A}}^+ \cdot \xi') \cdot \vec{\mathbf{A}} \cdot \xi' \text{ in } \mathbf{f4b}.$$

In any case, the energy balance (5.4.2) assumes

$$\frac{1}{2} e^{-2\nu\mu' t}(e^{2\nu\mu' t}(\vec{\mathbf{A}}^\mp \cdot \xi') \cdot \vec{\mathbf{A}} \cdot \xi')_t$$
$$= \frac{1}{2}((\vec{\mathbf{A}}^\mp \cdot \xi') \cdot \vec{\mathbf{A}} \cdot \xi')_t + \nu\mu'(\vec{\mathbf{A}}^\mp \cdot \xi') \cdot \vec{\mathbf{A}} \cdot \xi'$$
$$\leq \frac{1}{2}((\vec{\mathbf{A}}^\mp \cdot \xi') \cdot \vec{\mathbf{A}} \cdot \xi')_t + \nu(\vec{\mathbf{A}}^\mp \cdot \xi') \cdot \vec{\mathbf{B}} \cdot \xi' = 0,$$

which means that

$$e^{2\nu\mu' t}\left(\vec{\mathbf{A}}^\mp \cdot \xi'\right) \cdot \vec{\mathbf{A}} \cdot \xi' \leq e^{2\nu\mu' t}\left(\vec{\mathbf{A}}^\mp \cdot \xi'\right) \cdot \vec{\mathbf{A}} \cdot \xi'\Big|_{t=0} = \left(\vec{\mathbf{A}}^\mp \cdot \xi_0'\right) \cdot \vec{\mathbf{A}} \cdot \xi_0'$$

and leads to

$$(\lambda_3 - \lambda_2)\lambda_1 |\xi'|^2 e^{2\nu\mu' t} \leq e^{2\nu\mu' t}(\vec{\mathbf{A}}^- \cdot \xi') \cdot \vec{\mathbf{A}} \cdot \xi'$$
$$\leq (\vec{\mathbf{A}}^- \cdot \xi_0') \cdot \vec{\mathbf{A}} \cdot \xi_0' \leq (\lambda_3 - \lambda_1)\lambda_2 |\xi_0'|^2$$

and

$$(\lambda_2 - \lambda_1)\lambda_2 |\xi'|^2 e^{2\nu\mu' t} \leq e^{2\nu\mu' t}(\vec{\mathbf{A}}^+ \cdot \xi') \cdot \vec{\mathbf{A}} \cdot \xi'$$
$$\leq (\vec{\mathbf{A}}^+ \cdot \xi'_0) \cdot \vec{\mathbf{A}} \cdot \xi'_0 \leq (\lambda_3 - \lambda_1)\lambda_3 |\xi'_0|^2;$$

so,

$$|\xi'|^2 \leq \kappa |\xi'_0|^2 e^{-2\nu\mu' t}.$$

Returning to the energy equation (5.2.2) with the amendment (5.4.2), we may conclude that

$$
\begin{aligned}
&\frac{1}{2}e^{-2\nu\mu t}(e^{2\nu\mu t}\xi \cdot \vec{\mathbf{A}} \cdot \xi)_t \\
&= \frac{1}{2}(\xi \cdot \vec{\mathbf{A}} \cdot \xi)_t + \nu\mu\xi \cdot \vec{\mathbf{A}} \cdot \xi \leq \frac{1}{2}(\xi \cdot \vec{\mathbf{A}} \cdot \xi)_t + \nu\xi \cdot \vec{\mathbf{B}} \cdot \xi \\
&= \boldsymbol{\Omega}_* \cdot \xi' \times \vec{\mathbf{A}} \cdot \xi' \leq |\boldsymbol{\Omega}_*||\xi'||\vec{\mathbf{A}} \cdot \xi'| \leq |\boldsymbol{\Omega}_*||\xi'|^2\lambda_3 \\
&\leq |\xi_0'|^2\kappa\lambda_3|\boldsymbol{\Omega}_*|e^{-2\nu\mu' t},
\end{aligned}
$$

which results in (5.4.3), concluding the proof. $\square$

PART 3

Conservation Laws

Chapter 6

Fluid and Gas Momentums

6.1. Pressure force

Thus, the Archimedean torque (287–212 BC) had come into existence after the Pythagorean theorem (570–490 BC) to supply the surrounding space of always rotating inertia orts

$$\mathbf{i}, \mathbf{j}, \mathbf{k} = \mathbf{i}^t, \mathbf{j}^t, \mathbf{k}^t = \mathbf{i}(t), \mathbf{j}(t), \mathbf{k}(t), \quad t \geq 0,$$

hence, with the *screw* "$\times$" to be the cross product

$$\mathbf{j} \times \mathbf{k} = -\mathbf{k} \times \mathbf{j} = \mathbf{i}, \quad \mathbf{k} \times \mathbf{i} = -\mathbf{i} \times \mathbf{k} = \mathbf{j},$$
$$\mathbf{i} \times \mathbf{j} = -\mathbf{j} \times \mathbf{i} = \mathbf{k}, \quad \mathbf{i} \times \mathbf{i} = \mathbf{j} \times \mathbf{j} = \mathbf{k} \times \mathbf{k} = \mathbf{0}$$

in addition to the necessary *convolution* "$\cdot$" as the point product

$$\mathbf{i} \cdot \mathbf{j} = \mathbf{j} \cdot \mathbf{i} = \mathbf{j} \cdot \mathbf{k} = \mathbf{k} \cdot \mathbf{j} = \mathbf{k} \cdot \mathbf{i} = \mathbf{i} \cdot \mathbf{k} = 0 \quad \text{and}$$
$$\mathbf{i} \cdot \mathbf{i} = \mathbf{j} \cdot \mathbf{j} = \mathbf{k} \cdot \mathbf{k} = 1.$$

This was long before René Descartes decomposed (1637) the position of a point in space, or the radius-vector

$$\mathbf{r} = x\mathbf{i} + y\mathbf{j} + z\mathbf{k}, \quad -\infty < x, y, z < \infty,$$

into coordinate proportions of orts $\mathbf{i}, \mathbf{j}, \mathbf{k}$, which was subsequently found to agree with partial derivatives as

$$\mathbf{r}_x = \partial_x \mathbf{r} = \partial \mathbf{r}/\partial x = \mathbf{i}, \quad \mathbf{r}_y = \mathbf{j}, \quad \text{and} \quad \mathbf{r}_z = \mathbf{k}$$

to give rise to the *gradient* and *coordinate strain*,

$$\nabla = \mathbf{i}\partial_x + \mathbf{j}\partial_y + \mathbf{k}\partial_z \quad \text{and} \quad \partial_{\mathbf{r}} = \partial_x \mathbf{i} + \partial_y \mathbf{j} + \partial_z \mathbf{k},$$

$$v_{\mathbf{r}} \cdot \mathbf{u} = \mathbf{u} \cdot \nabla v,$$

Newton determined the *relative velocity*,

$$\mathbf{u} = u\mathbf{i} + v\mathbf{j} + w\mathbf{k} = \mathbf{r}_t = \lim_{t' \to t+0} \frac{\mathbf{r}' - \mathbf{r}}{t' - t} \quad (t' > t),$$

$$\text{for } \mathbf{i}, \mathbf{j}, \mathbf{k} = \mathbf{i}^0, \mathbf{j}^0, \mathbf{k}^0,$$

ant *acceleration* $\mathbf{u}_t$ of an *isolated* particle $\mathbf{r}$ as of a point in the void, and Euler found out the *convective velocity*

$$\mathbf{u} \cdot \nabla \mathbf{r} = u\mathbf{r}_x + v\mathbf{r}_y + w\mathbf{r}_z = u\mathbf{i} + v\mathbf{j} + w\mathbf{k} = \mathbf{u},$$

and *acceleration*

$$u\mathbf{i}_t + v\mathbf{j}_t + w\mathbf{k}_t = u\mathbf{u}_x + v\mathbf{u}_y + w\mathbf{u}_z = \mathbf{u} \cdot \nabla \mathbf{u} = \mathbf{u}_{\mathbf{r}} \cdot \mathbf{u},$$

produced with deformation rates of orts

$$\mathbf{i}_t = \mathbf{u}_x, \quad \mathbf{j}_t = \mathbf{u}_y \quad \text{and} \quad \mathbf{k}_t = \mathbf{u}_z,$$

and matrix of fluid deformations

$$\mathbf{u}_{\mathbf{r}} = \mathbf{u}_x \mathbf{i} + \mathbf{u}_y \mathbf{j} + \mathbf{u}_z \mathbf{k}, \quad \text{or} \quad \nabla \mathbf{u} = \mathbf{i}\mathbf{u}_x + \mathbf{j}\mathbf{u}_y + \mathbf{k}\mathbf{u}_z,$$

to make up the *full acceleration*

$$d_t \mathbf{u} = \frac{d\mathbf{u}}{dt} = \mathbf{u}_t + \mathbf{u} \cdot \nabla \mathbf{u}$$

produced in water or air with Archimedean *buoyancy*, or *buoyant force* **A** as follows.

Contrary to the torque, the force **A** was found accidentally and for reasons that are far from scientific. It is known only that the city Syracuse (located in the southeast corner of the Sicily island in the Mediterranean Sea) where Archimedes lived was a state governed by the Greek king Hiero the Second whose close relative was the astronomer Phidias, the father of Archimedes. Therefore, it is not surprising that one day, Hiero asked him to resolve some doubts and to establish the true share of silver in his new golden crown.

Currently, this problem could be resolved easily in any jewelry pawnshop. One needs to just drop a chemical reagent of the desired sample, say, the 750th or 585th test of gold, onto the clean metal surface of a jewelry product and one can conclude that the weight of pure gold in it does not make up the required 0.750 or 0.585 part of the total weight of its alloy with silver if an undesirable chemical reaction such as reagent boiling takes place.

Of course, Archimedes did not have the desired test. However, the absence of such a test should not have embarrassed him since silver is lighter than gold, and therefore an alloy of these two metals would be larger in volume than an ingot of pure gold (of the 999th test) equal in weight to the former.

As to the volume of a rigid body of any form, it has long been known to measure it with the amount of fluid displaced by the body. The difference in the volumes of two equal weights (of the gold–silver alloy and the ingot of pure gold) would help to determine the required share of silver in the alloy.

Meanwhile, according to the legend, Archimedes examined this seemingly simple solution for a long time. The reason was that the weight mysteriously lost by an immersed body had to be not only explained but also measured. The weight reduction, or the buoyant force **A**, was found finally by Archimedes to be equal to the weight of the liquid *dislodged* (or pushed out) by the immersed body, which made up subsequently the first equation of fluid dynamics.

But where does the force **A** come from?

The centuries-old experience of housewives tells us about the nature of buoyancy. To pick up a light-weight garbage from the flour in a bowl, one needs to just shake the bowl constantly. Then the required force **A** is created to push out garbage particles from the deep to the surface.

In other words, the flour when shaken is similar to water owing to the small-scale pulsations of its constituents. Like the visible particles of the shaken flour, the invisible molecules of water are also constantly pulsating and forcing some far larger (and distinguishable in a microscope) particles of pollen to move when suspended in water, which the Scotland botanist Robert Brown first noted in 1827

[Brown, 1828]. It is the molecular pulsations that deliver the pressure to water in the form of the buoyant force **A** of Archimedes.

As to the pressure of the surrounding air, or the *atmosphere*, as is well known, its buoyant force **A** was first measured by Evangelista Torricelli and Vincenzo Viviani, who found out that the air, like water, presses down although with a lesser force: "We all are living submerged on the bottom of an air ocean" (*Noi viviamo sommersi nel fondo d'un pelago d'aria*), as Torricelli wrote down on June 11, 1644, to his friend mathematician (and the future Cardinal) Michelangelo Ricci to tell him about what he had discovered in fact. It was the *mechanical identity* of water and air that was the very beginning of the future mechanics of continuous media, or *fluid dynamics* [Batchelor, 1967].

In comparison with water, the atmospheric pressure force was found to be small but sufficient to hold a column of mercury with a height of about 760 millimeters inside a sealed-up vertical, empty (without air) glass tube of Torricelli's *barometer*, which scale we still use.

After repeating the experiments of Torricelli and Viviani, Blaise Pascal found finally [Pascal, 1648] an unexpected *explanation for the buoyant force* **A** in water or air: it is directed upwards since the medium presses downwards. In other words, a continuous medium pushes its volumes in all directions without a change, or the corresponding pressure stress $\vec{\mathbf{p}}$ is *homogeneous* and *isotropic*: $\vec{\mathbf{p}} = p\vec{\mathbf{e}}$.

Accounting for this, we can compute the force **A**.

For this purpose, let us approximate a small part of a boundary ∂V of a volume V around a point $P \in \partial V$ by the piece of plane that touches ∂V at P and equals the part by the area dS. Then, we direct a unit *external normal*, or a vector (an arrow) of length 1 from the point P beyond the volume V and take its proportion of dS to obtain the familiar area vector:

$$d\mathbf{S} = \mathbf{ii} \cdot d\mathbf{S} + \mathbf{jj} \cdot d\mathbf{S} + \mathbf{kk} \cdot d\mathbf{S}, \quad dS = \sqrt{d\mathbf{S} \cdot d\mathbf{S}}.$$

According to the above reasons for Brownian motion and Pascal's explanation of the buoyant force, molecules of V produce the *isotropic pressure field*

$$\vec{\mathbf{p}} = p\vec{\mathbf{e}}, \quad \vec{\mathbf{e}} = \mathbf{ii} + \mathbf{jj} + \mathbf{kk} = \begin{pmatrix} 1 & 0 & 0 \\ 0 & 1 & 0 \\ 0 & 0 & 1 \end{pmatrix},$$

in which the *stresses* $p\mathbf{i}$, $p\mathbf{j}$, and $p\mathbf{k}$, as the forces pushing the areas $\mathbf{i} \cdot d\mathbf{S}$, $\mathbf{j} \cdot d\mathbf{S}$, and $\mathbf{k} \cdot d\mathbf{S}$, respectively, act from within the volume V on a small area dS of the vector $d\mathbf{S}$ in the point P with the force

$$p\vec{\mathbf{e}} \cdot d\mathbf{S} = pd\mathbf{S}.$$

At the same moment, to keep the equilibrium of the area dS, all the nearest molecules of the surrounding medium beyond the volume V push the molecules into V with the *reactive* (or opposite) force in the form of a surface, or*contact element*

$$d\mathbf{A} = -pd\mathbf{S}.$$

So, as the sum

$$\mathbf{A} = \int_{\partial V} d\mathbf{A} = -\int_{\partial V} pd\mathbf{S}$$

of such reactive elements is a Riemannian integral taking over all the boundary points $P \in \partial V$, the required buoyant force of Archimedes $\mathbf{A}$ comes into existence as in Fig. 6.1.1, to be reduced in future to *volume elements*

$$-(\nabla p)dV = -(\nabla \cdot p\vec{\mathbf{e}})dV$$

by the divergence theorem (3.4.1):

$$\mathbf{A} = -\int_{\partial V} pd\mathbf{S} = -\int_{V} (\nabla p)dV. \tag{6.1.1}$$

A solid body V immersed in the medium, while displacing its volume V, also removes its molecular pressure p. In this case, of the above two forces, $-\mathbf{A}$ and $\mathbf{A}$, acting on the boundary ∂V from inside and outside the volume V there remains only one force, $\mathbf{A}$. And reducing to the difference $p_+ - p_-$ of the largest and least pressures

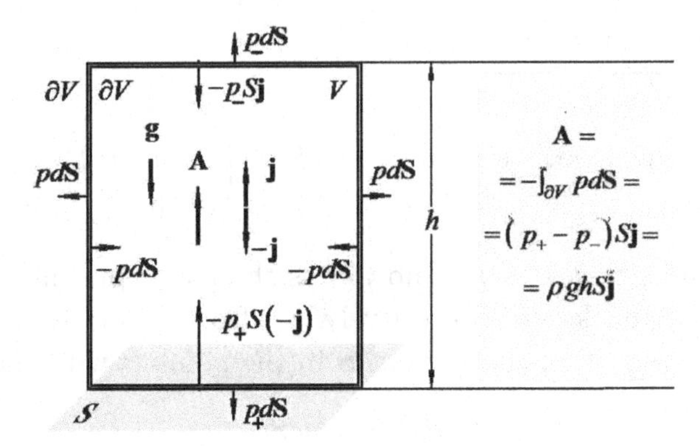

Figure 6.1.1. In the field $\mathbf{g} = -g\mathbf{j}$ of the earth's acceleration of the value g, the Archimedean force $\mathbf{A}$ diminishes the weight of a bar V of height h and area S by the weight of air of density ρ extruded by the bar.

on ∂V, p_+ and p_-, the latter pushes out the body V in the direction of the least pressure p_- (as in Fig. 6.1.1).

Thus, the buoyant force does not relate to gravity or other known forces. It exists by itself, highlighting fluid mechanics as a separate branch of physics.

6.2. Euler equations

Now, let us turn to examine another generalization by Euler of the dynamic law of Newton (4.2.2) as an alternative to the top (4.2.3).

For an inhomogeneous velocity field

$$\mathbf{u} = \mathbf{u}(t, \mathbf{r}),$$

of a *continuous medium* (which has the pressure p), Euler supposes the intuitively evident *mass conservation*

$$\frac{d}{dt}\int_V dM = \int_V \frac{d}{dt}dM = 0, \quad \text{for any small volume } V = V^t,$$

i.e.

$$\frac{d}{dt}\rho dV = \frac{d\rho}{dt}dV + \rho\frac{d}{dt}dV = \left(\frac{d\rho}{dt} + \rho\nabla\cdot\mathbf{u}\right)dV = 0,$$

or, with identities

$$\frac{d\rho}{dt} + \rho\nabla\cdot\mathbf{u} = \rho_t + \mathbf{u}\cdot\nabla\rho + \rho\nabla\cdot\mathbf{u}$$
$$= \rho_t + \nabla\cdot\rho\mathbf{u}\left(\frac{d}{dt} = \partial_t + \mathbf{u}\cdot\nabla\right),$$

the *continuity equation*

$$\rho_t + \nabla\cdot\rho\mathbf{u} = 0 \quad \left(\text{or } \frac{d}{dt}\int_V dM = 0\right) \tag{6.2.1}$$

and adds *pressure force* elements

$$-\nabla p dV$$

to *body force* elements

$$\rho\mathbf{g}dV$$

in Newton's law (4.2.2)

$$\frac{d}{dt}\int_V \rho\mathbf{u}dV = \mathbf{A} + \int_V \rho\mathbf{g}dV = \int_V (-\nabla p + \rho\mathbf{g})dV \tag{6.2.2}$$

to obtain, with identities

$$\frac{d}{dt}\int_V \rho\mathbf{u}dV = \int_V \frac{d}{dt}\rho\mathbf{u}dV = \int_V \rho\frac{d\mathbf{u}}{dt}dV \quad \left(\frac{d}{dt}\rho dV = 0\right),$$

the fluid *momentum equation*

$$\rho\frac{d\mathbf{u}}{dt} = -\nabla p + \rho\mathbf{g},$$

or

$$\rho\mathbf{u}_t + \rho\mathbf{u}\cdot\nabla\mathbf{u} = -\nabla p + \rho\mathbf{g} \tag{6.2.3}$$

that, due to (6.2.1), can be rewritten with

$$\rho\mathbf{u}_t + \rho\mathbf{u}\cdot\nabla\mathbf{u} = (\rho\mathbf{u})_t + \nabla\cdot\rho\mathbf{u}\mathbf{u} \quad \text{and} \quad \nabla p = \nabla\cdot p\vec{\mathbf{e}}$$

in the *divergence form* as

$$(\rho\mathbf{u})_t + \nabla\cdot(\rho\mathbf{u}\mathbf{u} + p\vec{\mathbf{e}}) = \rho\mathbf{g} \tag{6.2.4}$$

to make up a system of *Euler's hydrodynamic equations* [Euler, 1757]:

$$\begin{pmatrix} \rho \\ \rho \mathbf{u} \end{pmatrix}_t + \nabla \cdot \begin{pmatrix} \rho \mathbf{u} + \\ \rho \mathbf{u}\mathbf{u} + p\vec{\mathbf{e}} \end{pmatrix} = \begin{pmatrix} 0 \\ \rho \mathbf{g} \end{pmatrix}. \tag{6.2.5}$$

6.3. Vorticity equation

Due to the identity (3.3.3), or

$$\mathbf{u} \cdot \nabla \mathbf{u} = \nabla \frac{|\mathbf{u}|^2}{2} + \omega \times \mathbf{u} \quad \text{for } |\mathbf{u}|^2 = \mathbf{u} \cdot \mathbf{u} \quad \text{and} \quad \omega = \nabla \times \mathbf{u},$$

the momentum equation (6.2.3) takes the corresponding *Gromeka–Lamb form*

$$\mathbf{u}_t + \mathbf{u} \cdot \nabla \mathbf{u} = \mathbf{u}_t + \omega \times \mathbf{u} + \nabla \frac{|\mathbf{u}|^2}{2} = -\frac{1}{\rho} \nabla p + \mathbf{g} \tag{6.3.1}$$

that, for an *incompressible fluid*, or when

$$\rho = \text{const} > 0, \quad \text{hence,} \quad \nabla \cdot \mathbf{u} = 0 \quad (\text{since } \rho_t + \nabla \cdot \rho \mathbf{u} = 0) \tag{6.3.2}$$

leads to either the *Bernoulli integral* (1738)

$$\frac{|\mathbf{u}|^2}{2} + \frac{p}{\rho} = \mathit{const\ for\ } \mathbf{u}_t = \omega = \mathbf{g} = \mathbf{0} \tag{6.3.3}$$

or the Helmholtz *vorticity equation* [Helmholtz, 1858],

$$\omega_t + \mathbf{u} \cdot \nabla \omega - \omega \cdot \nabla \mathbf{u} = \nabla \times \mathbf{g}, \tag{6.3.4}$$

with the *vector commutator*

$$\mathbf{u} \cdot \nabla \omega - \omega \cdot \nabla \mathbf{u} = \nabla \times (\omega \times \mathbf{u}) \quad (\text{for } \nabla \cdot \omega = \nabla \cdot \mathbf{u} = 0). \tag{6.3.5}$$

For a *compressible fluid* ($0 < \rho \neq \text{const}$), the Gromeka–Lamb form (6.3.1) results in the vorticity equation of Friedman [1922]:

$$\omega_t + \mathbf{u} \cdot \nabla \omega - \omega \cdot \nabla \mathbf{u} = \frac{1}{\rho^2} \nabla \rho \times \nabla p + \nabla \times \mathbf{g}$$

$$\times \left(\frac{1}{\rho^2} \nabla \rho \times \nabla p = -\nabla \times \left(\frac{1}{\rho} \nabla p \right) \right).$$

6.4. Vortex law

The force **A** can be created by the motion of a body in a medium, the lifting power of Zhukovsky. To do this, it is required to design a wing with a well-thought-out geometry and accelerate it on the Earth to a difference $p_+ - p_-$ sufficient to lift the wing together with the aircraft.

The same force **A** acquires another form when the medium rotates. In this case, it follows the rule we will refer to as the *law of vortex.* According to this law, the local pressure always falls towards the vortex center. Such a pressure drop force **A** proves to be sometimes enormous when directed to the axis of rotation. For example, in such a spontaneous atmospheric vortex as a *tornado*, it is capable of drawing a loaded wagon into the funnel and twist a railroad track into a serpentine. Thanks to **A**, it becomes possible to separate uranium isotopes U-235 and U-238 in the gas mixture of uranium hexafluoride that is rotated by the rotor (cylinder) of an industrial centrifuge. Here, the force **A** pushes lighter molecules of U-235 faster than the heavier ones (of U-238). In the era preceding tea bags, it could be observed everywhere. So, while stirring tea with a spoon in a cup with tea leaves floating at the bottom, one could be surprised at how, contrary to the expected run-up to the walls (as in a merry-go-round), tea leaves accumulate near the center of rotation. Replacing the cup with tea by a glass bottle with small pieces of paper suspended in it, you can see how they regularly accumulate at the central axis of rotation evenly over the entire height of the rotating medium, without any additional movement along the axis.

In the case of a *stationary* ($\rho_t = 0$ and $\mathbf{u}_t = \mathbf{0}$) *two-dimensional* ($w = \mathbf{k} \cdot \mathbf{u} = 0$) *incompressible* ($\rho = \text{const} > 0$) *continuous* (in the sense of (4.2.1)) medium, its *plane-parallel* velocity field

$$\mathbf{u} = \mathbf{i}u(x, y) + \mathbf{j}v(x, y)$$

proves to be *solenoidal:*

$$u_x + v_y = 0 \quad (\text{or } \nabla \cdot \mathbf{u} = 0).$$

Then, when *free* of body forces ($\mathbf{g} = \mathbf{0}$), the continuous medium is subjected to the Euler equations (4.2.2) that take the form

$$uu_x + vu_y = -\frac{p_x}{\rho} \quad \text{and} \quad uv_x + vv_y = -\frac{p_y}{\rho}.$$

As is usual, additionally, the medium is taken to be as an *analytical* one whose velocity components $u(x, y)$ and $v(x, y)$ by definition are decomposed into convergent series of coordinate degrees $1, x, y, x^2, xy, y^2, \ldots$ in any small neighbourhood N of every point (x_0, y_0) in a flow region.

Now, following [Troshkin, 1995], let us turn to the *vortex center* $x_0 = y_0 = 0$ as a *stagnation point*, $\mathbf{u}(0, 0) = \mathbf{0}$,

$$u = ax + by + o(r) \quad \text{and} \quad v = cx + dy + o(r)$$

$$\text{for} \quad r = \sqrt{x^2 + y^2} \quad \text{and} \quad a, b, c, d = \text{const}.$$

For an incompressible medium, we have

$$d = -a \quad (u_x + v_y = 0).$$

To be a *real*, or *structurally stable* stagnation point (in which the pattern of stream lines, by definition, remains topologically unchanged for any small disturbances of u, v), the vortex center should have the positive *nondegeneracy*

$$e = \begin{vmatrix} a & b \\ c & -a \end{vmatrix} = -a^2 - bc > 0,$$

as in the ordinary case of a medium rotating *clockwise* or *counter-clockwise* around the centre $(0, 0)$:

$$u = y \quad \text{and} \quad v = -x \quad \text{or} \quad u = -y \quad \text{and}$$

$$v = -x, \; e = \begin{vmatrix} 0 & 1 \\ -1 & 0 \end{vmatrix} = \begin{vmatrix} 0 & -1 \\ 1 & 0 \end{vmatrix} = 1,$$

respectively.

Due to the Euler momentum equations

$$p_x = -\rho(uu_x + vu_y) = \rho ex + o(r) \quad \text{and}$$
$$p_y = -\rho(uv_x + vv_y) = \rho ey + o(r),$$

the pressure p falls toward the *vortex* centre $(0, 0)$ where it is lesser than that taken in any other point near $(0, 0)$, so rotating tea leaves do crowd around the center of rotation.

6.5. Cyclones and anticyclones

Another sign of nondegeneracy

$$e < 0$$

in the preceding case determines the remaining real stagnation point $(0, 0)$ in an incompressible fluid to be the familiar *saddle* point, or *anti-vortex* center where p is more than the p taken in any other point near $(0, 0)$ (due to the same momentum equations as above).

In the alternative case of a stationary two-dimensional free *compressible* medium ($\rho \neq$ const), the flow of *impulse density* $\tilde{\mathbf{u}} = \rho\mathbf{u}$, with components

$$\tilde{u} = \rho u \quad \text{and} \quad \tilde{v} = \rho v$$

is *solenoidal*

$$\tilde{u}_x + \tilde{v}_y = 0 \quad \text{(due to (4.1.1))},$$

and the Euler equations (4.2.2) take the form

$$\tilde{u}u_x + \tilde{v}u_y = -p_x \quad u \quad \tilde{u}v_x + \tilde{v}v_y = -p_y,$$

and, for the function

$$h = \ln\frac{\rho_0}{\rho}, \quad \rho_0 = \rho(0, 0),$$

reduce to the relations

$$\tilde{u}\tilde{u}_x + \tilde{v}\tilde{u}_y + \tilde{u}\tilde{u}h_x + \tilde{u}\tilde{v}h_y = -\rho p_x \quad \text{and}$$
$$\tilde{u}\tilde{v}_x + \tilde{v}\tilde{v}_y + \tilde{u}\tilde{v}h_x + \tilde{v}\tilde{v}h_y = -\rho p_y.$$

As in the preceding case, we have that

$$\tilde{u} = \tilde{a}x + \tilde{b}y + o(r), \quad \tilde{v} = \tilde{c}x - \tilde{a}y + o(r),$$
$$r = \sqrt{x^2 + y^2}, \;\; \tilde{a}, \tilde{b}, \tilde{c} = \text{const},$$

and

$$\tilde{u}\tilde{u},\; \tilde{u}v,\; \tilde{v}\tilde{v},\; \tilde{u}\tilde{u}h_x + \tilde{u}\tilde{v}h_y,\; \tilde{u}\tilde{v}h_x + \tilde{v}\tilde{v}h_y = o(r),$$

so

$$\tilde{\Delta}_0 = \begin{vmatrix} \tilde{a} & \tilde{b} \\ \tilde{c} & -\tilde{a} \end{vmatrix} > 0, \quad \text{or} \quad \tilde{\Delta}_0 < 0,$$

and

$$\nabla p = \frac{\tilde{\Delta}_0}{\rho}\mathbf{r} + \frac{1}{\rho}\mathbf{o}(r),$$

i.e. the pressure drops to the vortex center (or rises to the saddle point) of $\tilde{\mathbf{u}} = \rho\mathbf{u}$.

And since the pressure always rises to the center of the anticyclone, the latter cannot be the center of the flow vortex.

Chapter 7

Viscosity

7.1. Strain–stress dependence

As follows from the above, for either dry sand or pure snow to be a material of separate particles supplied with a mass density ρ submitted to the *mass conservation*, or continuity equation (3.4.2), a *dynamical equilibrium*, or *balance*, can be possible only in its initial *Newtonian form* (4.2.2). Contrary to sand and snow, for a water or an air as for a similar dense substance possessing, however, a molecular pressure p, and hence, forming a *continuous medium* provided with the same balance taken already in the further *Eulerian form* (6.2.2), there is needed the *Archimedean force* $\mathbf{A}$ from (6.1.1).

However, despite keeping birds flying on the Earth, the force $\mathbf{A}$ could not save them from the cosmic debris that continues to burn regularly and, as we shall see below (in the next paragraph), ultimately in the upper atmosphere due to another product of molecular motion, the *dynamic viscosity*

$$\mu = \mu(t, \mathbf{r}) > 0. \tag{7.1.1}$$

In other words, to restore the *dynamic balance*

$$\frac{d}{dt}\int_V \rho \mathbf{u} dV = \int_V \rho \mathbf{g} dV + \mathbf{A} + \mathbf{B} \quad \text{for } V = V^t$$

$$\text{where } \frac{d}{dt}\int_V \rho \mathbf{u} dV = \int_V \frac{d}{dt}\rho \mathbf{u} dV = \int_V ((\rho \mathbf{u})_t + \nabla \cdot \rho \mathbf{u}\mathbf{u}) dV, \tag{7.1.2}$$

in addition to the Archimedean *contact* (or surface) *force*

$$\mathbf{A} = -\int_{\partial V} p\vec{\mathbf{e}} \cdot d\mathbf{S} = -\int_V (\nabla p) dV \quad (\nabla \cdot p\vec{\mathbf{e}} = \nabla p)$$

of *pressure* p, with *stress* $-p\vec{\mathbf{e}}$, and *volume density* $-\nabla p$ independent of flow *strain* $\mathbf{u_r}$, the same invisible small-scale molecular motion creates the *friction force*

$$\mathbf{B} = \int_{\partial V} \mu\vec{\mathbf{b}} \cdot d\mathbf{S} = \int_V (\nabla \cdot \mu\vec{\mathbf{b}}) dV \tag{7.1.3}$$

of *viscosity* μ, with *stress* $\mu\vec{\mathbf{b}}$, *volume density* $\nabla \cdot \mu\vec{\mathbf{b}}$ and *structure* $\vec{\mathbf{b}}$.

The structure $\vec{\mathbf{b}}$ was first guessed as

$$\vec{\mathbf{b}} = u_y(\mathbf{ji} + \mathbf{ij}) \quad \text{for } \mathbf{u} = u\mathbf{i} \quad \text{and} \quad u_z = u_x = 0$$

by Newton [1687], changed to

$$\vec{\mathbf{b}} = \vec{\tau} \quad \text{and} \quad \vec{\tau} = \mathbf{u_r} + \nabla\mathbf{u} \quad \text{for } C = \nabla \cdot \mathbf{u} = 0 \ (\rho = \text{const} > 0)$$

by Navier [1823] when revealing the *shear property*

$$\text{Spur}(\vec{\mathbf{b}}) = \mathbf{i} \cdot \vec{\mathbf{b}} \cdot \mathbf{i} + \mathbf{j} \cdot \vec{\mathbf{b}} \cdot \mathbf{j} + \mathbf{k} \cdot \vec{\mathbf{b}} \cdot \mathbf{k} = 0,$$

and finally completed as

$$\vec{\mathbf{b}} = \vec{\tau} - \frac{2}{3}C\vec{\mathbf{e}}, \ C = \nabla \cdot \mathbf{u}, \quad \text{for any } \mathbf{u} = \mathbf{u}(t, \mathbf{r})$$

by Stokes [1845] until the *second*, or *volume viscosity*, or proportion $\varsigma\mu$ appears, for liquids and gases, in

$$\vec{\mathbf{b}} = \vec{\tau} - \left(\frac{2}{3} - \varsigma\right) C\vec{\mathbf{e}}, \quad C = \nabla \cdot \mathbf{u}, \ \varsigma = \text{const} \geq 0, \tag{7.1.4}$$

to produce

$$\text{Spur}(\vec{\mathbf{b}}) = \text{Spur}(\vec{\tau}) - \left(\frac{2}{3} - \varsigma\right) C\,\text{Spur}(\vec{\mathbf{e}}) = 2C - \left(\frac{2}{3} - \varsigma\right) 3C = 3\varsigma C$$

and is accounted in the dumping of shock waves [Landau, 1959].

Thus, due to pressure p and viscosity μ, the force

$$\mathbf{A} + \mathbf{B} = -\int_{\partial V} \vec{\mathbf{p}} \cdot d\mathbf{S}$$

consists of contact elements $-\vec{\mathbf{p}} \cdot d\mathbf{S}$ produced with the *fluid stress*

$$\begin{aligned} &\vec{\mathbf{p}} = p\vec{\mathbf{e}} - \mu\vec{\mathbf{b}} \quad \text{and} \quad \vec{\mathbf{b}} = \vec{\tau} - \left(\frac{2}{3} - \varsigma\right) C\vec{\mathbf{e}}, \\ \text{or} \quad &\vec{\mathbf{p}} = P\vec{\mathbf{e}} - \mu\vec{\tau} \quad \text{and} \quad P = p + \left(\frac{2}{3} - \varsigma\right) \mu C, \end{aligned} \tag{7.1.5}$$

of *actions* as vector columns or rays

$$\begin{aligned} \mathbf{p}^x &= p\mathbf{i} - \mu\mathbf{b}^x = P\mathbf{i} - \mu\tau^x, \quad \tau^x = \mathbf{u}_x + \nabla u, \\ \mathbf{p}^y &= p\mathbf{j} - \mu\mathbf{b}^y = P\mathbf{j} - \mu\tau^y, \quad \tau^y = \mathbf{u}_y + \nabla v, \\ \mathbf{p}^z &= p\mathbf{k} - \mu\mathbf{b}^z = P\mathbf{k} - \mu\tau^z, \quad \tau^z = \mathbf{u}_z + \nabla w, \end{aligned} \tag{7.1.6}$$

of the symmetric matrix

$$\vec{\mathbf{p}} = \mathbf{p}^x\mathbf{i} + \mathbf{p}^y\mathbf{j} + \mathbf{p}^z\mathbf{k} = \mathbf{i}\mathbf{p}^x + \mathbf{j}\mathbf{p}^y + \mathbf{k}\mathbf{p}^z = \vec{\mathbf{p}}^*, \quad \text{or} \quad \vec{\mathbf{p}} \cdot \mathbf{u} = \mathbf{u} \cdot \vec{\mathbf{p}},$$

since

$$\begin{aligned} &\mathbf{u}_x\mathbf{i} + \mathbf{u}_y\mathbf{j} + \mathbf{u}_z\mathbf{k} = \mathbf{i}\nabla u + \mathbf{j}\nabla v + \mathbf{k}\nabla w \quad \text{and} \\ &\nabla u\mathbf{i} + \nabla v\mathbf{j} + \nabla w\mathbf{k} = \mathbf{i}\mathbf{u}_x + \mathbf{j}\mathbf{u}_y + \mathbf{k}\mathbf{u}_z, \end{aligned}$$

hence, $\vec{\tau} = \vec{\tau}^*$, so $\vec{\mathbf{p}} = \vec{\mathbf{p}}^*$.

The divergence theorem

$$\int_{\partial V} \vec{\mathbf{p}} \cdot d\mathbf{S} = \int_V (\nabla \cdot \vec{\mathbf{p}}) dV \quad \text{for any small } V = V^t$$

reduces the dynamic balance (7.1.2) to the *momentum equilibrium*, or *Navier–Stokes equations*, while accounting for the mass conservation

(6.2.1),

$$\rho\mathbf{u}_t + \rho\mathbf{u}\cdot\nabla\mathbf{u} = \rho\mathbf{g} - \nabla\cdot\vec{\mathbf{p}}, \quad \text{or}$$
$$(\rho\mathbf{u})_t + \nabla\cdot(\rho\mathbf{u}\mathbf{u} + \vec{\mathbf{p}}) = \rho\mathbf{g}, t > 0, \qquad (7.1.7)$$

or, to the *dynamic equilibrium* as the coupled mass and momentum equations:

$$\begin{pmatrix}\rho\\ \rho\mathbf{u}\end{pmatrix}_t + \nabla\cdot\begin{pmatrix}\rho\mathbf{u}\\ \rho\mathbf{u}\mathbf{u} - \mu\vec{\tau} + ((2/3-\varsigma)\mu\nabla\cdot\mathbf{u} + p)\vec{\mathbf{e}}\end{pmatrix} = \begin{pmatrix}0\\ \rho\mathbf{g}\end{pmatrix},$$
$$\vec{\tau} = \mathbf{u}_\mathbf{r} + \nabla\mathbf{u} \quad (\varsigma = \text{const} \geq 0). \qquad (7.1.8)$$

With identities

$$\nabla\cdot\vec{\tau} = \Delta\mathbf{u} \quad \text{and} \quad \nabla\times(\omega\times\mathbf{u}) = \mathbf{u}\cdot\nabla\omega - \omega\cdot\nabla\mathbf{u} \ \text{for}\ \nabla\cdot\mathbf{u} = 0,$$
$$\nabla\times\Delta\mathbf{u} = \Delta\omega \quad \text{for} \quad \omega = \nabla\times\mathbf{u},$$

and the Gromeka–Lamb form (6.3.1), for a classical *incompressible viscous Newtonian medium* ($\rho, \mu = \text{const} > 0$) of *kinematic viscosity* μ/ρ, the equilibrium (7.1.8) assumes evidently the vector *vorticity equation*

$$\omega_t - \nu\Delta\omega + \mathbf{u}\cdot\nabla\omega - \omega\cdot\nabla\mathbf{u} = \nabla\times\mathbf{g}. \qquad (7.1.9)$$

Meanwhile, further clarification of the fluid force $\mathbf{A}+\mathbf{B}$ is closely related to more (as with pitch, clay, or plasticine) or less (as with paint, yogurt, or blood) *viscous–plastic* media with their own *viscous rheology of Bingham* [1922], as a non-constant (or a *non-Newtonian*) dependence of viscosity on shear strain:

$$\mu = \mu(\vec{\tau}).$$

On the other hand, it is known by now that at least for an incompressible fluid ($\rho = \text{const} > 0$) in any fixed spatial bounded domain $V = V^0$, $t \geq 0$, with a piecewise smooth boundary ∂V, hence, with a compact *closure* $\bar{V} = V \cup \partial V$, the fluid motion (7.1.8) with (7.1.5) is *uniquely determined* by every smooth initial data

$\mathbf{u}_0 = \mathbf{u}_0(\mathbf{r})$ consistent with the boundary ones, $\mathbf{u}_* = \mathbf{u}_*(t, \mathbf{r})$,

$$\mathbf{u}|_{t=0} = \mathbf{u}_0 \text{ in } \bar{V} \quad \text{and} \quad \mathbf{u}|_{\partial V} = \mathbf{u}_* \text{ for all } t \geq 0, \text{ with}$$
$$\mathbf{u}_0|_{\partial V} = \mathbf{u}_*|_{t=0},$$

only for the *rheology of Ladyzhenskaya* [1969]:

$$\mu(\vec{\tau}) = \eta(\tau), \; \tau = \|\vec{\tau}\| = \sqrt{|\tau^x|^2 + |\tau^y|^2 + |\tau^z|^2},$$
$$\text{for } \nabla \cdot \mathbf{u} = 0,$$
$$\text{with } \frac{d\eta}{d\tau} > 0, \; a_- \tau^b \leq \eta(\tau) \leq a_+ \tau^b, \; a_\mp, b = \text{const},$$
$$a_+ \geq a_- > 0 \quad \text{and} \quad b \geq \tfrac{1}{2}.$$

7.2. Power identity

As is usual, when convolved with velocity, the momentum equilibrium (7.1.7) gives us the necessary *power identity*

$$\rho \mathbf{u} \cdot \mathbf{u}_t + \rho \mathbf{u} \cdot (\mathbf{u} \cdot \nabla \mathbf{u}) = \rho \mathbf{u} \cdot \mathbf{g} - \mathbf{u} \cdot (\nabla \cdot \vec{\mathbf{p}}). \tag{7.2.1}$$

In doing so, like the body force $\rho \mathbf{g}$ per volume dV that produces the work per second, or the ordinary mechanic power $\rho \mathbf{u} \cdot \mathbf{g}$, the corresponding fluid force $-\nabla \cdot \vec{\mathbf{p}}$ in (7.2.1) produces the *fluid power* $-\mathbf{u} \cdot (\nabla \cdot \vec{\mathbf{p}})$.

When added to *momentum flow* $\nabla \cdot (\vec{\mathbf{p}} \cdot \mathbf{u})$, while accounting for scalar product (1.3.2) and norm (1.3.3), or

$$\vec{\tau} \cdot \vec{\mathbf{p}} = \tau^x \cdot \mathbf{p}^x + \tau^y \cdot \mathbf{p}^y + \tau^z \cdot \mathbf{p}^z \quad \text{and}$$
$$\|\vec{\tau}\|^2 = \tau^x \cdot \tau^x + \tau^y \cdot \tau^y + \tau^z \cdot \tau^z,$$

symmetry

$$\mathbf{i} \cdot \mathbf{p}^y = \mathbf{j} \cdot \mathbf{p}^x, \quad \mathbf{j} \cdot \mathbf{p}^z = \mathbf{k} \cdot \mathbf{p}^y \quad \text{and} \quad \mathbf{k} \cdot \mathbf{p}^x = \mathbf{i} \cdot \mathbf{p}^z$$

and necessary identities

$$
\begin{aligned}
&(\nabla u)\cdot \mathbf{p}^x + (\nabla v)\cdot \mathbf{p}^y + (\nabla w)\cdot \mathbf{p}^z \\
&= \begin{pmatrix} u_x\mathbf{i} \\ +u_y\mathbf{j} \\ +u_z\mathbf{k} \end{pmatrix} \cdot \mathbf{p}^x \begin{pmatrix} +v_x\mathbf{i} \\ +v_y\mathbf{j} \\ +v_z\mathbf{k} \end{pmatrix} \cdot \mathbf{p}^y \begin{pmatrix} +w_x\mathbf{i} \\ +w_y\mathbf{j} \\ +w_z\mathbf{k} \end{pmatrix} \cdot \mathbf{p}^z \\
&= \begin{pmatrix} u_x\mathbf{i} \\ +v_x\mathbf{j} \\ +w_x\mathbf{k} \end{pmatrix} \cdot \mathbf{p}^x \begin{pmatrix} +u_y\mathbf{i} \\ +v_y\mathbf{j} \\ +w_y\mathbf{k} \end{pmatrix} \cdot \mathbf{p}^y \begin{pmatrix} +u_z\mathbf{i} \\ +v_z\mathbf{j} \\ +w_z\mathbf{k} \end{pmatrix} \cdot \mathbf{p}^z \\
&= \mathbf{u}_x \cdot \mathbf{p}^x + \mathbf{u}_y \cdot \mathbf{p}^y + \mathbf{u}_z \cdot \mathbf{p}^z
\end{aligned}
$$

and

$$
\begin{aligned}
\nabla\cdot(\vec{\mathbf{p}}\cdot\mathbf{u}) - \mathbf{u}\cdot(\nabla\cdot\vec{\mathbf{p}}) &= \nabla\cdot(u\mathbf{p}^x + v\mathbf{p}^y + w\mathbf{p}^z) \\
&\quad - u\nabla\cdot\mathbf{p}^x - v\nabla\cdot\mathbf{p}^y - w\nabla\cdot\mathbf{p}^z \\
&= (\nabla u)\cdot\mathbf{p}^x + (\nabla v)\cdot\mathbf{p}^y + (\nabla w)\cdot\mathbf{p}^z = \frac{1}{2}\vec{\tau}\cdots\vec{\mathbf{p}},
\end{aligned}
$$

it leads to what can be called the *mixing power*

$$
\nabla\cdot(\vec{\mathbf{p}}\cdot\mathbf{u}) - \mathbf{u}\cdot(\nabla\cdot\vec{\mathbf{p}}) = \frac{1}{2}\vec{\tau}\cdot\vec{\mathbf{p}},
$$

with the corresponding *mixing power identity*

$$
-\mathbf{u}\cdot(\nabla\cdot\vec{\mathbf{p}}) = \frac{1}{2}\vec{\tau}\cdot\vec{\mathbf{p}} - \nabla\cdot(\vec{\mathbf{p}}\cdot\mathbf{u}) \quad \text{whatever } \vec{\mathbf{p}} = \vec{\mathbf{p}}_*. \tag{7.2.2}
$$

As a consequence, for a *specific* (per mass ρdV) *living force* (kinetic energy)

$$
K = \frac{|\mathbf{u}|^2}{2}, \quad |\mathbf{u}|^2 = \mathbf{u}\cdot\mathbf{u},
$$

with mass conservation (6.2.1),

$$
\rho\mathbf{u}\cdot\mathbf{u}_t + \rho\mathbf{u}\cdot(\mathbf{u}\cdot\nabla\mathbf{u}) = \rho K_t + \rho\mathbf{u}\cdot\nabla K = (\rho K)_t + \nabla\cdot\rho K\mathbf{u},
$$

the power identity (7.2.1) takes the form

$$(\rho K)_t + \nabla\cdot(\rho K\mathbf{u} + \vec{\mathbf{p}}\cdot\mathbf{u}) - \rho\mathbf{u}\cdot\mathbf{g} = \frac{1}{2}\vec{\tau}\cdot\vec{\mathbf{p}} \quad \text{for any } \vec{\mathbf{p}} = \vec{\mathbf{p}}_*, \tag{7.2.3}$$

or, for (7.1.5),

$$(\rho K)_t + \nabla\cdot(\rho K\mathbf{u} + \vec{\mathbf{p}}\cdot\mathbf{u}) - \rho\mathbf{u}\cdot\mathbf{g} = \frac{P}{2}\vec{\tau}\cdot\vec{\mathbf{e}} - \frac{\mu}{2}\vec{\tau}\cdot\vec{\tau},$$

with

$$\vec{\tau}\cdot\vec{\mathbf{e}} = Sp(\vec{\tau}) = 2\nabla\cdot\mathbf{u} = 2C, \quad \vec{\tau}\cdot\vec{\tau} = \|\vec{\tau}\|^2,$$

$$P = p + \sigma\mu C \quad \text{and} \quad \sigma = \frac{2}{3} - \varsigma,$$

the form

$$(\rho K)_t + \nabla\cdot(\rho K\mathbf{u} + \vec{\mathbf{p}}\cdot\mathbf{u}) - \rho\mathbf{u}\cdot\mathbf{g} = PC - \frac{\mu}{2}\|\vec{\tau}\|^2$$

$$= pC + \left(\frac{2}{3} - \zeta\right)\mu C^2 - \mu\frac{\|\vec{\tau}\|^2}{2},$$

or, with (2.7.7) as

$$D = \frac{\|\vec{\tau}\|^2}{2} = |\omega|^2 + 2C^2 + 2\nabla\cdot\mathbf{D}, \quad \omega = \nabla\times\mathbf{u}, \quad C = \nabla\cdot\mathbf{u},$$

$$\mathbf{D} = \mathbf{u}\cdot\nabla\mathbf{u} - C\mathbf{u}, \quad \mathbf{u}\cdot\nabla\mathbf{u} = \omega\times\mathbf{u} + \nabla K,$$

the form

$$(\rho K)_t + \nabla\cdot(\rho K\mathbf{u} + \vec{\mathbf{p}}\cdot\mathbf{u}) - \rho\mathbf{u}\cdot\mathbf{g}$$

$$= pC - \left(\frac{4}{3} + \varsigma\right)\mu C^2 - \mu|\omega|^2 - 2\mu\nabla\cdot\mathbf{D},$$

or, with

$$\mu\nabla\cdot\mathbf{D} = \nabla\cdot\mu\mathbf{D} - \mathbf{D}\cdot\nabla\mu \quad \text{and} \quad \mu = \rho\nu,$$

the *final form*

$$(\rho K)_t + \nabla \cdot (\rho K \mathbf{u} + \mathbf{E}) - \rho \mathbf{u} \cdot \mathbf{g} = pC - \dot{q}_f \quad \text{for } K = \frac{|\mathbf{u}|^2}{2},$$

$$\mathbf{E} = \vec{\mathbf{p}} \cdot \mathbf{u} + 2\mu \mathbf{D}, \quad \dot{q}_f = \left(\frac{4}{3} + \zeta\right) \mu C^2 + \mu |\omega|^2 - 2\mathbf{D} \cdot \nabla \mu,$$

$$\mathbf{D} = \mathbf{u} \cdot \nabla \mathbf{u} - C\mathbf{u} = \nabla K + \omega \times \mathbf{u} - C\mathbf{u}, \quad \varsigma = \text{const} \geq 0, \tag{7.2.4}$$

for any smooth *kinematic viscosity*

$$\nu = \frac{\mu}{\rho} = \nu(t, \mathbf{r}) > 0. \tag{7.2.5}$$

7.3. Axial symmetry

In addition, it is worth noting that the equilibrium (7.1.8) is not restricted by the system of Cartesian coordinates x, y, z and orts

$$\mathbf{i} = \mathbf{j} \times \mathbf{k}, \quad \mathbf{j} = \mathbf{k} \times \mathbf{i} \quad u \quad \mathbf{k} = \mathbf{i} \times \mathbf{j}, \quad \mathbf{i} \cdot \mathbf{j} = \mathbf{j} \cdot \mathbf{k} = \mathbf{k} \cdot \mathbf{i} = 0,$$
$$\mathbf{i} \cdot \mathbf{i} = \mathbf{j} \cdot \mathbf{j} = \mathbf{k} \cdot \mathbf{k} = 1.$$

Therefore, it is natural to try equations (7.1.8) in *cylindrical coordinates*

$$r = \sqrt{x^2 + y^2}, \quad \theta = \operatorname{arctg} \frac{y}{x}, \quad \text{or} \quad x = r \cos \theta,$$
$$y = r \sin \theta \quad (\text{and the same } z),$$

or *polar orts*

$$\mathbf{I} = \mathbf{I}(\theta) = \mathbf{i} \cos \theta + \mathbf{j} \sin \theta,$$
$$\mathbf{J} = \mathbf{J}(\theta) = \mathbf{j} \cos \theta - \mathbf{i} \sin \theta \quad (\text{and the same } \mathbf{k}),$$

reversible,

$$\mathbf{i} = \mathbf{I} \cos \theta - \mathbf{J} \sin \theta, \quad \mathbf{j} = \mathbf{J} \cos \theta + \mathbf{I} \sin \theta$$
$$(\text{with} \quad \mathbf{i}_\theta = \mathbf{j}_\theta = \mathbf{k}_\theta = \mathbf{0}),$$

satisfying equations

$$\mathbf{I}_\theta = \mathbf{J} \quad \text{and} \quad \mathbf{J}_\theta = -\mathbf{I},$$

and conserving both initial properties of orts,

$$\mathbf{I} = \mathbf{J} \times \mathbf{k}, \quad \mathbf{J} = \mathbf{k} \times \mathbf{I}, \quad \mathbf{k} = \mathbf{I} \times \mathbf{J},$$

$$\mathbf{I} \cdot \mathbf{J} = \mathbf{J} \cdot \mathbf{k} = \mathbf{k} \cdot \mathbf{I} = 0 \quad \text{and} \quad \mathbf{I} \cdot \mathbf{I} = \mathbf{J} \cdot \mathbf{J} = \mathbf{k} \cdot \mathbf{k} = 1,$$

and the sum of direct squares

$$\mathbf{II} + \mathbf{JJ} = \mathbf{ii} + \mathbf{jj},$$

where components

$$\partial_x = \cos\theta\partial_r - \frac{\sin\theta}{r}\partial_\theta \quad \text{and} \quad \partial_y = \sin\theta\partial_r + \frac{\cos\theta}{r}\partial_\theta,$$

of gradient

$$\nabla = \mathbf{i}\partial_x + \mathbf{j}\partial_y + \mathbf{k}\partial_z = \mathbf{I}\partial_r + \mathbf{k}\partial_z + \mathbf{J}\frac{\partial_\theta}{r}$$

are replaced for

$$\partial_r = \cos\theta\partial_x + \sin\theta\partial_y \quad \text{and} \quad \frac{\partial_\theta}{r} = \cos\theta\partial_y - \sin\theta\partial_y.$$

Besides, in this case we have the following radius-vector:

$$\mathbf{r} = x\mathbf{i} + y\mathbf{j} + z\mathbf{k} = r\mathbf{I}(\theta) + z\mathbf{k},$$

body acceleration

$$\mathbf{g} = g^x\mathbf{i} + g^y\mathbf{j} + g^z\mathbf{k} = g^r\mathbf{I} + g^z\mathbf{k} + g^\theta\mathbf{J}$$

and fluid velocity

$$\mathbf{u} = \mathbf{r}_t = u\mathbf{i} + v\mathbf{j} + w\mathbf{k} = \hat{u}\mathbf{I} + \hat{v}\mathbf{J} + w\mathbf{k},$$

of polar components as *radial*, *azimuthal*, and *axial* velocities,

$$\hat{u} = r_t, \quad \hat{v} = r\theta_t \quad \text{and} \quad w = z_t$$

returning the orthogonal ones

$$u = \hat{u}\cos\theta - \hat{v}\sin\theta \quad \text{and} \quad v = \hat{u}\sin\theta + \hat{v}\cos\theta.$$

Together with mass density $\rho > 0$, pressure p, and viscosity $\mu > 0$, all the polar functions φ in hand are assumed to be *axially symmetric*, or independent of polar angle θ:

$$\varphi_\theta = 0 \quad \text{for } \varphi = \rho, p, \mu, \hat{u}, \hat{v}, w, g^{r,z,\theta}. \tag{7.3.1}$$

As a consequence, we obtain the *polar* derivative of velocity

$$\mathbf{u}_\theta = \hat{u}\mathbf{I}_\theta + \hat{v}\mathbf{J}_\theta = \hat{u}\mathbf{J} - \hat{v}\mathbf{I} = \mathbf{k} \times \mathbf{u},$$

mass flow $\rho\mathbf{u}$ volume density

$$\nabla \cdot \rho\mathbf{u} = \begin{pmatrix} \mathbf{I}\partial_r + \mathbf{k}\partial_z \\ +\mathbf{J}\dfrac{\partial_\theta}{r} \end{pmatrix} \cdot \begin{pmatrix} \mathbf{I}\rho\hat{u} \\ +\mathbf{k}\rho w \\ +\mathbf{J}\rho\hat{v} \end{pmatrix} = \begin{pmatrix} (\rho\hat{u})_r \\ +(\rho w)_z \end{pmatrix}$$

$$+ \frac{1}{r}\mathbf{J} \cdot \begin{pmatrix} \mathbf{J}\rho\hat{u} \\ -\mathbf{I}\rho\hat{v} \end{pmatrix} = \frac{(r\rho\hat{u})_r}{r} + (\rho w)_z,$$

with the continuity equation (6.2.1), or

$$\rho_t + \nabla \cdot \rho\mathbf{u} = \rho_t + \frac{1}{r}(r\rho\hat{u})_r + (\rho w)_z = \rho_t + \hat{u}\rho_r + w\rho_z + C\rho = 0$$

$$\text{where } C = \nabla \cdot \mathbf{u} = \tfrac{(r\hat{u})_r}{r} + w_z, \tag{7.3.2}$$

momentum flow $\rho\mathbf{u}\mathbf{u}$ volume density, while accounting for $\mathbf{u}_\theta = \mathbf{k}\times\mathbf{u}$,

$$\nabla \cdot \rho\mathbf{u}\mathbf{u} = \begin{pmatrix} \mathbf{k} \cdot (\rho\mathbf{u}\mathbf{u})_z \\ +\mathbf{I} \cdot (\rho\mathbf{u}\mathbf{u})_r \\ +\dfrac{1}{r}\mathbf{J} \cdot (\rho\mathbf{u}\mathbf{u})_\theta \end{pmatrix} = \begin{pmatrix} (\rho\mathbf{k} \cdot \mathbf{u}\mathbf{u})_z \\ +(\rho\mathbf{I} \cdot \mathbf{u}\mathbf{u})_r \\ +\dfrac{\rho}{r}\mathbf{J} \cdot \mathbf{u}_\theta\mathbf{u} + \tfrac{\rho}{r}\mathbf{J} \cdot \mathbf{u}\mathbf{u}_\theta \end{pmatrix}$$

$$= \begin{pmatrix} (\rho\hat{u}\mathbf{u})_r + \dfrac{\rho\hat{u}\mathbf{u}}{r} \\ +(\rho w\mathbf{u})_z \\ +\dfrac{\rho\hat{v}}{r}(\hat{u}\mathbf{J} - \hat{v}\mathbf{I}) \end{pmatrix},$$

or

$$\nabla \cdot \rho\mathbf{uu} = \begin{pmatrix} \dfrac{(r\rho\hat{u}\mathbf{u})_r}{r} \\ +(\rho w\mathbf{u})_z \\ +\dfrac{\rho\hat{v}}{r}(\hat{u}\mathbf{J} - \hat{v}\mathbf{I}) \end{pmatrix} = \mathbf{I}\begin{pmatrix} \dfrac{(r\rho\hat{u}\hat{u})_r}{r} \\ +(\rho\hat{u}w)_z \\ -\dfrac{\rho\hat{v}^2}{r} \end{pmatrix}$$

$$+\mathbf{J}\begin{pmatrix} \dfrac{(r\rho\hat{v}\hat{u})_r}{r} \\ +(\rho\hat{v}w)_z \\ +\dfrac{\rho\hat{v}\hat{u}}{r} \end{pmatrix} + \begin{pmatrix} \dfrac{(r\rho w\hat{u})_r}{r} \\ +(\rho ww)_z \end{pmatrix}\mathbf{k},$$

velocity gradient

$$\nabla\mathbf{u} = \begin{pmatrix} \mathbf{I}\partial_r + \mathbf{k}\partial_z \\ +\mathbf{J}\dfrac{\partial_\theta}{r} \end{pmatrix}\begin{pmatrix} \hat{u}\mathbf{I} \\ +\hat{v}\mathbf{J} \\ +w\mathbf{k} \end{pmatrix} = \begin{pmatrix} \hat{u}_r\mathbf{II} + \hat{v}_r\mathbf{IJ} + w_r\mathbf{Ik} \\ -\dfrac{\hat{v}}{r}\mathbf{JI} + \dfrac{\hat{u}}{r}\mathbf{JJ} \\ +\hat{u}_z\mathbf{kI} + \hat{v}_z\mathbf{kJ} + w_z\mathbf{kk} \end{pmatrix},$$

strain

$$\mathbf{u_r} = (\nabla\mathbf{u})^* = \begin{pmatrix} \hat{u}_r\mathbf{II} - \dfrac{\hat{v}}{r}\mathbf{IJ} + \hat{u}_z\mathbf{Ik} \\ +\hat{v}_r\mathbf{JI} + \dfrac{\hat{u}}{r}\mathbf{JJ} + \hat{v}_z\mathbf{Jk} \\ +w_r\mathbf{kI} + w_z\mathbf{kk} \end{pmatrix}, \quad \text{with}$$

$$\hat{v}_r + \frac{\hat{v}}{r} = \frac{(r\hat{v})_r}{r} \quad \text{and} \quad \hat{v}_r - \frac{\hat{v}}{r} = r\left(\frac{\hat{v}}{r}\right)_r,$$

torsion

$$\vec{\omega} = \mathbf{u_r} - \nabla\mathbf{u} = \begin{pmatrix} -\dfrac{(r\hat{v})_r}{r}\mathbf{IJ} + (\hat{u}_z - w_r)\mathbf{Ik} \\ +\dfrac{(r\hat{v})_r}{r}\mathbf{JI} + \hat{v}_z\mathbf{Jk} \\ (w_r - \hat{u}_z)\mathbf{kI} - \hat{v}_z\mathbf{kJ} \end{pmatrix} = \begin{pmatrix} \mathbf{II} \\ +\mathbf{JJ} \\ +\mathbf{kk} \end{pmatrix}$$

$$\times\begin{pmatrix} \mathbf{k}\dfrac{(r\hat{v})_r}{r} + \\ \mathbf{J}(\hat{u}_z - w_r) \\ -\mathbf{I}\hat{v}_z \end{pmatrix} = \vec{\mathbf{e}} \times \omega,$$

of vorticity

$$\omega = -\mathbf{I}\hat{v}_z + \mathbf{J}(\hat{u}_z - w_r) + \mathbf{k}\frac{(r\hat{v})_r}{r} \tag{7.3.3}$$

and shear

$$\vec{\tau} = \mathbf{u_r} + \nabla\mathbf{u} = \begin{pmatrix} 2\hat{u}_r\mathbf{II} + r\left(\frac{\hat{v}}{r}\right)_r \mathbf{IJ} + (\hat{u}_z + w_r)\,\mathbf{Ik} \\ +r\left(\frac{\hat{v}}{r}\right)_r \mathbf{JI} + \frac{2\hat{u}}{r}\mathbf{JJ} + \hat{v}_z\mathbf{Jk} \\ \\ +(\hat{u}_z + w_r)\,\mathbf{kI} + \hat{v}_z\mathbf{kJ} + 2w_z\mathbf{kk} \end{pmatrix}$$

$$= \begin{pmatrix} \mathbf{I}\tau^r \\ +\mathbf{J}\tau^\theta \\ +\mathbf{k}\tau^z \end{pmatrix} = \begin{pmatrix} \tau^r\mathbf{I} \\ +\tau^\theta\mathbf{J} \\ +\tau^z\mathbf{k} \end{pmatrix},$$

with *viscous actions*

$$\mu\tau^r = \begin{pmatrix} 2\mu\hat{u}_r\mathbf{I} \\ +\mu r\left(\frac{\hat{v}}{r}\right)_r \mathbf{J} \\ +\mu\,(\hat{u}_z + w_r)\,\mathbf{k} \end{pmatrix}, \quad \mu\tau^\theta = \begin{pmatrix} \mu r\left(\frac{\hat{v}}{r}\right)_r \mathbf{I} + \frac{2\mu\hat{u}}{r}\mathbf{J} \\ +\mu\hat{v}_z\mathbf{k} \end{pmatrix},$$

$$\mu\tau^z = \begin{pmatrix} \mu\,(\hat{u}_z + w_r)\,\mathbf{I} \\ +\mu\hat{v}_z\mathbf{J} \\ +2\mu w_z\mathbf{k} \end{pmatrix},$$

and *force density*

$$\nabla\cdot\mu\vec{\tau} = \begin{pmatrix} \mathbf{I}\partial_r \\ +\mathbf{k}\partial_z \end{pmatrix}\cdot\begin{pmatrix} \mathbf{I}\mu\tau^r + \mathbf{J}\mu\tau^\theta \\ +\mathbf{k}\mu\tau^z \end{pmatrix} + \frac{1}{r}\mathbf{J}\cdot\begin{pmatrix} \mathbf{J}\mu\tau^r \\ -\mathbf{I}\mu\tau^\theta \end{pmatrix}$$

$$= \frac{(r\mu\tau^r)_r}{r} + (\mu\tau^z)_z,$$

or

$$\nabla \cdot \mu\vec{\tau} = \begin{pmatrix} 2\frac{(\mu r\hat{u}_r)_r}{r} \\ +\left(\mu\left(\hat{u}_z + w_r\right)\right)_z \end{pmatrix} \mathbf{I} + \begin{pmatrix} \frac{1}{r}\left(\mu r^2\left(\frac{\hat{v}}{r}\right)_r\right)_r \\ +\left(\mu\hat{v}_z\right)_z \end{pmatrix} \mathbf{J}$$

$$+ \begin{pmatrix} \frac{\left(\mu r\left(\hat{u}_z + w_r\right)\right)_r}{r} \\ +2\left(\mu w_z\right)_z \end{pmatrix} \mathbf{k},$$

$$\text{with } \frac{1}{r}\left(\mu r^2\left(\frac{\hat{v}}{r}\right)_r\right)_r = \mu\hat{v}_{rr} \quad \text{for } \mu = \text{const},$$

and deformation measure, or dissipation

$$D = \frac{\|\vec{\tau}\|^2}{2} = \frac{|\tau^r|^2 + |\tau^\theta|^2 + |\tau^z|^2}{2}$$

$$= 2\hat{u}_r^2 + 2\left(\frac{\hat{u}}{r}\right)^2 + 2w_z^2 + \left(r\left(\frac{\hat{v}}{r}\right)_r\right)^2 + \hat{v}_z^2 + (\hat{u}_z + w_r)^2.$$

So, the dynamic equilibrium

$$\begin{pmatrix} \rho \\ \rho\mathbf{u} \end{pmatrix}_t + \nabla \cdot \begin{pmatrix} \rho\mathbf{u} \\ \rho\mathbf{u}\mathbf{u} - \mu\vec{\tau} + \left(\left(2/3 - \varsigma\right)\mu\nabla\cdot\mathbf{u} + p\right)\vec{\mathbf{e}} \end{pmatrix}$$

$$= \begin{pmatrix} 0 \\ \rho\mathbf{g} \end{pmatrix}, \quad \vec{\tau} = \mathbf{u_r} + \nabla\mathbf{u}(\varsigma = \text{const} \geq 0). \tag{7.1.8}$$

for the classical *strain–stress dependence*

$$\begin{aligned} &\vec{\mathbf{p}} = p\vec{\mathbf{e}} - \mu\vec{\mathbf{b}} \quad \text{and} \quad \vec{\mathbf{b}} = \vec{\tau} - \left(\frac{2}{3} - \varsigma\right) C\vec{\mathbf{e}}, \\ \text{or} \quad &\vec{\mathbf{p}} = P\vec{\mathbf{e}} - \mu\vec{\tau} \quad \text{and} \quad P = p + \left(\frac{2}{3} - \varsigma\right)\mu C, \end{aligned} \tag{7.1.5}$$

takes the form

$$\rho_t + \hat{u}\rho_r + w\rho_z + C\rho = 0, \quad C = \frac{(r\hat{u})_r}{r} + w_z,$$

$$\begin{pmatrix} (\rho\hat{u})_t \\ +\dfrac{(r\rho\hat{u}\hat{u})_r}{r} \\ +(\rho\hat{u}w)_z \\ -\dfrac{\rho\hat{v}^2}{r} \\ -2\dfrac{(\mu r\hat{u}_r)_r}{r} \\ -(\mu(\hat{u}_z + w_r))_z \\ +P_r \end{pmatrix}\mathbf{I} + \begin{pmatrix} (\rho\hat{v})_t \\ +\dfrac{(r\rho\hat{v}\hat{u})_r}{r} \\ +(\rho\hat{v}w)_z \\ +\dfrac{\rho\hat{v}\hat{u}}{r} - \\ \dfrac{1}{r}\left(\mu r^2\left(\dfrac{\hat{v}}{r}\right)_r\right)_r \\ -(\mu\hat{v}_z)_z \end{pmatrix}\mathbf{J}$$

$$+\begin{pmatrix} (\rho w)_t \\ +\dfrac{(r\rho w\hat{u})_r}{r} \\ +(\rho ww)_z \\ -\dfrac{(\mu r(\hat{u}_z + w_r))_r}{r} \\ -2(\mu w_z)_z \\ +P_z \end{pmatrix}\mathbf{k} = \begin{pmatrix} g^r\mathbf{I}+ \\ g^\theta\mathbf{J}+ \\ g^z\mathbf{k} \end{pmatrix},$$

$$P = p + \left(\frac{2}{3} - \zeta\right)\mu C, \quad \zeta = \text{const} \geq 0. \tag{7.3.4}$$

7.4. Helical flow and viscosity integral

Without radial velocity $\hat{u} = 0$ and time and axial dependences $\partial_t = \partial_z = 0$, the corresponding stationary *helical flow*

$$\mathbf{u} = w(r)\mathbf{k} + \hat{v}(r)\mathbf{J} \text{ for any } \rho(r), \mu(r) > 0 \quad (\partial_t = \partial_z = \hat{u} = 0) \tag{7.4.1}$$

becomes *solenoidal*:

$$\nabla \cdot \mathbf{u} = C = \frac{1}{r}(r\hat{u})_r + w_z = 0 \text{ in (7.3.2), hence, } P = p \quad \text{in (7.1.5),}$$

so the equilibrium equations left in (7.3.4)

$$\left(p_r - \frac{\rho\hat{v}^2}{r}\right)\mathbf{I} - \frac{1}{r}\left(\mu r^2\left(\frac{\hat{v}}{r}\right)_r\right)_r \mathbf{J} + \left(p_z - \frac{(\mu r w_r)_r}{r}\right)\mathbf{k}$$
$$= g^r\mathbf{I} + g^\theta\mathbf{J} + g^z\mathbf{k} \tag{7.4.2}$$

admit the *viscosity integral*

$$\mu r^2\left(\frac{\hat{v}}{r}\right)_r + \int_r^a r' g^\theta\left(r'\right)dr' = \text{const}, \quad -\frac{1}{r}\left(\mu r^2\left(\frac{\hat{v}}{r}\right)_r\right)_r = g^\theta(r),$$
$$a = \text{const} > 0. \tag{7.4.3}$$

As a result, when treated with negligible body acceleration

$$\mathbf{g} = \mathbf{0}, \quad \text{or} \quad g^r = g^\theta = g^z = 0,$$

between the coaxial inner and outer cylinders,

$$\varepsilon a < r < a, \quad 0 < \varepsilon = \text{const} < 1\varepsilon a < r < a, \tag{7.4.4}$$

rotating and translating with constant azimuthal and axial velocities, $\hat{v}_\mp$ and $w_\mp$, longitudinal *pressure drop* $p_z < 0$ prescribed and *no-slip conditions*:

$$-p_z = b = \text{const} > 0, \quad \hat{v}|_{r=\varepsilon a, a} = \hat{v}_-, \hat{v}_+, \quad w|_{r=\varepsilon a, a} = w_-, w_+,$$
$$\left(\mu r^2\left(\frac{\hat{v}}{r}\right)_r\right)_r = 0, -\frac{(\mu r w_r)_r}{r} = -p_z, \quad \varepsilon a < r < a, \tag{7.4.5}$$

the corresponding free equilibrium (7.4.2) is realized by the pressure field

$$p = \text{const} - zb - \int_r^a \frac{\rho'\hat{v}'^2}{r'}dr' \quad \text{with } \rho' = \rho(r') \quad \text{and} \quad \hat{v}' = \hat{v}(r'), \tag{7.4.6}$$

and helical flow (7.4.1) of components

$$\hat{u} = 0, \quad \frac{\hat{v}}{r} = (1-\Omega)\frac{\hat{v}_+}{a} + \Omega\frac{\hat{v}_-}{\varepsilon a} \quad \text{and} \quad w = w^{(c)} + w^{(p)},$$

$$\text{where } w^{(c)} = w_+(1-\Phi) + w_-\Phi \quad \text{and} \quad w^{(p)} = (\Psi - \Phi)\frac{b}{2}\int_{\varepsilon a}^{a}\frac{rdr}{\mu}, \tag{7.4.7}$$

determined with *velocity factors*

$$\Omega = \frac{\int_r^a \frac{dr'}{\mu' r'^2}}{\int_{\varepsilon a}^a \frac{dr}{\mu r^2}}, \quad \Phi = \frac{\int_r^a \frac{dr'}{\mu' r'}}{\int_{\varepsilon a}^a \frac{dr}{\mu r}}, \quad \text{and}$$

$$\Psi = \frac{\int_r^a \frac{r'dr'}{\mu'}}{\int_{\varepsilon a}^a \frac{rdr}{\mu}}, \quad \text{for } \mu' = \mu(r'), \tag{7.4.8}$$

such that

$$F|_{r=\varepsilon a} = 1 > F(r) > F|_{r=a} = 0 \quad \text{and}$$
$$F_r < 0 \quad \text{for } \varepsilon a < r < a \quad \text{and} \quad F = \Omega, \Phi, \Psi,$$

with

$$\Omega_r = -\frac{1}{\mu r^2 \int_{\varepsilon a}^a \frac{dr}{\mu r^2}}, \Phi_r = -\frac{1}{\mu r \int_{\varepsilon a}^a \frac{dr}{\mu r}}, \quad \text{and} \quad \Psi_r = -\frac{1}{\mu \int_{\varepsilon a}^a \frac{rdr}{\mu}}.$$

The obtained formulae may be useful in solving such industrial problems as the extruding of a viscous–plastic medium [Händle, 2007; Händle *et al.*, 2015; Bizhanov *et al.*, 2014, 2015; Kurunov and Bizhanov, 2017].

7.5. Pressure head rate

In the case of a *Newtonian liquid* ($\rho, \mu = \text{const} > 0$), velocity factors (7.4.8) depend no more on the constant viscosity μ, and formulae (7.4.7) give us the familiar *pressure head* flow of Hagen–Poiseuille $w^{(p)}\mathbf{k}$ [Hagen, 1839; Poiseuille, 1840], the *shear* flows of Couette $w^{(c)}\mathbf{k}$ and Taylor–Couette $\hat{v}\mathbf{J}$ [Joseph, 1976], and their sum $w^{(p)}\mathbf{k}+w^{(c)}\mathbf{k}+$

$\hat{v}$**J** [Langlois and Deville, 2014] satisfying (7.4.5) for dimensionless azimuthal

$$\frac{\hat{v}}{\hat{v}_-} = Y + \frac{\hat{v}_+}{\hat{v}_-}(X - \varepsilon Y), \quad Y = \frac{\varepsilon}{1-\varepsilon^2}\left(\frac{1}{X} - X\right),$$
$$\varepsilon \le X = \frac{r}{a} \le 1 \ (\hat{v}_- \ne 0)$$

and

$$\frac{w^{(c)}}{w_p} = \delta_+ - \frac{\delta}{\ln(1/\varepsilon)}\ln X, \quad \delta = \delta_- - \delta_+, \quad \delta_\mp = \frac{w_\mp}{w_p}(\ln(1/\varepsilon) > 0),$$

at the *velocity deviation* δ, and axial

$$\frac{w^{(p)}}{w_p} = 1 - X^2 + \frac{1-\varepsilon^2}{\ln(1/\varepsilon)}\ln X \quad \text{where } w_p = \frac{a^2 b}{4\mu}$$

is the *rate* of the *pressure head* b that, together with boundary velocities $w_\mp$, or $\delta_\mp$, *transports* the liquid between cylinders in (7.4.5), with the *maximum velocity*

$$w_* = \max_{\varepsilon \le X \le 1} w(X) = w(X_*)$$

as follows.

Both the *transport velocity*

$$\frac{w}{w_p} = \frac{w^{(c)} + w^{(p)}}{w_p} = \delta_+ + Z, \quad Z = 1 - X^2 + E\ln X^2,$$
$$E = \frac{1-\varepsilon^2-\delta}{2\ln(1/\varepsilon)}, \tag{7.5.1}$$

and the shear strain *deformation* as a derivative

$$Z' = \frac{dZ}{dX} = \frac{2}{X}(E - X^2)$$

have a *threshold* $E = E(\varepsilon, \delta)$ such that

(i) for a *moderate* threshold $\varepsilon^2 < E < 1$, or velocity deviation

$$1 - \varepsilon^2 - 2\ln(1/\varepsilon) = \delta < \delta < \bar{\delta} = 1 - \varepsilon^2 - 2\varepsilon^2 \ln(1/\varepsilon),$$

the deformation Z' changes its sign in the flow domain, namely,

$$\frac{dZ}{dX} > (<)0 \quad \text{for } X < (>)\sqrt{E};$$

(ii) contrarily, for a *marginal* threshold

$$E \leq 0, \quad 0 < E \leq \varepsilon^2, \quad \text{or} \quad E \geq 1,$$

or deviation

$$\delta \geq 1 - \varepsilon^2, \quad \bar{\delta} \leq \delta < 1 - \varepsilon^2, \quad \text{or} \quad \delta \leq \delta,$$

the deformation Z' keeps its sign in the flow domain, namely,

$$\frac{dZ}{dX} < 0 \quad \text{for } \varepsilon < X < 1,$$

so

$$\frac{dZ}{dX} = 0 \quad \text{for } X = \varepsilon \text{ at } E = \varepsilon^2(\delta = \bar{\delta}) \quad \text{or for } X = 1$$
$$\text{at } E = 1(\delta = \underline{\delta}).$$

Excluding the shift of the outer cylinder ($\delta_+ = 0$), we find, from (7.5.1), that

$$w_* < w_p \quad \text{and} \quad X_* = \sqrt{E} \quad \text{for } \textbf{(i)} \quad \text{or}$$
$$w_* = w_p \quad \text{and} \quad X_* = \varepsilon, 1 \quad \text{for } \textbf{(ii)},$$

i.e. *the transporting velocity w cannot exceed the pressure head rate w_p*, which is illustrated in Fig. 7.5.1.

As to the displacement of the outer cylinder ($\delta_+ \neq 0$) in (7.5.1), it merely shifts *the maximum value*

$$w_* < w_+ + w_p \quad \text{and} \quad X_* = \sqrt{E} \text{ for } \textbf{(i)}$$
$$\text{or} \quad w_* = w_+ + w_p \quad \text{and} \quad X_* = \varepsilon, 1 \text{ for } \textbf{(ii)}$$

that depends not on w_-, as before.

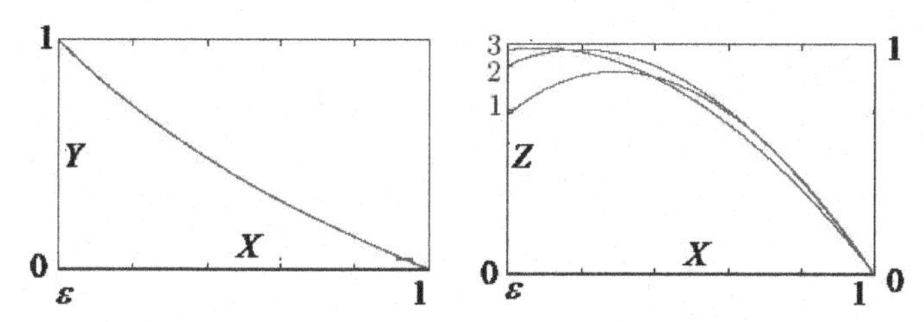

Figure 7.5.1. Dimensionless profiles of azimuthal (to the left) and axial (to the right) velocity components of helical flow for two moderate (1 and 2), as in (**i**), and one marginal (3), as in (**ii**), values of δ_- when $\delta_+ = 0$.

Finally, in the limiting case of an ordinary tube at rest when

$$\varepsilon \to +0 \quad \text{and} \quad \hat{v}_+ = w_+ = 0 \text{ in (7.4.5)},$$

we come to the Hagen–Poiseuille–Stokes flow [Hagen, 1839; Poiseuille, 1840; Stokes, 1845] $w\mathbf{k}$ with the parabolic velocity profile

$$w = w^{(p)} = w_p(1 - X^2). \tag{7.5.2}$$

Chapter 8

Heat Conductivity

8.1. Heat balance

Initially, *heat* was thought to be a fluid that flows from every warmed volume V through the boundary ∂V in the process of *heat loss* that was first described exactly by the mathematical physicist Jean–Baptiste Joseph Fourier [Fourier, 1822], who found that the volume V gives away heat per second in a quantity equal to the *boundary thermal flow*

$$-\int_{\partial V} \kappa \nabla T \cdot d\mathbf{S} = -\int_V \nabla \cdot (\kappa \nabla T) dV, \quad \kappa > 0, \tag{8.1.1}$$

of contact elements $-\kappa \nabla T \cdot d\mathbf{S}$ as a negative proportion of the temperature T gradient ∇T directed normally to the boundary ∂V from within the volume V and taken with the *coefficient of thermal conductivity* κ to be reduced integrally to the volume density $-\nabla \cdot (\kappa \nabla T)$ in (8.1.1) due to the divergence theorem.

In other words, the *Fourier law* (8.1.1) gives the required quantity of heat $-\nabla \cdot (\kappa \nabla T) dV$ lost by every volume element dV through boundary elements $-\kappa \nabla T \cdot d\mathbf{S}$.

As to the corresponding inverse process of *heat gain*, it is due mainly to the observations and experiences of the military engineer Carnot [1824] and the ship physician Mayer [1842], who have revealed what is now known as the *First Law of Thermodynamics*, or the *Heat Equation*, in which, when reduced by the thermal flow (8.1.1), the *heat*$\int_V \dot{q} dV$ obtained by every volume V not only increases the *internal heat* $\int_V e\rho dV$ at the rate of $\frac{d}{dt}\int_V e\rho dV$ but also makes the *pressure work*

$$\int_V p\frac{d}{dt}dV = \int_V pCdV, \quad C = \nabla\cdot\mathbf{u}, \tag{8.1.2}$$

that goes per second to the *expansion* or *compression* of every dV with the positive or negative rate $\frac{d}{dt}dV > 0$ or $\frac{d}{dt}dV < 0$, respectively:

$$\int_V \dot{q}dV + \int_{\partial V}\kappa\nabla T\cdot d\mathbf{S} = \frac{d}{dt}\int_V e\rho dV + \int_V p\frac{d}{dt}dV \text{ for any } V = V^t. \tag{8.1.3}$$

Accounting for (8.1.1) and the fact that the integral rate is the integral of rates in (3.3.2), or

$$\frac{d}{dt}\int_V e\rho dV = \int_V \frac{d}{dt}e\rho dV, \quad V = V^t,$$

for an arbitrarily small V, the heat equation (8.1.2) reduces to volume elements,

$$\dot{q}dV + \nabla\cdot(\kappa\nabla T)dV = \frac{d}{dt}e\rho dV + p\frac{d}{dt}dV,$$

and, with mass conservation and volume rate,

$$\frac{d}{dt}\rho dV = 0 \text{ and } \frac{d}{dt}dV = CdV, \quad C = \nabla\cdot\mathbf{u},$$

takes both the common *differential* and *divergence* forms

$$\rho e_t + \rho\mathbf{u}\cdot\nabla e - \nabla\cdot(\kappa\nabla T) = \dot{q} - pC \quad \left(\text{for } \frac{de}{dt} = e_t + \mathbf{u}\cdot\nabla e\right)$$

and

$$(\rho e)_t + \nabla\cdot(\rho e\mathbf{u} - \kappa\nabla T) = \dot{q} - pC \quad (\text{for } \rho_t + \nabla\cdot\rho\mathbf{u} = 0), \tag{8.1.4}$$

respectively.

8.2. The heat source in a fluid and a gas

In fact, the mass density of *internal heat* e proves to be proportional to the *absolute temperature* $T > 0$ of Kelvin,

$$e = c_v T, \quad c_v = \text{const} > 0, \tag{8.2.1}$$

after James Joule measured in 1850 the so-called *mechanical equivalent* of heat c_v [Joule, 1850], now referred to as the *specific heat capacity at a constant volume* [Landau and Lifshitz, 1959].

Joule's measurements of heating [Joule, 1850] dealt with either water or mercury mixing for constant density and viscosity,

$$\rho = \text{const} > 0 \text{ (hence, } C = \nabla \cdot \mathbf{u} = 0) \quad \text{and} \quad \mu = \text{const} > 0,$$

to exclude both pressure power (8.1.2) and any possible rheology,

$$\nabla \mu = \mathbf{0},$$

in a fixed volume $V = V^0$ bounded with walls that are not only rigid and no-slipping as

$$\mathbf{u}|_{\partial V} = \mathbf{0}, \quad \text{hence, } \mathbf{E}|_{\partial V} = \mathbf{0} \quad \text{in (7.3.5)} \quad \text{and}$$
$$\rho e \mathbf{u}|_{\partial V} = \mathbf{0} \quad \text{in (8.1.4)}$$

but also *adiabatic* (or thermos-isolated) to make negligible both the heat loss (8.1.1),

$$\int_{\partial V} \kappa \nabla T \cdot d\mathbf{S} = 0,$$

and *energy flow*

$$\int_{\partial V} \mathbf{E} \cdot d\mathbf{S} = \int_V (\nabla \cdot \mathbf{E}) dV = 0.$$

The main albeit invisible part of Joule's mixer "a" was a metal grid "b" rotated with a paddle wheel, as shown in Fig. 8.2.1.

When made with weights falling in the constant gravitation field $\mathbf{g} = -g\mathbf{j}$ of the Earth, the rotation produces such an intensive

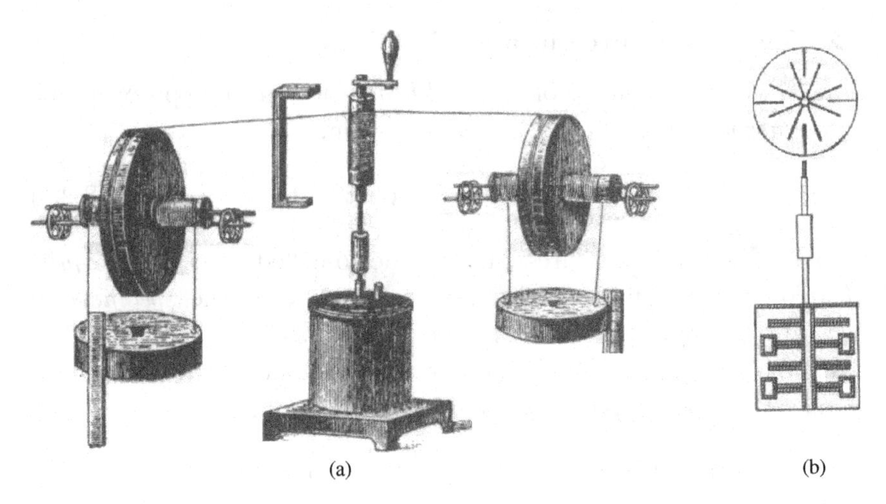

Figure 8.2.1. Joule's apparatus for heating (a) by mixing (b) [Joule, 1850].

vorticity field ω that it mixes a fluid and thereby excludes practically the work of the body force:

$$\int_V \rho \mathbf{u} \cdot \mathbf{g} dV = -\rho g \mathbf{j} \int_V v dV = 0.$$

As a result, the negative and positive rates of the living force and internal heat delivered by the power and heat equations, (7.3.5) and (8.1.4), are

$$\left(\int_V K \rho dV \right)_t = - \int_V \mu |\omega|^2 dV \quad \text{and} \quad \left(\int_V \rho e dV \right)_t$$
$$= \int_V \dot{q} dV, \quad V = V^0, \tag{8.2.2}$$

respectively.

According to the measurements of heating by mixing [Joule, 1850], the sum of rates, or the rate of the corresponding *full energy* $E = e + K$ per mass, reduces to zero,

$$\left(\int_V E \rho dV \right)_t = 0 \quad \text{for } E = e + K. \tag{8.2.3}$$

Then, in Joule's mixer (Fig. 8.2.1),

$$\frac{dT}{dt} = \frac{\mu}{c_v} \frac{\int_V |\omega|^2 dV}{\int_V \rho dV}, \quad \omega = \nabla \times \mathbf{u}, \quad \text{for } T = T(t),$$

which follows immediately from (8.2.1)–(8.2.3).

Thus, it is the *molecular motion* as invisible fluctuations of molecules whose tremors were first detected in 1827 by Robert Brown that initiates fluid dynamics by means of the specific pressure force $-\nabla p$ and at the same time concludes this part of physics with mixing power $\mu|\omega|^2$ producing the heat e at the temperature T and thereby expelling mechanics from its classical limits (Chapters 4–7) to *thermodynamics* (8.1).

We can conclude that

Theorem 8.2.1 (on the fluid source of heat). *Any arbitrarily small fixed fluid volume* $V = V^0$ *filled with vortex flow* $\mathbf{u}$, $\omega = \nabla \times \mathbf{u} \neq \mathbf{0}$, *proves to be a source of heat in the quantity*

$$\begin{aligned} \dot{q} = \dot{q}_f &= \left(\frac{4}{3} + \varsigma\right) \mu C^2 + \mu|\omega|^2 - 2\mathbf{D} \cdot \nabla\mu, \\ C &= \nabla \cdot \mathbf{u}, \quad \omega = \nabla \times \mathbf{u}, \\ \mathbf{D} &= \mathbf{u} \cdot \nabla \mathbf{u} - C\mathbf{u} = \nabla K + \omega \times \mathbf{u} - C\mathbf{u}, \\ K &= \frac{|\mathbf{u}|^2}{2} \quad (\varsigma = \text{const} \geq 0), \end{aligned} \tag{8.2.4}$$

expended per mass and per second in every particle $\mathbf{r}$, *namely, emitted as* $\dot{q}_f > 0$, *for a Newtonian medium, with* $\nabla\mu = \mathbf{0}$, *or, possibly, absorbed as* $\dot{q}_f \leq 0$, *for an elastic continuous one, with* $\nabla\mu \neq \mathbf{0}$, *in the heat equation* (8.1.4), *in the sense that when free in large, as* $\int_V \mathbf{u} \cdot \mathbf{g} dV = 0$ *(e.g., for* $\mathbf{g} = \mathbf{0}$*), and bounded by the walls* ∂V *that are not only impenetrable,* $\mathbf{u} \cdot d\mathbf{S}|_{\partial V} = \mathbf{0}$, *but also no-slipping,* $\mathbf{u} \cdot d\mathbf{S}|_{\partial V} = \mathbf{0}$, *and adiabatic,* $\nabla T \cdot d\mathbf{S}|_{\partial V} = \mathbf{0}$, *the volume* V *keeps the energy conservation* (8.2.3).

Proof. Really, adding the power identity (7.2.4), or

$$(\rho K)_t + \nabla \cdot (\rho K \mathbf{u} + \mathbf{E}) - \rho \mathbf{u} \cdot \mathbf{g} = pC - \dot{q}_f,$$

to the heat equation (8.1.4), or

$$(\rho e)_t + \nabla \cdot (\rho e \mathbf{u} - \kappa \nabla T) = \dot{q} - pC,$$

we come to the local *full energy E balance*, or the *energy equation*

$$\begin{aligned}
(\rho E)_t + \nabla \cdot \left(\rho E \mathbf{u} + \mathbf{E} - \frac{\kappa}{c_v}\nabla e\right) &= \rho \mathbf{u} \cdot \mathbf{g} + \dot{q} - \dot{q}_f,\\
E &= e + K,\\
\mathbf{E} &= \vec{\mathbf{p}} \cdot \mathbf{u} + 2\mu \mathbf{D}.
\end{aligned} \tag{8.2.5}$$

In our case, when the volume V is *conservative* as

$$\int_V \left(\nabla \cdot \begin{pmatrix} \rho E \mathbf{u} + \mathbf{E} \\ -\dfrac{\kappa}{c_v}\nabla e \end{pmatrix}\right) dV = \int_{\partial V} \begin{pmatrix} \rho E \mathbf{u} + \vec{\mathbf{p}} \cdot \mathbf{u} \\ +2\mu \mathbf{u} \cdot \nabla \mathbf{u} \\ -2\mu C \mathbf{u} - \kappa \nabla T \end{pmatrix} \cdot d\mathbf{S} = 0 \quad \text{and}$$

$$\int_V \mathbf{u} \cdot \mathbf{g} dV = 0,$$

we obtain

$$\left(\int_V E\rho dV\right)_t = \int_V \mathbf{u} \cdot \mathbf{g} dV + \int_V (\dot{q} - \dot{q}_f) dV = \int_V (\dot{q} - \dot{q}_f) dV;$$

so, for an arbitrary small V, equalities (8.2.3) and (8.2.4) are equivalent, which concludes the proof. □

Thus, the specification of the heat source $\dot{q}$ in (8.1.4) needs clarification by the sum $\dot{q}_e + \dot{q}_f$ of the *external* (or applied) heat $\dot{q}_e$ prescribed as a function of time t and point $\mathbf{r}$ and the fluid source $\dot{q}_f$:

$$\dot{q} = \dot{Q} + \dot{q}_f, \quad \dot{Q} = \dot{q}_e(t, \mathbf{r}). \tag{8.2.6}$$

8.3. The set of conservation laws

For the stress–strain dependence (7.1.5), internal heat (8.2.1), and full heat source (8.2.6), both the heat balance (8.1.4) and the dynamic equilibrium (7.1.8) give us the required full set of *conservation laws*

$$\begin{pmatrix} \rho \\ \rho\mathbf{u} \\ \rho e \end{pmatrix}_t + \nabla \cdot \begin{pmatrix} \rho\mathbf{u} \\ \rho\mathbf{u}\mathbf{u} + \vec{\mathbf{p}} \\ \rho e\mathbf{u} - (\kappa/c_v)\,\nabla e \end{pmatrix} = \begin{pmatrix} 0 \\ \rho\mathbf{g} \\ \dot{q} - pC \end{pmatrix}, \quad T = \frac{e}{c_v},$$

$$\dot{q} = \dot{q}_e + \dot{q}_f, \quad c_v = \text{const} > 0, \tag{8.3.1}$$

with (7.1.5), or

$$\vec{\mathbf{p}} = P\vec{\mathbf{e}} - \mu\vec{\tau}, \quad \vec{\tau} = \mathbf{u_r} + \nabla\mathbf{u}, \quad P = p + \mu\left(\frac{2}{3} - \zeta\right)C,$$

$$C = \nabla \cdot \mathbf{u}, \quad \zeta = \text{const} \geq 0, \tag{8.3.2}$$

for the *classical liquid* or *gas*, with the corresponding *equation of state* as

$$\begin{aligned} &\text{either} \quad \rho = \text{const} > 0 \ (\text{for } C = 0, \ \text{hence}, \ \nabla C = \mathbf{0}) \\ &\text{or } p = (c_p - c_v)\rho T = (\gamma - 1)\rho e \ (\nabla C \neq \mathbf{0}) \\ &\text{where } \gamma = \tfrac{c_p}{c_v} = \text{const} > 1 \end{aligned} \tag{8.3.3}$$

is the *gas constant* of Mayer [1842] prescribed for the specific heat capacities c_v and c_p taken at *constant volume* and *pressure*, respectively.

With the help of identities

$$\nabla \cdot \vec{\tau} = \nabla \cdot \mathbf{u_r} + \nabla \cdot \nabla\mathbf{u} = \nabla C + \Delta\mathbf{u}, \quad C = \nabla \cdot \mathbf{u},$$

and what could be referred to as the *redundant fluid force* $\mathbf{P}_\zeta$ in the volume density

$$-\nabla \cdot \vec{\mathbf{p}} = -\nabla p + \frac{\mu}{3}\nabla C + \mu\Delta\mathbf{u} + \mathbf{P}_\zeta,$$

$$\mathbf{P}_\zeta = \zeta\nabla\mu C - \frac{2}{3}C\nabla\mu + (\nabla\mu) \cdot \vec{\tau}, \tag{8.3.4}$$

to be trivial

$$\mathbf{P}_\zeta = \mathbf{0} \text{ when } \nabla\mu = \mathbf{0} \quad \text{and} \quad \zeta C = 0 \quad (\text{i.e.,} \quad \zeta = 0 \quad \text{or} \quad C = 0), \tag{8.3.5}$$

one can rewrite (8.3.1) as

$$\begin{pmatrix} \rho \\ \rho\mathbf{u} \\ \rho e \end{pmatrix}_t + \begin{pmatrix} \nabla\cdot\rho\mathbf{u} \\ \nabla\cdot\rho\mathbf{u}\mathbf{u} - \frac{\mu}{3}\nabla C - \mu\Delta\mathbf{u} + \nabla p \\ \nabla\cdot\rho e\mathbf{u} - \nabla\cdot(\kappa/c_v)\,\nabla e \end{pmatrix} = \begin{pmatrix} 0 \\ \rho\mathbf{g} + \mathbf{P}_\zeta \\ \dot{q} - pC \end{pmatrix},$$

$$T = \frac{e}{c_v}, \tag{8.3.6}$$

$$\dot{q} = \dot{q}_e + \dot{q}_f.$$

Accounting for the fluid heat source (8.2.4) and adding the power identity (7.2.4) to the heat equation (8.1.4), we have the full energy balance (8.2.5) for an *external heat source* $\dot{q}_e$ prescribed per mass and per second, which is equivalent to the heat equation (8.1.4).

With the full energy balance (8.3.7), the set of conservation laws (8.3.1) takes the form

$$\begin{pmatrix} \rho \\ \rho\mathbf{u} \\ \rho E \end{pmatrix}_t + \nabla\cdot\begin{pmatrix} \rho\mathbf{u} \\ \rho\mathbf{u}\mathbf{u} + \vec{\mathbf{p}} \\ \rho E\mathbf{u} + \mathbf{E} - (\kappa/c_v)\,\nabla e \end{pmatrix} = \begin{pmatrix} 0 \\ \rho\mathbf{g} \\ \rho\mathbf{u}\cdot\mathbf{g} + \dot{q} \end{pmatrix},$$

$$T = \frac{e}{c_v}, \tag{8.3.8}$$

$$\dot{q} = \dot{q}_e + \dot{q}_f,$$

for invariable *fluid stress* (8.3.2) and classical liquid or gas equations of state (8.3.3).

Chapter 9

A Centrifuge

9.1. A caught tornado

The set of laws (8.3.8) is illustrated using the known axially rotating cylinder as a *rotor* of a gas centrifuge [Kofman, 1941] in Fig. 9.1.1, where there is nothing but a gas mixture of *uranium hexafluoride*, UF_6 [Borisevich and Wood, 2000], rotated with the rate $\mathbf{u}$ while swirling and heating with the strain $\mathbf{u_r}$ at the *characteristic* velocity and temperature

$$v_* \approx 600\frac{\mathrm{m}}{\mathrm{s}} \quad \text{and} \quad T_* \approx 300\ \mathrm{K} = 273.15 + 26.85°\mathrm{C},$$

when approximated [Belotserkovskii *et al.*, 2011] by the ideal gas medium equation of state

$$p = (c_p - c_v)\rho T = (\gamma - 1)\rho e = \gamma R\rho T = c_s^2\rho, \quad \gamma = \frac{c_p}{c_v},$$

$$R = \frac{\gamma - 1}{\gamma}c_v, \quad c_s = \sqrt{\gamma RT},$$

with heat volume–constant capacity, its pressure–constant ratio, gas constant, speed of sound

$$c_v = 349.65\frac{m^2}{s^2 \cdot K}, \quad \gamma = 1.065,$$

$$R = 21.34\frac{m^2}{s^2 K}, \quad c_s = 83\frac{m}{s},$$

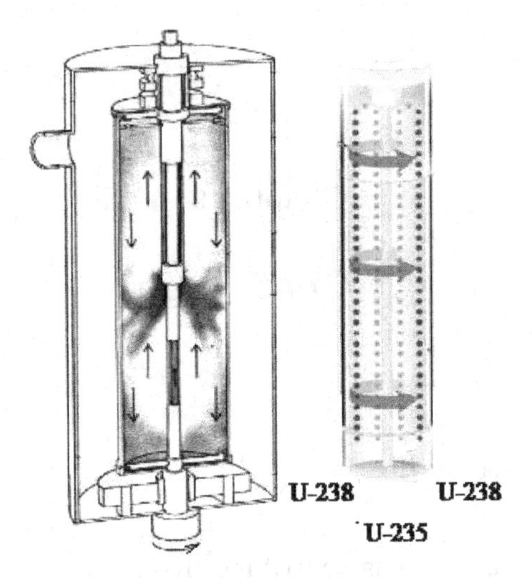

Figure 9.1.1. A centrifugal rotor as a tube (the inner cylinder to the left) with the vertical arrows of a counter-flow and the rigid-body rotation separating a gas mixture (to the right).

dynamic viscosity, heat conductivity

$$\mu = 17.05 \times 10^{-6} \frac{kg}{m \cdot s}, \qquad \kappa = 0.00783 \frac{kg \cdot m}{s^3 \cdot K},$$

and the related *Prandtl, Brinkman and Eckert numbers* [Mulliken, 1922]

$$\mathrm{Pr} = \frac{\gamma \mu}{\kappa / c_v} \simeq 0.81, \quad \mathrm{Br} = \frac{\mu v_*^2}{\kappa T_*} \simeq 2.61, \quad \text{and}$$

$$\mathrm{Ec} = \frac{v_*^2}{\gamma c_v T_*} \simeq 3.22. \ (\mathrm{Pr} \cdot \mathrm{Ec} = \mathrm{Br}).$$

As this takes place, the main isotope component U−235 (0.7114%) of UF_6 is required to be separated basically from U−238 [Borisevich and Wood 2000] with the help of the *vortex law* (of §6.4) that can be traced back to the previous century: when *tea leaves* are commonly stirred in a cup of tea, they unexpectedly begin to cluster near the center of rotation rather than fly apart to the walls of the cup as in a merry-go-round.

Due to the viscosity of and no-slip conditions on the side cylindrical rigid boundary, the gas mixture is unwound up to the hypersonic rates v_* to be seven or more times as much as the speed of sound c_s in the *centrifugal method* suggested first by Lindemann and Aston [1919] based on the above-mentioned vortex law that prescribes the pressure p to fall to the center of fluid rotation and thereby separate the lighter molecules (U−235) from the weightier ones (U−238) with the first addition ∇p in the fluid force density (8.3.4).

Then, Mulliken [1922] conceived the idea of increasing the rate of separation through the counter-flow circulation, although his opinion on how to organize such a circulation was initially pessimistic ("This increase would, however, hardly be worth the added difficulties" [Mulliken, 1922]). Later, Yu. B. Khariton proved that for the only rigid-body like rotation of a mixture (free of counter-flow), in the *tube*, or the rotor of a typical centrifuge [Kofman, 1941], centrifugal separation would have no advantages over the traditional gas diffusion method [Khariton, 1937]. Eventually, Urey [1939] yet came back to the idea of counter-flow and developed the evaporative centrifugal method of Mulliken in a tube ("Liquid would be pumped into the top of this tube, would flow to the bottom, where it would be boiled by radiant heat. The vapour would flow upward mostly through the center of the tube and escape at the top" [Urey, 1939]), as in Fig. 9.1.2.

The mechanics of centrifugal flow had been understood finally by Dirac, who recognized the fact that the counter-flow vortex in Fig. 9.1.2 has to take the place of the only rigid-body rotation in Fig. 9.1.1 to turn a rotating gas tube to what he referred to as a *self-fractionating centrifuge* [Dirac, 1946].

In other words, as per Dirac, the latter proved to be quite similar to the natural *tornado* caught in a gas tube to permanently separate the mixture (by the rigid-body rotation) and simultaneously renew the separation (by the counter-flow circulation).

After joining a research group of the first atomic project "Tube Alloys", UK, and finishing the related analysis in his subsequently declassified and published report BR-42 [Dirac, 1946], Dirac wrote

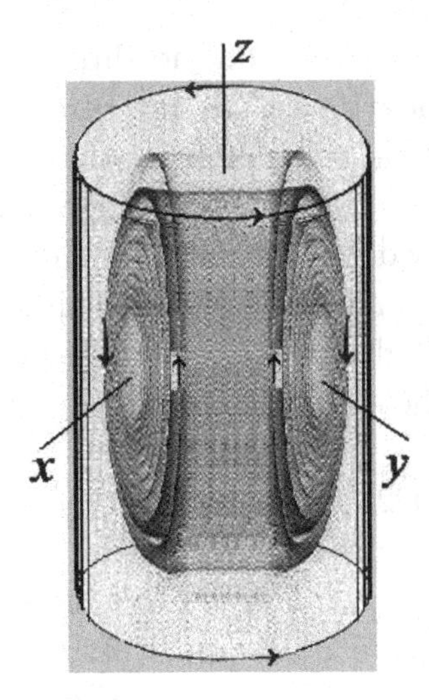

Figure 9.1.2. A counter-flow vortex.

therein to R. E. Peierls on 11 May 1942: "I have written up my work about the circulation in a self-fractionating centrifuge and have enclosed it forthwith. I have not done any calculations on the effect of temperature variation along the tube, but I believe the effect is important and someone should work it out" [Dirac, 1946].

Following [Troshkin, 2010a, 2010b; Belotserkovskii *et al.*, 2011], let us try to trace the effect mentioned.

9.2. Rigid-body and torsional rotations of a gas tube

We shall start with the rotation in Fig. 9.1.1.

For a gas medium in (8.3.3), with axial symmetry (7.3.1), polar laplacian

$$\Delta e = \nabla \cdot \nabla e = \begin{pmatrix} \mathbf{I}\partial_r + \mathbf{k}\partial_z \\ +\mathbf{J}\dfrac{\partial_\theta}{r} \end{pmatrix} \cdot \begin{pmatrix} \mathbf{I}e_r \\ +\mathbf{k}e_z \end{pmatrix} = \begin{pmatrix} e_{rr} \\ +e_{zz} \end{pmatrix} + \frac{\mathbf{J} \cdot \mathbf{J}e_r}{r}$$

$$= \frac{(re_r)_r}{r} + e_{zz}, \quad e = e(t, r, z),$$

and relevant identities

$$\nabla \cdot \rho e\mathbf{u} = \begin{pmatrix} \mathbf{I}\partial_r \\ +\mathbf{k}\partial_z \end{pmatrix} \cdot \begin{pmatrix} \mathbf{I}\rho e\hat{u} \\ +\mathbf{J}\rho e\hat{v} \\ +\mathbf{k}\rho e w \end{pmatrix} + \frac{1}{r}\mathbf{J} \cdot \begin{pmatrix} \mathbf{I}_\theta \rho e\hat{u} \\ +\mathbf{J}_\theta \rho e\hat{v} \end{pmatrix}$$

$$= \frac{(r\rho e\hat{u})_r}{r} + (\rho e w)_z$$

in (8.3.1) and

$$\frac{1}{r}\left(r^2\left(\frac{\hat{v}}{r}\right)_r\right)_r = \frac{1}{r}\left(r^2\left(\frac{\hat{v}_r}{r} - \frac{\hat{v}}{r^2}\right)\right)_r = \frac{1}{r}(r\hat{v}_r - \hat{v})_r = \hat{v}_{rr}$$

in (7.3.2), the system of conservation laws (8.3.1)–(8.3.3) takes the form of equations

$$\rho_t + \hat{u}\rho_r + w\rho_z = -C\rho, \quad C = \frac{(r\hat{u})_r}{r} + w_z,$$

$$(\rho\hat{u})_t + \frac{(r\rho\hat{u}\hat{u})_r}{r} + (\rho\hat{u}w)_z - \frac{\rho\hat{v}^2}{r} - 2\frac{\mu\,(r\hat{u}_r)_r}{r}$$
$$- \mu\,(\hat{u}_z + w_r)_z + P_r = g^r,$$

$$(\rho\hat{v})_t + \frac{(r\rho\hat{v}\hat{u})_r}{r} + (\rho\hat{v}w)_z + \frac{\rho\hat{v}\hat{u}}{r} - \mu\hat{v}_{rr} - \mu\hat{v}_{zz} = g^\theta,$$

$$(\rho w)_t + \frac{(r\rho w\hat{u})_r}{r} + (\rho w w)_z - \frac{\mu\,(r\,(\hat{u}_z + w_r))_r}{r} - 2\mu w_{zz} + P_z = g^z,$$

$$\text{and} \quad (\rho e)_t + \frac{(r\rho e\hat{u})_r}{r} + (\rho e w)_z - \kappa\Delta T = \dot{q}_e + \dot{q}_f - pC, T = \frac{e}{c_v},$$

for

$$\dot{q}_e = \dot{q}_e(t,z,r), \quad \dot{q}_f = \left(\frac{4}{3} + \zeta\right)\mu C^2 + \mu\,|\omega|^2,$$

$$T = \frac{e}{c_v}, \quad P = p + \left(\frac{2}{3} - \zeta\right)\mu C, \quad \zeta = \text{const} \geq 0,$$

$$p = (\gamma - 1)\rho e, \quad \gamma = \frac{c_p}{c_v} > 1,$$

$$\text{and} \quad c_v, c_p, \kappa, \mu = \text{const} > 0,$$

or, accounting for (7.3.3) and enstrophy

$$|\omega|^2 = \hat{v}_z^2 + (\hat{u}_z - w_r)^2 + \left(\frac{(r\hat{v})_r}{r}\right)^2,$$

equations

$$\rho_t + \frac{(r\rho\hat{u})_r + (r\rho w)_z}{r} = 0, \quad p = (\gamma - 1)\rho e, \quad \gamma = \frac{c_p}{c_v} > 1,$$

$$(\rho\hat{u})_t + \frac{(r\rho\hat{u}\hat{u})_r + (r\rho\hat{u}w)_z}{r} - \frac{\rho\hat{v}^2}{r} - \mu\left(\Delta\hat{u} + \frac{\hat{u}_r}{r}\right)$$
$$+ \left(p - \left(\frac{1}{3} + \zeta\right)\mu C\right)_r = g^r,$$

$$(\rho\hat{v})_t + \frac{(r\rho\hat{v}\hat{u})_r + (r\rho\hat{v}w)_z}{r} + \frac{\rho\hat{v}\hat{u}}{r} - \mu\left(\Delta\hat{v} - \frac{\hat{v}_r}{r}\right) = g^\theta,$$

$$(\rho w)_t + \frac{(r\rho w\hat{u})_r + (r\rho ww)_z}{r} - \mu\Delta w + \left(p - \left(\frac{1}{3} + \zeta\right)\mu C\right)_z = g^z,$$

$$(\rho e)_t + \frac{(r\rho e\hat{u})_r + (r\rho ew)_z}{r} - \kappa\Delta T = \dot{q}_e + \dot{q}_f - pC, \quad e = c_v T,$$

$$C = w_z + \frac{(r\hat{u})_r}{r}, \quad \Delta T = T_{zz} + \frac{(rT)_r}{r},$$

$$\dot{q}_f = \left(\frac{4}{3} + \zeta\right)\mu C^2 + \mu|\omega|^2, \tag{9.2.1}$$

to keep the mass–momentum–heat balance for a Newtonian gas medium in motion.

For negligible external heart source and body accelerations

$$\dot{q}_e = g^r = g^\theta = g^z = 0,$$

in the rotor of *aspect ratio* α,

$$V: 0 < r < a, 0 < z < h = \alpha a, \quad \alpha = \frac{h}{a}, \quad a, h = \text{const} > 0,$$

in Fig. 9.2.1, what we have first is the *rotation*

$$\mathbf{u} = \hat{v}(\bar{z}, \bar{r})\mathbf{J} \ (w = \hat{u} = C = 0), \quad \bar{z} = \frac{z}{h}, \quad \bar{r} = \frac{r}{a} = \alpha\frac{r}{h},$$

balanced with equations (9.2.1) reduced to relations

$$p = (\gamma - 1)\rho e, \quad p_{\bar{r}} = \frac{\rho \hat{v}^2}{\bar{r}} \text{ and } p_{\bar{z}} = 0, \text{ or } (\rho \hat{v}^2)_{\bar{z}} = \bar{r} p_{\bar{r}\bar{z}} = \bar{r} p_{\bar{r}\bar{z}} = 0,$$

$$\Delta \hat{v} - \frac{\hat{v}_r}{r} = \frac{\hat{v}_{\bar{z}\bar{z}} + \alpha^2 \hat{v}_{\bar{r}\bar{r}}}{h^2} = 0,$$

$$-h^2 \Delta T = -T_{\bar{z}\bar{z}} - \alpha^2 \frac{(\bar{r} T_{\bar{r}})_{\bar{r}}}{\bar{r}} = \frac{h^2 \dot{q}_f}{\kappa}$$

$$= \frac{\mu}{\kappa} \left(\hat{v}_{\bar{z}}^2 + \alpha^2 \left(\frac{(\bar{r}\hat{v})_{\bar{r}}}{\bar{r}} \right)^2 \right) \text{ in V}$$

for *rigid-body* boundary conditions

$$\begin{gathered} \rho|_{\bar{r}=1} = \rho_*, \quad \hat{v}|_{\bar{r}=0} = 0, \quad \hat{v}|_{\bar{r}=1} = v_* \quad \text{and} \\ T|_{\bar{r}=1} = T_*, \rho_*, \\ v_*, T_* = \text{const} > 0, \end{gathered} \tag{9.2.2}$$

and the *axial* and *radial factors*, m, ξ, ζ and N, η, n, of density, velocity and temperature,

$$\begin{gathered} \rho = \rho_* m(\bar{z}) N(\bar{r}), \ \hat{v} = v_* \xi(\bar{z}) \eta(\bar{r}) \quad \text{and} \quad T = T_* \text{Br} \zeta(\bar{z}) n(\bar{r}) \\ \text{for} \quad p = p_* m \zeta N n, \quad p_* = (\gamma - 1) c_v \rho_* T_* \text{Br} \quad \text{and} \quad \text{Br} \simeq (2.61), \end{gathered} \tag{9.2.3}$$

to satisfy the necessary system of ordinary differential equations

$$\begin{gathered} (m\zeta)_{\bar{z}} = (m\xi^2)_{\bar{z}} = \xi \eta_{\bar{r}\bar{r}} + \xi_{\bar{z}\bar{z}} \eta = 0, \\ \zeta_{\bar{z}\bar{z}} n + \alpha^2 \zeta \frac{(\bar{r} n_{\bar{r}})_{\bar{r}}}{\bar{r}} + \xi_{\bar{z}}^2 \eta^2 + \alpha^2 \xi^2 \left(\frac{(\bar{r}\eta)_{\bar{r}}}{\bar{r}} \right)^2 = 0, \end{gathered} \tag{9.2.4}$$

$$\text{and} \quad \zeta m (Nn)_{\bar{r}} = \xi^2 m \Gamma \frac{\eta^2}{n \bar{r}} N n, \quad \Gamma = \frac{\gamma \text{Ec}}{\gamma - 1} \simeq 52.7, 0 < \bar{r} < 1.$$

with proper restrictions

$$\begin{gathered} \eta|_{\bar{r}=0} = 0, \quad \eta|_{\bar{r}=1} = n|_{\bar{r}=1} = N|_{\bar{r}=1} = 1 \quad \text{and} \\ \eta, n, N > 0 \quad \text{for} \quad 0 < \bar{r} < 1, \end{gathered} \tag{9.2.5}$$

which with identities

$$\zeta_{\bar{z}\bar{z}}n + \alpha^2\zeta\frac{(\bar{r}n_{\bar{r}})_{\bar{r}}}{\bar{r}} + \xi_{\bar{z}}^2\eta^2 + \alpha^2\xi^2\left(\frac{(\bar{r}\eta)_{\bar{r}}}{\bar{r}}\right)^2$$

$$= \alpha^2\frac{(\bar{r}n_{\bar{r}})_{\bar{r}}}{\bar{r}} + \alpha^2\left(\frac{(\bar{r}\eta)_{\bar{r}}}{\bar{r}}\right)^2 = 0$$

for

$$m = \xi = \zeta = 1, \quad \eta(\bar{r}) = \bar{r}, \quad n(\bar{r}) = 2 - \bar{r}^2$$

$$\text{and} \quad Nn = e^{-\Gamma\int_{\bar{r}}^{1}\frac{\eta^2(\bar{r})d\bar{r}}{n(\bar{r})\bar{r}}} = n^{-\frac{\Gamma}{2}} \tag{9.2.6}$$

gives us the required solution (9.2.3) for the rigid-body rotation in Fig. 9.1.1 that provides a double increase in temperature and the $2^{26.35} > 85000000$-fold pressure drop $2^{-26.35}$ to rarify the gas up to the vacuum around the axis $\bar{r} = 0$ compared to the boundary $\bar{r} = 1$ as in Fig. 9.2.1.

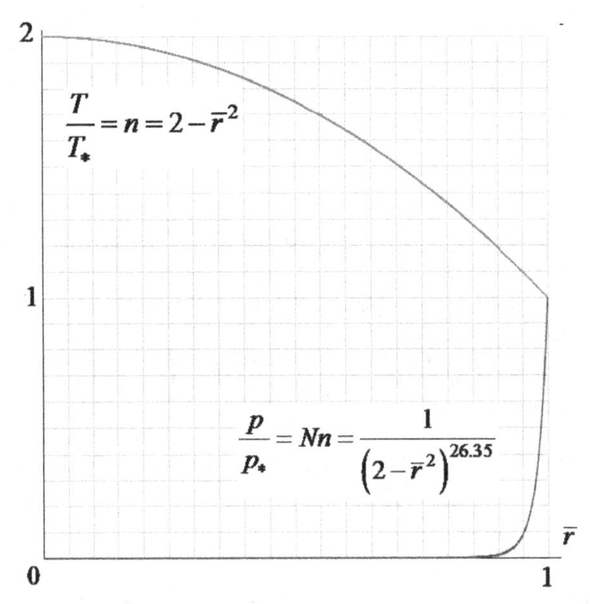

Figure 9.2.1. The Internal bottom pressure drop and the upper temperature increase of rigid-body rotation.

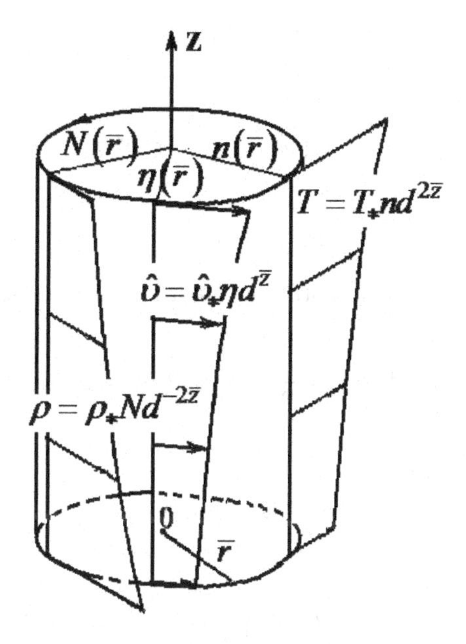

Figure 9.2.2. Torsional rotation.

Further replacement [Troshkin, 2010a, 2010b] of constant rigid-body factors $m = \xi = \zeta = 1$ and boundary conditions (9.2.2) by the variable *torsional* ones in (9.2.3) caused by the difference in azimuthal velocities

$$\hat{\upsilon}|_{\substack{\bar{r}=1\\ \bar{z}=0}} = \hat{\upsilon}_* \quad \text{and} \quad \hat{\upsilon}|_{\substack{\bar{r}=1\\ \bar{z}=1}} = d\hat{\upsilon}_*, \quad \text{for } d = \text{const} > 1,$$

depicted in Fig. 9.2.2,
for

$$m = d^{-2\bar{z}}, \quad \xi = d^{\bar{z}} \quad \text{and} \quad \zeta = d^{2\bar{z}} = \xi^2,$$

and

$$m\zeta = m\xi^2 = 1, \quad \xi_{\bar{z}} = \delta\xi, \quad \zeta_{\bar{z}} = 2\delta\zeta \quad \text{for} \quad \delta = \ln d,$$

with

$$\rho|_{\bar{r}=1} = \rho_* m(\bar{z}), \quad \hat{\upsilon}|_{\bar{r}=0} = 0, \quad \hat{\upsilon}|_{\bar{r}=1} = \upsilon_*\xi(\bar{z}) \quad \text{and}$$
$$T|_{\bar{r}=1} = T_*\zeta(\bar{z}), \tag{9.2.7}$$

changes no restrictions (9.2.5) for (9.2.4), or for

$$\eta_{\bar{r}\bar{r}} + \delta^2 \eta = 0 \quad \text{and} \quad 4\delta^2 n + \alpha^2 \frac{(\bar{r} n_{\bar{r}})_{\bar{r}}}{\bar{r}} + \delta^2 \eta^2 + \alpha^2 \left(\frac{(\bar{r}\eta)_{\bar{r}}}{\bar{r}} \right)^2 = 0, \tag{9.2.8}$$

but supplies the torsional rotation with new radial factors

$$\eta = \sin \delta \bar{r} \quad \text{for } \delta = \frac{\pi}{2} = 1,570796\ldots \ (\sin \delta = 1),$$

and n in (9.2.8) for the case of respectively the *short rotor*, with aspect ratio $\alpha < \delta/2 = \pi 4 < 1$, so that

$$-\frac{1}{\bar{r}}(\bar{r} n_{\bar{r}})_{\bar{r}} - \beta^2 n = f = (\beta\eta)^2 + \left(\frac{1}{\bar{r}} (\bar{r}\eta)_{\bar{r}} \right)^2, \quad n > 0, \quad 0 < \bar{r} < 1,$$
$$n|_{\bar{r}=1} = 1 \tag{9.2.9}$$
$$\text{for } \beta = \frac{\delta}{\alpha} = \frac{\pi}{2\alpha} \geq 2, \text{ or } \alpha \leq \frac{\pi}{4}, \text{ i.e. } h \leq \frac{\pi}{4} a < a,$$

for the positive source of heat

$$f = (\beta\eta)^2 + \left(\frac{(\bar{r}\eta)_{\bar{r}}}{\bar{r}} \right)^2 > 0, \quad 0 < \bar{r} < 1 \ (\eta = \sin \delta \bar{r}).$$

In doing so, the temperature $n = 2 - \bar{r}^2$ of the rigid body rotation $\eta = \bar{r}$ would reduce f by a constant

$$2\beta^2 = (\beta\bar{r})^2 + \left(\frac{1}{\bar{r}} \left(\bar{r}^2\right)_{\bar{r}} \right)^2 + \frac{1}{\bar{r}} \left(\bar{r} \left(2 - \bar{r}^2\right)_{\bar{r}} \right)_{\bar{r}} + \beta^2 (2 - \bar{r}^2)$$

degenerating in the case without torsion: $\beta = 0$.

Consequently, to hold the heat balance in the torsional rotation $\eta = \sin \delta \bar{r}$, there is required the additional *enlargement* ϕ of the temperature increase

$$n = 2 - \bar{r}^2 + \phi,$$

coming from the heat source

$$\psi = \frac{1}{\bar{r}}(\bar{r}(2 - \bar{r}^2)_{\bar{r}})_{\bar{r}} + \beta^2(2 - \bar{r}^2) + f = \beta^2(2 - \bar{r}^2) - 4 + f,$$

that proves to be positive in the short rotor,

$$\psi = \beta^2(2 - \bar{r}^2) - 4 + f > \beta^2 - 4 > 0, \quad 0 < \bar{r} < 1,$$

to provide the necessary sign of ϕ due to the maximum principle [Landis, 1997]:

$$\phi(0 < \bar{r} < 1) > 0 \text{ if } -\frac{(\bar{r}\phi_{\bar{r}})_{\bar{r}}}{\bar{r}} - \beta^2\phi = \psi > 0, \ 0 < \bar{r} < 1,$$

$$\phi|_{\bar{r}=1} = 0, \quad \text{and} \quad \phi|_{\bar{r}=0} \geq 0.$$

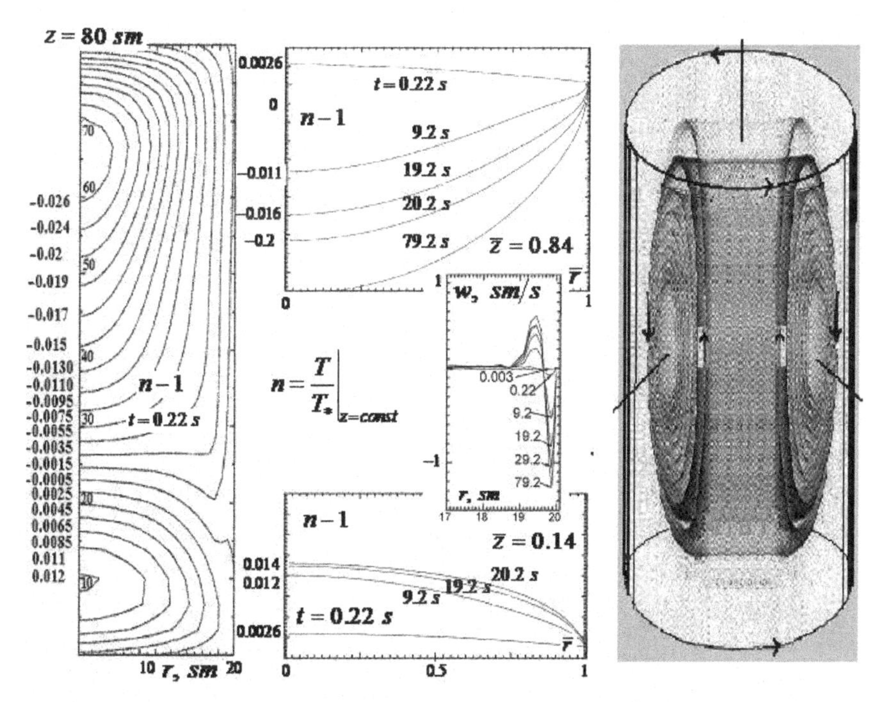

Figure 9.3.1. Temperature separation in DNS of (9.2.1) [Belotserkovskii *et al.*, 2011].

9.4. Temperature separation in a counter-flow vortex

In the finite-volume DNS (direct numerical simulation) of conservation laws (9.2.1) in a centrifugal rotor [Troshkin, 2010a, 2010b; Belotserkovskii *et al.*, 2011], the counter-flow vortex comes into existence from the initial torsional rotation in Fig. 9.2.2 after two-tenths of a second together with a zonal separation of temperature as shown in Fig. 9.3.1 to the left.

However, over time, the counter-flow circulation heats the bottom and cools the upper layers of the initial tube in Fig. 9.2.2, and the temperature is equalized, which is illustrated in the middle of Fig. 9.3.1.

PART 4

Turbulence

Chapter 10

Stress of Pulsations

10.1. Vortex cascades of instabilities

To become a new force originating in both natural experiments [Reynolds, 1883; Comte-Bellot, 1969] and direct numerical simulations of fluid dynamic equilibrium (7.1.8) [Belotserkovskii *et al.*, 2011b], turbulence comes up with vortex cascades of hydrodynamic instabilities, as in Fig. 10.2.1, or in Fig. 10.1.2, resulting in the energy spectrum balance between large- and small-scale eddies in the form of Kolmogorov–Obukhov's law of $-5/3$ [Kolmogorov, 1941; Obukhov, 1941] in Fig. 10.1.3.

10.2. Turbulence as a turbulent force

Thus, when taken as an ordinary open connected set $V = V^t$ in a surrounding space of points $\mathbf{r} = x\mathbf{i} + y\mathbf{j} + z\mathbf{k}$ arbitrarily moving not only with the flow of *velocity* $\mathbf{u} = \mathbf{u}(t, \mathbf{r}) = u\mathbf{i} + v\mathbf{j} + w\mathbf{k}$ but also with the flow of *strain*, or *fluid deformation* $\mathbf{u_r} = \mathbf{u}_x\mathbf{i} + \mathbf{u}_y\mathbf{j} + \mathbf{u}_z\mathbf{k}$, including its *conjugate*, or *gradient* $\nabla\mathbf{u} = \mathbf{i}\mathbf{u}_x + \mathbf{j}\mathbf{u}_y + \mathbf{k}\mathbf{u}_z$, as well as *shear* $\vec{\tau} = \mathbf{u_r} + \nabla\mathbf{u}$, *pressing*, or *bending* $\vec{\sigma} = u_x\mathbf{ii} + v_y\mathbf{jj} + w_z\mathbf{kk}$, *push* $(\nabla \cdot \mathbf{u})\vec{e}$, *shift* $\vec{\gamma} = \vec{\tau} - \vec{\sigma}$, and *torsion* $\vec{\omega} = \mathbf{u_r} - \nabla\mathbf{u}$, every spatial domain V becomes a *smooth substance*.

The latter turns into a *continuous medium* when the conventional *momentum rate* $(\rho\mathbf{u})_t + \nabla \cdot \rho\mathbf{uu}$ per volume dV balances a mass force $\rho\mathbf{g}$ applied to $\mathbf{r}$ (as in the initial Newton's dynamic law) only when increased by the additional *fluid momentum rate* $\nabla \cdot \vec{\mathbf{p}}$ (or reduced by the corresponding *fluid force* $-\nabla \cdot \vec{\mathbf{p}}$) of a *stress* $\vec{\mathbf{p}}$ determined

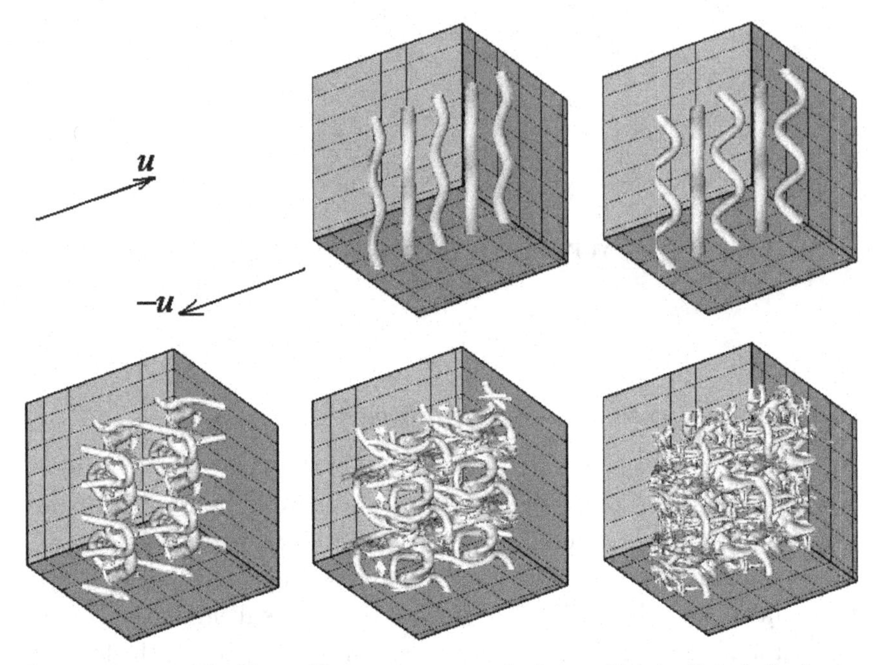

Figure 10.1.1. The Konuykhov vortex cascade due to Kelvin–Helmholtz instability [Betolserkovskii *et al.*, 2011b].

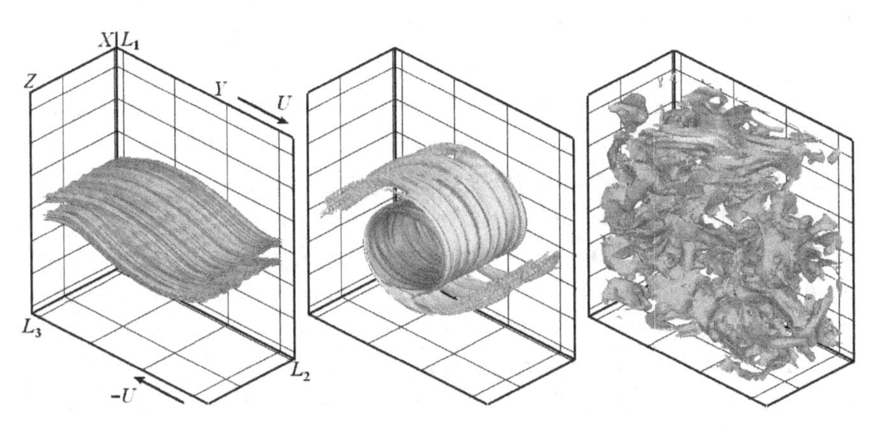

Figure 10.1.2. The Fortova vortex cascade due to Kelvin–Helmholtz instability [Belotserkovskii *et al.*, 2011b].

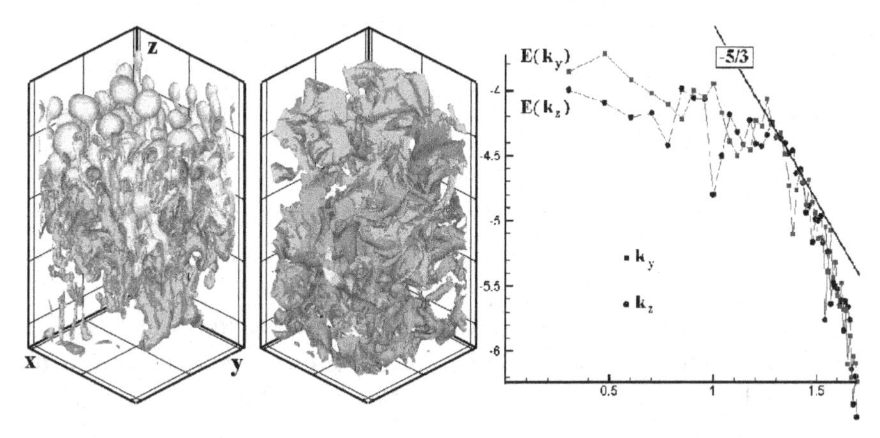

Figure 10.1.3. Rayleigh–Taylor instability [Belotserkovskii *et al.*, 2011] with the specific turbulent energy dependence $E = (\langle u'^2\rangle + \langle v'^2\rangle + \langle w'^2\rangle)/2$ on the longitudinal and transverse wave numbers k_y and k_z in the inertial Kolmogorov–Obukhov's part of $-59^\circ \simeq \mathrm{arctg}(-5/3)$ [Kolmogorov, 1941; Obukhov, 1941].

first by the pressure p as $\vec{\mathbf{p}} = p\vec{\mathbf{e}}$ (the Euler equations) and then by the dynamic and volume viscosities, $\mu > 0$ and $\varsigma\mu$, $\varsigma = \text{const} \geq 0$, as $\vec{\mathbf{p}} = p\vec{\mathbf{e}} - \mu\vec{\mathbf{b}}$ for $\vec{\mathbf{b}} = \tilde{\tau} - (\frac{2}{3} - \varsigma)(\nabla \cdot \mathbf{u})\vec{\mathbf{e}}$ (the Navier–Stokes equations) that leads to the *pressure force* $\mathbf{A} = -\int_{\partial V} p\vec{\mathbf{e}} \cdot d\mathbf{S}$ and its *viscous amendment* $\mathbf{B} = \int_{\partial V} \mu\vec{\mathbf{b}} \cdot d\mathbf{S}$ to make the required balance in the *dynamic equilibrium* (7.1.2):

$$\frac{d}{dt}\int_V \rho\mathbf{u}dV = \int_V \rho\mathbf{g}dV + \mathbf{A} + \mathbf{B}.$$

All the remaining "live" of the medium in hand proved to be "calm" or "not calm" depending on the eternal problem of this kind on how its flow $\mathbf{u}$ proves *to be or not to be stable* to keep the equilibrium in question.

As with Hamlet, the answer to this question had come insensibly, but inevitably, and hence, naturally in the form of the required *third force* $\mathbf{C} = \int_{\partial V} \rho\vec{\mathbf{c}} \cdot d\mathbf{S}$, determined directly by the *turbulence* $\vec{\mathbf{c}}$ as the *correlation matrix* of the *velocity pulsation* for the *time averaging*,

$$\vec{\mathbf{c}} = \overline{\mathbf{u}'\mathbf{u}'} \text{ of } \mathbf{u}' = \mathbf{u} - \bar{\mathbf{u}}, \quad \text{for } \bar{\mathbf{u}} = \lim_{T\to\infty} \frac{1}{T}\int_0^T \mathbf{u}(t+s,\mathbf{r})ds,$$

It seemed, however, to be certain, if the eddies were owing to one particular cause, that integration would show the birth of eddies to depend upon some definite value of—

$$\frac{c\rho U}{\mu}$$

Figure 10.2.1. How the number Re was guessed [Reynolds, 1883].

measured immediately through the *anemometer*, as in [Comte-Bellot, 1969], and delivered finally with the initial dynamic equilibrium to be valued now for both *instant* and *averaged* velocities, $\mathbf{u}$ and $\bar{\mathbf{u}}$, respectively.

In 1883, the turbulence had come with instabilities deforming the parabolic velocity profile (7.5.2) with eddies spontaneously originating in a tube of diameter c when a dimensionless parameter Re of kinematic viscosity μ/ρ and characteristic length c and velocity $U = 2\int_0^{c/2} w dr/c$ was increased [Reynolds, 1883].

Regardless of any parameter, four years later [Thomson, 1887], an alternative treatment of the phenomenon in hand as a *physical definition* of turbulence had come into existence where this instability development proved to be a *force* $\mathbf{C}$ of *turbulent stress* $\rho\vec{\mathbf{c}}$ of some new *structure* $\vec{\mathbf{c}}$ in the momentum equilibrium of Euler (6.2.2), or (6.2.4) as

$$\frac{d}{dt}\int_V \rho\bar{\mathbf{u}}dV = \int_V \rho\mathbf{g}dV + \bar{\mathbf{A}} + \mathbf{C}, \quad \bar{\mathbf{A}} = -\int_{\partial V} \bar{\mathbf{p}}\vec{\mathbf{e}}\cdot d\mathbf{S},$$

$$\mathbf{C} = -\int_{\partial V} \rho\vec{\mathbf{c}}\cdot d\mathbf{S},$$

or

$$(\rho\bar{\mathbf{u}})_t + \nabla\cdot\overline{\rho\mathbf{u}\mathbf{u} + p\vec{\mathbf{e}}} = \rho\mathbf{g}, \quad \text{i.e.}$$

$$(\rho\bar{\mathbf{u}})_t + \nabla\cdot(\rho\bar{\mathbf{u}}\bar{\mathbf{u}} + \bar{\mathbf{p}}\vec{\mathbf{e}} + \rho\vec{\mathbf{c}}) = \rho\mathbf{g}, \tag{10.2.1}$$

when *averaged*, i.e. considered for proper velocity $\bar{\mathbf{u}} = \bar{\mathbf{u}}\mathbf{i} + \bar{\upsilon}\mathbf{j} + \bar{\mathbf{w}}\mathbf{k}$ and pressure $\bar{\mathbf{p}}$, respectively, such that

$$\bar{\bar{\mathbf{p}}} = \bar{\mathbf{p}} \quad \text{and} \quad \overline{\alpha p + \beta q} = \alpha\bar{\mathbf{p}} + \beta\bar{\mathbf{q}} \quad \text{for any } \alpha, \beta = \text{const},$$

with values $\bar{\mathbf{u}}, \bar{\upsilon}, \bar{\mathbf{w}}$ and $\bar{\mathbf{p}}$ including their *pulsations* u', υ', w' and p' in *instantaneous*

$$\mathbf{u} = \bar{\mathbf{u}} + \mathbf{u}', \quad \overline{\mathbf{u}'} = \mathbf{0}, \quad \text{and} \quad p = \bar{\mathbf{p}} + p', \quad \overline{p'} = 0,$$

supposed to be smooth in [Thomson, 1887] and measured subsequently by *anemometers* and *barometers*, respectively, as in [Comte-Bellot, 1969].

The density at that was taken to be *non-pulsating* as

$$\rho = \bar{\rho}(\rho' = 0),$$

to provide the required absence of mass sources to continuity equation (6.2.1), or to the momentum flow $\rho\mathbf{u}'$ to be *solenoidal*:

$$\rho_t + \nabla \cdot \rho\mathbf{u} = 0 \text{ is } \rho_t + \nabla \cdot \rho\bar{\mathbf{u}} = 0 \quad \text{and} \quad \nabla \cdot \rho\mathbf{u}' = 0. \qquad (10.2.2)$$

The same averaging of Euler equation (6.2.4), accounting for

$$\overline{\mathbf{u}\mathbf{u}} = \bar{\mathbf{u}}\bar{\mathbf{u}} + \bar{\mathbf{u}}\overline{\mathbf{u}'} + \overline{\mathbf{u}'}\bar{\mathbf{u}} + \overline{\mathbf{u}'\mathbf{u}'} = \bar{\mathbf{u}}\bar{\mathbf{u}} + \overline{\mathbf{u}'\mathbf{u}'},$$

in [Thomson, 1887] had led (6.2.4) to (10.2.1), to reveal $\vec{\mathbf{c}}$ as the *correlation matrix*

$$\vec{\mathbf{c}} = \overline{\mathbf{u}'\mathbf{u}'} = \mathbf{i}\overline{u'\mathbf{u}'} + \mathbf{j}\overline{\upsilon'\mathbf{u}'} + \mathbf{k}\overline{w'\mathbf{u}'} = \overline{\mathbf{u}'u'}\mathbf{i} + \overline{\mathbf{u}'\upsilon'}\mathbf{j} + \overline{\mathbf{u}'w'}\mathbf{k}. \qquad (10.2.3)$$

After seven years [Reynolds, 1894], the same force $\mathbf{C}$ was found in the Navier–Stokes equations (7.1.2), or (7.1.7) when averaged as (6.2.4):

$$\frac{d}{dt}\int_V \rho\bar{\mathbf{u}}dV = \int_V \rho\mathbf{g}dV + \bar{\mathbf{A}} + \bar{\mathbf{B}} + \mathbf{C}, \quad \bar{\mathbf{A}} = -\int_{\partial V} \bar{\mathbf{p}}d\mathbf{S},$$

$$\mathbf{C} = -\int_{\partial V} \vec{\mathbf{c}} \cdot d\mathbf{S}, \quad \vec{\mathbf{c}} = \overline{\mathbf{u}'\mathbf{u}'},$$

$$\bar{\mathbf{B}} = -\int_{\partial V} (\bar{\bar{\mathrm{p}}} - \bar{\mathrm{p}}\vec{\mathrm{e}}) \cdot d\mathbf{S} = \int_{\partial V} \mu\bar{\bar{\tau}} \cdot d\mathbf{S} - \left(\frac{2}{3} - \varsigma\right) \int_{\partial V} \mu\bar{\mathbf{C}} d\mathbf{S},$$

$$\bar{\mathbf{C}} = \nabla \cdot \bar{\mathbf{u}}, \quad \text{for } \mu = \bar{\mu},$$

or

$$\mathbf{f} = (\rho\bar{\mathbf{u}})_t + \nabla \cdot (\rho\overline{\bar{\mathbf{u}}\bar{\mathbf{u}}} + \bar{\mathbf{P}}\vec{\mathrm{e}} - \mu\bar{\vec{\tau}} + \rho\vec{\mathbf{c}}) - \rho\mathbf{g} = \mathbf{0} \tag{10.2.4}$$

for

$$\bar{\mathbf{P}} = \bar{\mathrm{p}} + \left(\frac{2}{3} - \varsigma\right) \mu\bar{\mathbf{C}}, \quad \bar{\mathbf{C}} = \nabla \cdot \bar{\mathbf{u}}, \quad \bar{\bar{\tau}} = \bar{\mathbf{u}}_{\mathbf{r}} + \nabla\bar{\mathbf{u}} \text{ and } \vec{\mathbf{c}} = \overline{\mathbf{u}'\mathbf{u}'}.$$

10.3. Generation, relaxation, and diffusion

There is something more [Troshkin, 1989, 1990a, 1990b, 1992b, 1993, 1995] in the definition of turbulence with the correlation matrix (10.2.3), as in [Thomson, 1887], than that with the proper force $-\nabla \cdot \rho\vec{\mathbf{c}}$ applied to volume dV in the momentum equilibrium (10.2.1), namely, the interaction existing between averaged flow $\bar{\mathbf{u}}$ and correlation $\vec{\mathbf{c}}$ known to be the *generation*

$$\vec{\mathbf{c}} \cdot \nabla\bar{\mathbf{u}} + \bar{\mathbf{u}}_{\mathbf{r}} \cdot \vec{\mathbf{c}} \tag{10.3.1}$$

of correlation $\vec{\mathbf{c}}$, that is required for the average momentum equilibrium (10.2.1) to *approximate* the instantaneous one (6.2.4) at the following physical *second order closure* for turbulent momentum equilibrium (10.2.1) for $\mu = 0$ [Thomson, 1887] or (10.2.4) for $\mu > 0$ [Frost and Moulden, 1977; Zubarev *et al.*, 1992].

To obtain the required closure for $\mu = 0$, let us note first that the equilibrium (10.2.1) becomes evidently *equivalent* to the momentum one (6.2.4) when additionally subtracted from the latter to lead to the *pulsating momentum equilibrium*

$$(\rho\mathbf{u}')_t + \nabla \cdot (\rho\mathbf{u}\mathbf{u} + p\vec{\mathrm{e}} - \rho\overline{\bar{\mathbf{u}}\bar{\mathbf{u}}} - \bar{\mathrm{p}}\vec{\mathrm{e}} - \rho\vec{\mathbf{c}}) = \mathbf{0},$$

or

$$(\rho\mathbf{u}')_t + \nabla \cdot (\rho\bar{\mathbf{u}}\mathbf{u}' + \rho\mathbf{u}'\bar{\mathbf{u}} + p'\vec{\mathrm{e}}) = \mathbf{0}(\mathbf{u}\mathbf{u} - \bar{\mathbf{u}}\bar{\mathbf{u}} - \vec{\mathbf{c}} = \bar{\mathbf{u}}\mathbf{u}' + \mathbf{u}'\bar{\mathbf{u}}), \tag{10.3.2}$$

that together with (10.2.1) is equivalent to (6.2.4) as to the sum of average (10.2.1) and pulsation (10.3.2).

Then, let us apply to (10.3.2) the following procedure of *second order averaging*. Multiplying directly (10.3.2) by velocity pulsation $\mathbf{u}'$ to the left and the right, then adding the two products,

$$\mathbf{u}'(\rho\mathbf{u}')_t + (\rho\mathbf{u}')_t\mathbf{u}' + \mathbf{u}'\nabla\cdot\begin{pmatrix}\rho\bar{\mathbf{u}}\mathbf{u}' + \rho\mathbf{u}'\bar{\mathbf{u}}\\ +p'\vec{\mathbf{e}}\end{pmatrix}$$
$$+\left(\nabla\cdot\begin{pmatrix}\rho\bar{\mathbf{u}}\mathbf{u}' + \rho\mathbf{u}'\bar{\mathbf{u}}\\ +p'\vec{\mathbf{e}}\end{pmatrix}\right)\mathbf{u}' = \vec{\mathbf{0}},$$

further, using mutual conjugation of strain $\bar{\mathbf{u}}_\mathbf{r}$ and gradient $\nabla\bar{\mathbf{u}}$,

$$\mathbf{u}'\cdot\nabla\bar{\mathbf{u}} = \bar{\mathbf{u}}_\mathbf{r}\cdot\mathbf{u}',$$

continuity Equations (10.2.2) and immediately verified identities

$$\mathbf{u}'(\rho\mathbf{u}')_t + (\rho\mathbf{u}')_t\mathbf{u}' + \mathbf{u}'\nabla\cdot\rho\bar{\mathbf{u}}\mathbf{u}' + (\nabla\cdot\rho\bar{\mathbf{u}}\mathbf{u}')\mathbf{u}'$$
$$= (\rho\mathbf{u}'\mathbf{u}')_t + \nabla\cdot\rho\bar{\mathbf{u}}\mathbf{u}'\mathbf{u}',$$

and

$$\mathbf{u}'\nabla\cdot\rho\mathbf{u}'\bar{\mathbf{u}} + (\nabla\cdot\rho\mathbf{u}'\bar{\mathbf{u}})\mathbf{u}' = \mathbf{u}'(\rho\mathbf{u}'\cdot\nabla\bar{\mathbf{u}}) + (\rho\mathbf{u}'\cdot\nabla\bar{\mathbf{u}})\mathbf{u}'$$
$$= \rho\mathbf{u}'\mathbf{u}'\cdot\nabla\bar{\mathbf{u}} + \rho\bar{\mathbf{u}}_\mathbf{r}\cdot\mathbf{u}'\mathbf{u}',$$

and

$$\mathbf{u}'\nabla\cdot p'\vec{\mathbf{e}} + (\nabla\cdot p'\vec{\mathbf{e}})\mathbf{u}' = \nabla\cdot p'(\mathbf{u}'\vec{\mathbf{e}} + \vec{\mathbf{e}}\mathbf{u}')$$
$$-2p'\vec{\mathbf{e}}\nabla\cdot\mathbf{u}'\left(\nabla\cdot\mathbf{u}' = -\frac{1}{\rho}\mathbf{u}'\cdot\nabla\rho\right),$$

and, finally, averaging the resulting sum, we obtain the requred *second-order amendment*

$$(\rho\vec{\mathbf{c}})_t + \nabla\cdot\rho\bar{\mathbf{u}}\vec{\mathbf{c}} + \rho(\vec{\mathbf{c}}\cdot\nabla\bar{\mathbf{u}} + \bar{\mathbf{u}}_\mathbf{r}\cdot\vec{\mathbf{c}}) = \rho\vec{\mathbf{R}}_{p'\mathbf{u}'},$$

for the averaged momentum equilibrium (10.2.4), with *relaxation*

$$-\rho\vec{\mathbf{R}}_{p'\mathbf{u}'} = \nabla\cdot(\overline{\vec{\mathbf{e}}p'\mathbf{u}' + p'\mathbf{u}'\vec{\mathbf{e}}}) + \left(\frac{2}{\rho}\overline{p'\mathbf{u}'}\cdot\nabla\rho\right)\vec{\mathbf{e}}$$

$$(\nabla = \vec{\mathbf{e}}\cdot\nabla = \nabla\cdot\vec{\mathbf{e}}). \tag{10.3.3}$$

For $\mu > 0$, the same averaging applied to the pulsating component

$$(\rho\mathbf{u}')_t + \nabla\cdot(\rho\bar{\mathbf{u}}\mathbf{u}' + \rho\mathbf{u}'\bar{\mathbf{u}} + P'\vec{\mathbf{e}} - \mu\vec{\tau}') = \mathbf{0},$$

$$P' = p' + \sigma\mu C', \quad C' = \nabla\cdot\mathbf{u}', \quad \vec{\tau}' = \mathbf{u}'_{\mathbf{r}} + \nabla\mathbf{u}'$$

of the momentum equilibrium (7.1.7), with identities

$$\begin{aligned}\nabla\cdot\mathbf{u}'\mu\vec{\tau}' &= \partial_x\mathbf{u}'\mu\vec{\tau}' + \cdots = (\nabla\cdot\mathbf{u}')\mu\vec{\tau}' + \mathbf{u}'\partial_x\mu\vec{\tau}' + \cdots \\ &= (\nabla\cdot\mathbf{u}')\mu\vec{\tau}' + \mathbf{u}'\cdot\nabla\mu\vec{\tau}', \\ \nabla\cdot\mu\vec{\tau}'\mathbf{u}' &= \partial_x\mu\tau'^x\mathbf{u}' + \cdots = (\partial_x\mu\tau'^x)\mathbf{u}' + \mu\tau'^x\partial_x\mathbf{u}' + \cdots \\ &= (\nabla\cdot\mu\vec{\tau}')\mathbf{u}' + \mu\vec{\tau}'(\nabla\cdot\mathbf{u}')\end{aligned}$$

and

$$\begin{aligned}&\mathbf{u}'\nabla\cdot\mu\vec{\tau}' + (\nabla\cdot\mu\vec{\tau}')\mathbf{u}' - \nabla\cdot\mu(\mathbf{u}'\vec{\tau}' + \vec{\tau}'\mathbf{u}') \\ &\quad = -2(\nabla\cdot\mathbf{u}')\mu\vec{\tau}' = 2\frac{\mu}{\rho}\vec{\tau}'\mathbf{u}'\cdot\nabla\rho,\end{aligned}$$

leads to the resulting *second-order amendment:*

$$(\rho\vec{\mathbf{c}})_t + \nabla\cdot\rho\bar{\mathbf{u}}\vec{\mathbf{c}} + \rho(\vec{\mathbf{c}}\cdot\nabla\bar{\mathbf{u}} + \bar{\mathbf{u}}_{\mathbf{r}}\cdot\vec{\mathbf{c}}) = \rho\vec{\mathbf{R}}_{p'\mathbf{u}'} + \rho\vec{\mathbf{D}}_{\mu\vec{\tau}'}, \tag{10.3.4}$$

for the averaged momentum equilibrium (10.2.4) with generation (10.3.1), relaxation (10.3.3), and *diffusion*

$$\begin{aligned}\rho\vec{\mathbf{D}}_{\mu\vec{\tau}'} &= \overline{\mathbf{u}'\nabla\cdot\mu\vec{\tau}' + (\nabla\cdot\mu\vec{\tau}')\mathbf{u}'} \\ &= \nabla\cdot\mu(\overline{\mathbf{u}'\vec{\tau}'} + \overline{\vec{\tau}'\mathbf{u}'}) - 2\frac{\mu}{\rho}\overline{\vec{\tau}'\mathbf{u}'}\cdot\nabla\rho,\end{aligned} \tag{10.3.5}$$

the last two of them being approximated by the relevant physical processes to constitute the so-called *second order closure* for *RANS* (short for Reynolds Averaged Navier–Stokes system) [Frost and Moulden, 1977] or, equivalently, by the *diffusion–relaxation approximation* [Zubarev *et al.*, 1992]:

$$\vec{\mathbf{R}}_{P'\mathbf{u}'} + \vec{\mathbf{D}}_{\mu\vec{\tau}'} = \nabla \cdot (\nu + \Gamma)\nabla\vec{\mathbf{c}} - \frac{\aleph}{\nu+\gamma}(\vec{\mathbf{c}} - c^2\vec{\mathbf{e}}), \qquad (10.3.6)$$

where

$$c = c(\mathbf{u}') = \sqrt{\frac{Sp(\vec{\mathbf{c}})}{3}} \quad \text{and} \quad Sp(\vec{\mathbf{c}}) = \overline{u'u'} + \overline{v'v'} + \overline{w'w'},$$

and

$$\aleph = c^2 + \frac{\nu}{\gamma}c^2 + \nu\Lambda(\nu+\gamma), \quad \text{or} \quad \frac{\aleph}{\nu+\gamma} = \nu\Lambda + \frac{c^2}{\gamma} \quad \text{for } \nu = \frac{\mu}{\rho},$$
$$\Gamma, \Lambda, \gamma > 0, \qquad (10.3.7)$$

10.4. Laminar–turbulent transition

After the first evidence in a tube [Reynolds, 1883], both further natural experiments and direct numerical simulations of momentum equilibrium ((6.2.4) or (7.1.7)), [Comte-Bellot, 1969; Frost and Moulden, 1977; Belotserkovskii, *et al.*, 2011], in a *plane channel* with no-slipping walls, reveal that, even for one-dimensional averaged flow

$$\bar{\mathbf{u}} = \bar{\mathbf{u}}(y)\mathbf{i}, \quad |y| < h, \quad \text{with } \bar{\mathbf{u}}\big|_{|y|=h} = 0,$$

pulsations $\mathbf{u}'$ inevitably prove to be three-dimensional while producing the *heterogeneous turbulence* $\vec{\mathbf{c}} = \overline{\mathbf{u}'\mathbf{u}'}$, with *even* parts of double specific energy of pulsations,

$$\overline{w'w'} = \overline{u'u'} = \overline{v'v'}\,(y) = \overline{v'v'}\,(-y) \quad \text{of}$$
$$c^2 = \frac{\overline{u'u'} + \overline{v'v'} + \overline{w'w'}}{3} = \overline{v'v'},$$

and the *odd* correlation component submitted to boundary conditions [Comte-Bellot, 1969]:

$$\overline{u'v'}\,(y) = \mathbf{i} \cdot \vec{\mathbf{c}} \cdot \mathbf{j} = -\overline{u'v'}\,(-y) \quad (\text{hence, } \left.\overline{u'v'}\right|_{y=0} = 0) \text{ and}$$
$$\left.\overline{u'v'}\right|_{y=h} = 0. \tag{10.4.1}$$

As to the above-mentioned further live of flow $\mathbf{u} = \bar{\mathbf{u}} + \mathbf{u}'$ when supplied with molecular pressure p, but constantly *disturbed* by pulsations $\mathbf{u}'$ as fluctuations, or deviations of $\mathbf{u}$ from a *basic*, or undisturbed averaged flow $\bar{\mathbf{u}} = \mathbf{a}$, it will depend on whether it turns out to be deprived of them in time as $\mathbf{u}' \to \mathbf{0}$ to remain stable, or $\bar{\mathbf{u}} = \mathbf{a}$, with the *laminar* (or trivial) *correlation* $\vec{\mathbf{c}} = \vec{\mathbf{0}}$, or, contrarily, if not so, on whether pulsations $\mathbf{u}'$ are sufficiently strong and, what is most important, numerous to create a *turbulent* one $\vec{\mathbf{c}} \neq \vec{\mathbf{0}}$, which develops a force $-\nabla \cdot \rho\vec{\mathbf{c}}$ capable of making a *transition* in (10.2.4) to be a branch, or a *bifurcation* of the initial *laminar flow* $\bar{\mathbf{u}} = \mathbf{a}$ ($\vec{\mathbf{c}} = \vec{\mathbf{0}}$), to the proper *turbulent* one $\bar{\mathbf{u}} = \mathbf{b}$ ($\vec{\mathbf{c}} \neq \vec{\mathbf{0}}$), both resolving the momentum equilibrium $\mathbf{f} = \mathbf{0}$ in (10.2.4) with the same body force $\rho\mathbf{g}$.

As is found in [Zubarev *et al.*, 1992], such a transition can be really possible in a plane channel when in addition to the diffusion and relaxation processes in (10.3.6), the turbulence $\vec{\mathbf{c}} \neq \vec{\mathbf{0}}$ proves to be *heterogeneous by Nevzglyadov* [1960] to be definite as

$$\vec{\mathbf{c}} - c^2\vec{\mathbf{e}} = -Fc^2\vec{\bar{\tau}} = -Fc^2(\bar{\mathbf{u}}_{\mathbf{r}} + \nabla\bar{\mathbf{u}}), \quad \rho, \mu, F = \text{const} > 0,$$

or, more precisely, to be related linearly as

$$\vec{\mathbf{c}} - c^2\vec{\mathbf{e}} = -Fc^2(\mathbf{a}_{\mathbf{r}} + \nabla\mathbf{a}),$$

to the strain $\vec{\bar{\tau}}$ of the averaged velocity of the Hagen–Poiseuille flow $\bar{\mathbf{u}} = \mathbf{a}$, or the Stokes velocity $\bar{\mathbf{u}}^{(p)}$ [Stokes, 1845],

$$\bar{\mathbf{u}} = \mathbf{a} = \bar{\mathbf{u}}^{(p)}\mathbf{i}, \quad \bar{\mathbf{u}}^{(p)} = \frac{b}{2\mu}(h^2 - y^2), \quad -\bar{\mathbf{p}}_x = b = \text{const} > 0,$$

satisfying the Navier–Stokes equations (10.2.4) as

$$\mathbf{f} = -\mu\left(\bar{\mathbf{u}}^{(p)}_{yy} + \frac{b}{\mu}\right)\mathbf{i} = \mathbf{0} \text{ for } \mathbf{g} = \mathbf{0} \quad \text{and} \quad \vec{\mathbf{c}} = \vec{\mathbf{0}},$$

so,

$$\begin{aligned}\overline{u'v'} &= \mathbf{i}\cdot(\vec{\mathbf{c}} - c^2\vec{\mathbf{e}})\cdot\mathbf{j} = -Fc^2\mathbf{i}\cdot(\mathbf{a}_\mathbf{r} + \nabla\mathbf{a})\cdot\mathbf{j} \\ &= -F\overline{v'v'}\bar{\mathbf{u}}_y^{(p)} = \frac{Fb}{\mu}\overline{v'v'}y,\end{aligned}$$

or

$$\overline{v'v'} = d^2h\left|\frac{\overline{u'v'}}{y}\right| \quad \text{for } d = \sqrt{\frac{\mu}{Fbh}} = \text{const} > 0. \tag{10.4.2}$$

The corresponding turbulent flow

$$\bar{\mathbf{u}} = \mathbf{b} = \bar{\mathbf{u}}\mathbf{i}, \quad \bar{\mathbf{u}} = \frac{b}{2\mu}(h^2 - y^2) - \frac{1}{\nu}\int_y^h \overline{u'v'}(y')dy',$$

$$\bar{\mathbf{u}}_y = \frac{\overline{u'v'}}{\nu} - \frac{by}{\mu},$$

is produced by the same equilibrium (10.2.4), however, with $\overline{u'v'} \neq 0$, or

$$\mathbf{f} = -\mu\nabla\cdot\vec{\vec{\tau}} + \rho\nabla\cdot\vec{\mathbf{c}} + \bar{\mathbf{p}}_x\mathbf{i} = -\mu\left(\bar{\mathbf{u}}_{yy} - \frac{1}{\nu}\overline{u'v'}_y + \frac{b}{\mu}\right)\mathbf{i} = \mathbf{0}.$$

$$|y| < h, \quad \bar{\mathbf{u}}|_{y=0,h} = 0,$$

in which the correlation $\overline{u'v'} \neq 0$ has to satisfy the relevant projection of the second-order amendment (10.3.5) to (10.2.4) with approximation (10.3.6):

$$\mathbf{i}\cdot(\vec{\mathbf{c}}_t + \bar{\mathbf{u}}\cdot\nabla\vec{\mathbf{c}} + \vec{\mathbf{c}}\cdot\nabla\bar{\mathbf{u}} + \bar{\mathbf{u}}_\mathbf{r}\cdot\vec{\mathbf{c}} = \vec{\mathbf{R}}_{p'\mathbf{u}'} + \vec{\mathbf{D}}_{\mu\vec{\tau}'})\cdot\mathbf{j}.$$

As a result, with identities

$$\begin{aligned}&\mathbf{i}\cdot(\vec{\mathbf{c}}_t + \bar{\mathbf{u}}\cdot\nabla\vec{\mathbf{c}} + \vec{\mathbf{c}}\cdot\nabla\bar{\mathbf{u}} + \bar{\mathbf{u}}_\mathbf{r}\cdot\vec{\mathbf{c}} = \vec{\mathbf{c}}\cdot\nabla\bar{\mathbf{u}} + \bar{\mathbf{u}}_\mathbf{r}\cdot\vec{\mathbf{c}})\cdot\mathbf{j} \\ &\quad = \bar{\mathbf{u}}_y\mathbf{j}\cdot\overline{\mathbf{u}'\mathbf{u}'}\cdot\mathbf{j} = \bar{\mathbf{u}}_y\overline{v'v'},\end{aligned}$$

(10.3.6) and (10.3.7), or

$$\mathbf{i}\cdot(\vec{\mathbf{R}}_{p'\mathbf{u}'} + \vec{\mathbf{D}}_{\mu\vec{\tau}'})\cdot\mathbf{j} = ((\nu + \Gamma)\overline{u'v'}_y)_y - \left(\nu\Lambda + \frac{\overline{v'v'}}{\gamma}\right)\overline{u'v'},$$

and accounting for (10.4.1) and (10.4.2), we come to the following *correlation boundary value problem:*

$$-((\nu+\Gamma)\overline{u'v'}_y)_y + \nu\Lambda\overline{u'v'} + \frac{d^2h}{y}|\overline{u'v'}|\left(\bar{\mathbf{u}}_y + \frac{\overline{u'v'}}{\gamma}\right) = 0,$$

$$0 < y < h, \quad \overline{u'v'}\big|_{y=0,h} = 0.$$

Then, following [Zubarev *et al.*, 1992] and putting

$$\frac{\Gamma}{U_p h} = \frac{\gamma}{U_p h} = \varepsilon\kappa = \gamma_0(1-Y^2), \ \ \gamma_0 = \text{const} > 0, \ \ 0 < Y = \frac{y}{h} < 1,$$

with the characteristic velocity, the bifurcation parameter, and the Reynolds number

$$U_p = \sqrt{\frac{bh}{\rho}}, \quad \varepsilon = \frac{\nu}{U_p h} \quad \text{and} \quad \mathrm{Re} = \frac{1}{\nu}\int_{-h}^{h} \bar{\mathbf{u}}_p dy = \frac{2}{3\varepsilon^2},$$

respectively, and dimensionless constants,

$$\alpha = \Lambda h^2 = 2000, \quad \beta = \left(\frac{d}{\varepsilon}\right)^2 = \left(\frac{d}{\gamma_0\bar{\varepsilon}}\right)^2$$

$$= \frac{3d^2}{2}\mathrm{Re}, \quad d = 1, \quad \bar{\varepsilon} = \frac{\varepsilon}{\gamma_0}, \quad \gamma_0 = 0.105,$$

velocity and correlation,

$$U = \frac{\bar{\mathbf{u}}}{U_P} = \frac{1-Y^2}{2\varepsilon} - \frac{1}{\varepsilon}\int_Y^1 \tau(Y')dY' \quad \text{and} \quad \tau = \frac{\overline{u'v'}}{U_p^2},$$

we may reduce the correlation boundary value problem to

$$-((1+\kappa)\tau_Y)_Y + \alpha\tau + \beta\frac{|\tau|}{Y}\left(\varepsilon U_Y + \frac{\tau}{\kappa}\right) = 0,$$

$$0 < Y < 1, \quad \tau|_{Y=0,1} = 0,$$

or

$$-((1+\kappa)\tau_Y)_Y + \alpha\tau = \beta\,|\tau|\left(1 - \frac{\tau}{\varsigma}\right),$$

$$0 < Y < 1, \quad \tau|_{Y=0,1} = 0,$$

with

$$\varsigma = \frac{\kappa Y}{\kappa + 1}, \quad \kappa = \frac{1 - Y^2}{\bar{\varepsilon}}, \quad 0 < Y < 1, \quad \varsigma|_{Y=0,1} = 0.$$

or to

$$-((1+\kappa)\tau_Y)_Y + \alpha\tau = \beta\,|\tau|\left(1 - \frac{\tau}{\varsigma}\right), \quad 0 < Y < 1, \ \ \tau|_{Y=0,1} = 0,$$

$$\text{or} \qquad (10.4.3)$$

$$-\tau_{\bar{Y}\bar{Y}} + \bar{\alpha}\tau = \bar{\beta}\,|\tau|\left(1 - \frac{\tau}{\varsigma}\right), \quad 0 < \bar{\mathbf{Y}} < 1, \ \ \tau|_{\bar{\mathbf{Y}}=0,1} = 0,$$

with

$$\bar{\alpha} = \alpha H^2 (1+\kappa) \quad \text{and} \quad \bar{\beta} = \beta H^2 (1+\kappa)$$

where the *scheme constant*

$$H = \int_0^1 \frac{dY}{1+\kappa} = \frac{\bar{\varepsilon}}{2\sqrt{1+\bar{\varepsilon}}} \ln \frac{\sqrt{1+\bar{\varepsilon}}+1}{\sqrt{1+\bar{\varepsilon}}-1}$$

is depicted in Fig. 10.4.1, for either

$$\bar{\mathbf{Y}} = \frac{1}{H}\int_0^Y \frac{dY'}{1+\kappa(Y')} = \frac{\bar{\varepsilon}}{2H\sqrt{1+\bar{\varepsilon}}} \ln \frac{\sqrt{1+\bar{\varepsilon}}+Y}{\sqrt{1+\bar{\varepsilon}}-Y}, \quad 0 \le Y \le 1$$

$$\left(d\bar{\mathbf{Y}} = \frac{1}{H}\frac{dY}{1+\kappa}\right),$$

or

$$Y = \sqrt{1+\bar{\varepsilon}}\,\mathrm{th}\frac{\bar{\mathbf{Y}}H\sqrt{1+\bar{\varepsilon}}}{\bar{\varepsilon}}, \quad 0 \le \bar{\mathbf{Y}} \le 1 \left(\mathrm{th}z = \frac{e^z - e^{-z}}{e^z + e^{-z}}\right).$$

Meanwhile, as follows from the evident identity

$$-\int_0^1 \tau_{\bar{\mathbf{Y}}\bar{\mathbf{Y}}}\varphi d\bar{\mathbf{Y}} = \int_0^1 \tau_{\bar{\mathbf{Y}}}\varphi_{\bar{\mathbf{Y}}} d\bar{\mathbf{Y}} \quad \forall\, \varphi|_{\bar{\mathbf{Y}}=0,1} = 0$$

$$(\text{for any smooth } \varphi \quad \text{with} \quad \varphi|_{\bar{\mathbf{Y}}=0,1} = 0),$$

every solution τ of relaxation problem (10.4.3) proves to a *critical point* of proper *functional*

$$E(\tau) = \int_0^1 \left(\frac{1}{2}\tau_{\bar{\mathbf{Y}}}^2 + A(\tau)\right) d\bar{\mathbf{Y}}, \quad \tau|_{\bar{\mathbf{Y}}=0,1} = 0, \qquad (10.4.4)$$

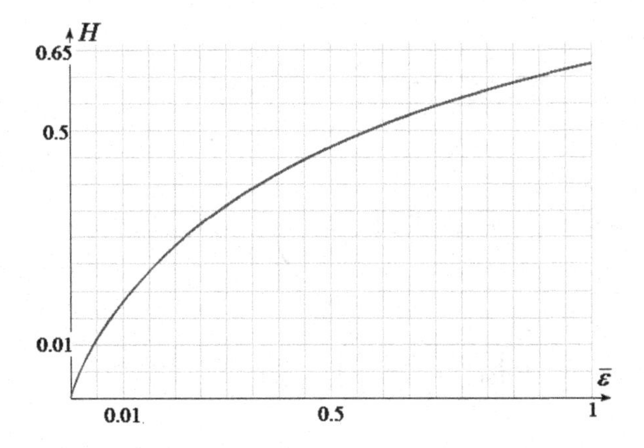

Figure 10.4.1. The scheme constant $H(\bar{\varepsilon})$.

with such a *potential*

$$A(\tau) = \frac{\bar{\alpha}\tau^2}{2} - \frac{\bar{\beta}|\tau|\tau}{2} + \frac{\bar{\beta}|\tau|^3}{3\varsigma}, \quad -\infty < \tau < \infty, \tag{10.4.5}$$

as in Fig. 10.4.2, to degenerate, as

$$(E'(\tau), \varphi) = 0 \quad \forall\, \varphi|_{\bar{\mathbf{Y}}=0,1} = 0,$$

the *variational differential*

$$(E'(\tau), \varphi) = \int_0^1 (\tau_{\bar{\mathbf{Y}}}\varphi_{\bar{\mathbf{Y}}} + A_\tau \varphi)d\bar{\mathbf{Y}} \quad \text{for } A_\tau = \bar{\alpha}\tau - \bar{\beta}|\tau|\left(1 - \frac{\tau}{\varsigma}\right).$$

At that, in addition to the *lowest points*

$$0 \quad \text{for} \quad \alpha \geq \beta \quad \text{and} \quad \frac{\bar{\beta} - \bar{\alpha}}{\bar{\beta}}\varsigma = \frac{\beta - \alpha}{\beta}\varsigma \text{ for } \beta > \alpha$$

of potential $A(\tau)$ in Fig. 10.4.2, both the *boundedness from below* of

$$A(\tau) \geq A\left(\frac{\beta - \alpha}{\beta}\varsigma(Y)\right) = -\frac{(\beta - \alpha)^3 H^2}{6\bar{\varepsilon}\beta^2} f(Y),$$

$$f(Y) = \frac{(1 - Y^2)^2 Y^2}{1 + \bar{\varepsilon} - Y^2}, \quad 0 < Y < 1,$$

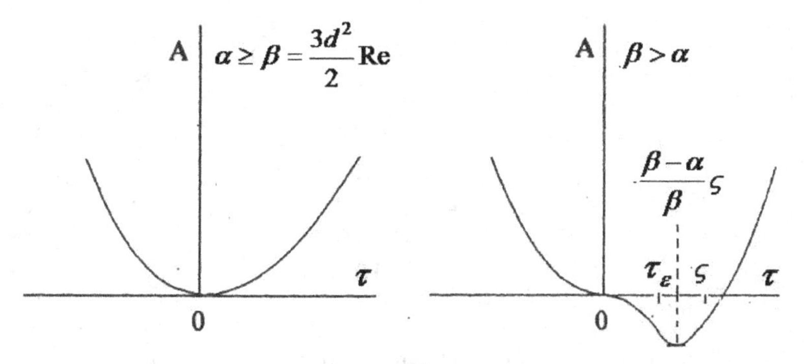

Figure 10.4.2. The lowest points 0 and $\frac{\beta-\alpha}{\beta}\varsigma$ of $A(\tau)$ and the minimum point τ_ε of $E(\tau)$.

and the Poincare–Steklov *compact imbedding*

$$\|\varphi\|_1 = \sqrt{\int_0^1 \varphi_{\bar{\mathbf{Y}}}^2 d\bar{\mathbf{Y}}} \geq \pi \sqrt{\int_0^1 \varphi^2 d\bar{\mathbf{Y}}} = \pi \|\varphi\|_0 \quad \forall\, \varphi|_{\bar{\mathbf{Y}}=0,1} = 0,$$
$$\pi = 3.14159\ldots$$

(to be an equality for $\varphi = \sin \pi \bar{\mathbf{Y}}$) whereby in every sequence $\tau^{(n)}$ of smooth functions $\tau^{(n)}|_{\bar{\mathbf{Y}}=0,1} = 0$ *bounded* in the *strong* norm $\|\tau^{(n)}\|_1 \leq \text{const}$, $n = 1, 2, \ldots$, one can find a sub-sequence $\tau^{(n_k)}$, $k = 1, 2, \ldots$, *fundamental* in the *weak* norm as $\|\tau^{(n_{k+p})} - \tau^{(n_k)}\|_0 \to 0$ for $k, p \to \infty$ $(n_k, n_{k+p} \to \infty, p = 1, 2, \ldots)$, provide the functional

$$E(\tau) \geq -\frac{(\beta - \alpha)^3 H}{6\bar{\varepsilon}\beta^2 (1 + \kappa)} \int_0^1 f(Y)\, dY \quad \forall\, \tau|_{\bar{\mathbf{Y}}=0,1} = 0$$

(as bounded from below) with a *minimum point* τ_ε, $\|\tau_\varepsilon - \tau_{n_k}\|_0 \to 0$, $k \to \infty$,

$$E(\tau) \geq E(\tau_\varepsilon) \quad (\text{hence, } (E'(\tau_\varepsilon), \varphi) = 0 \quad \forall\, \varphi|_{\bar{\mathbf{Y}}=0,1} = 0),$$

to be a unique non-trivial solution of (10.4.3) [Zubarev *et al.*, 1992] localized between the lowest points of potential $A(\tau)$ for sufficiently

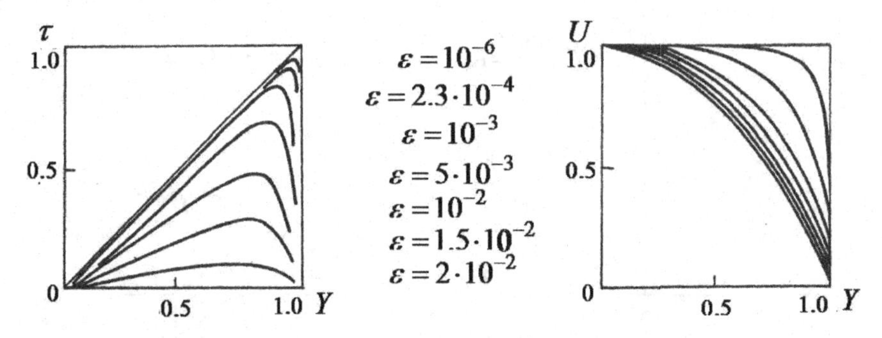

Figure 10.4.3. Both supercritical channel correlations $\tau = \tau_\varepsilon$ and velocities U.

large $\beta > \alpha$,

$$0 < \tau = \tau_\varepsilon < \frac{\beta - \alpha}{\beta}\varsigma \quad (\text{with } -\tau_{\bar{Y}\bar{Y}} > 0), 0 < Y < 1, \ \text{for } \alpha < \beta,$$

as in Fig. 10.4.2, or for the corresponding *supercritical* values of the bifurcation parameter or the Reynolds number,

$$\varepsilon < \varepsilon_*, \quad \text{or} \quad \bar{\varepsilon} < \bar{\varepsilon}_* = \frac{\varepsilon_*}{\gamma_0}, \quad \text{and} \quad \text{Re} > \text{Re}_* = \frac{2}{3\varepsilon_*^2},$$

shown in Fig. 10.4.3.

The corresponding *critical values*

$$\text{Re}_* \approx 1380 \quad (\text{for } \alpha = 2000, d = 1 \quad \text{and} \quad \gamma_0 = 0.105),$$

delivered by the least eigenvalue

$$\lambda_* = \beta_* - \alpha, \quad \beta_* = \left(\frac{d}{\varepsilon_*}\right)^2 = \frac{3d^2}{2}\text{Re}_*,$$

of the linearized (10.4.3), or the corresponding *spectral problem*

$$-((1+\kappa)\varphi_Y)_Y = \lambda\varphi, \quad \lambda = \lambda(\bar{\varepsilon}), \quad \bar{\varepsilon}\kappa = 1 - Y^2,$$
$$\varphi \neq 0, \quad 0 < Y < 1, \quad \varphi|_{Y=0,1} = 0,$$

such that

$$\pi^2 = \inf_{\tau|_{Y=0,1}=0} \frac{\int_0^1 \tau_Y^2 dY}{\int_0^1 \tau^2 dY} < \lambda = \inf_{\tau|_{Y=0,1}=0} \frac{\int_0^1 (1+\kappa)\tau_Y^2 dY}{\int_0^1 \tau^2 dY} < \pi^2 \frac{1+\bar{\varepsilon}}{\bar{\varepsilon}}.$$

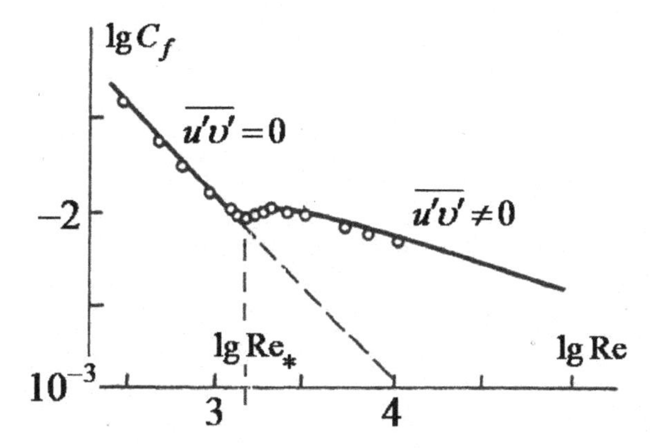

Figure 10.4.4. Resistance curve for a plane channel with experimental points [Patel and Head, 1969; Beavers and Sparrow, 1971].

is verified by the known plane channel *resistance curve* [Patel and Head, 1969; Beavers and Sparrow, 1971] of

$$C_f = \frac{-hp_x}{\frac{1}{2h}\int_{-h}^{h}\bar{\mathrm{u}}dy} = C_f\,(\mathrm{Re}) \quad (\text{or } C_f = 10^{\lg C_f} = C_f\,(\lg \mathrm{Re}))$$

with its *laminar* ($\tau = 0$) and *turbulent* ($\tau \neq 0$) branches in Fig. 10.4.4.

Chapter 11

Stress–Flow Interaction

11.1. Discrepancy angles of turbulent outflow

Other experiments [Abramovich, 1973; Loitsyanskii, 1966] evidence that when created with the help of turbulence $\vec{\mathbf{c}}$ in a channel $x < 0$, $|y| < h$, for rather large Reynolds numbers $Re \gg Re_*$, or small bifurcation parameters $\varepsilon \ll 1$, the velocity profile $\bar{u}$ becomes almost rectangular, as in Fig. 10.3.3, to form an outflow $\bar{\mathbf{u}}|_{x=0} = \bar{u}\mathbf{i}$ in which the profile $\bar{u}$ initiates the so-called *near-turbulent wake* in the outside space $x > 0$, without walls $|y| = h$ and with character *angles of discrepancy* $c_\pm$, as in Fig. 11.1.1.

These can be obtained with the help of the *characteristic equation* [Troshkin, 1989, 1990a, 1990b, 1992b, 1993, 1995]

$$(\varphi_t + \bar{\mathbf{u}} \cdot \nabla\varphi)^2 = \varphi_{\mathbf{r}} \cdot \vec{\mathbf{c}} \cdot \nabla\varphi \tag{11.1.1}$$

(derived below by momentum equilibrium (10.1.4) and second order amendment (10.2.4)) for a smooth *phase* $\varphi = \varphi(t, \mathbf{r})$ as follows.

Really, in our case (Fig. 11.1.1), Equation (10.1.2) takes the form

$$(\bar{u}^2 - \overline{u'u'})\varphi_x^2 - 2\overline{u'v'}\varphi_x\varphi_y - \overline{v'v'}\varphi_y^2 = 0,$$

or

$$(1 - \tau_{11})\kappa^2 + 2\tau_{12}\kappa - \tau_{22} = 0 \quad \text{for } \kappa = \kappa(x, y) = -\varphi_x/\varphi_y(\varphi_y \neq 0)$$

with coefficients

$$\tau_{11} = \overline{u'u'}/\bar{u}^2, \quad \tau_{12} = \overline{u'v'}/\bar{u}^2 \quad \text{and} \quad \tau_{22} = \overline{v'v'}/\bar{u}^2$$

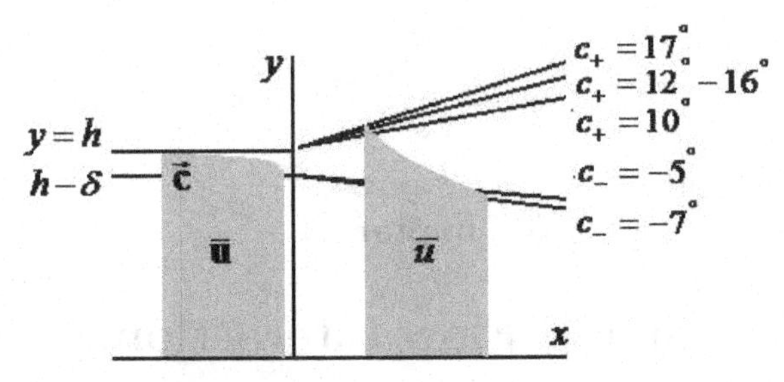

Figure 11.1.1. The outflow $\bar{\mathbf{u}} = \bar{u}\mathbf{i}$ formed by the channel turbulence $\vec{\mathbf{c}}$ and the near-wake velocity profile $\bar{u}$ with discrepancy angles $c_{\pm}$ measured as $c_{+} = 12^{\circ} - 16^{\circ}$ and $c_{-} = -7^{\circ}$ [Abramovich, 1973] or $c_{+} = 10^{\circ}$ and $c_{-} = -5^{\circ}$ [Loitsyanskii, 1966] and calculated from the inside data [Comte-Bellot, 1966] as $c_{+} = 17^{\circ}$ and $c_{-} = -5^{\circ}$ [Troshkin, 1990a].

measured in [Comte-Bellot, 1969] as approximately equal to

$$\tau_{12} = 0.4\sqrt{\tau_{11}}\sqrt{\tau_{22}}, \quad \tau_{11} = 2.5/(12)^2, \quad \tau_{22} = 1/(12)^2 \quad \text{at } y = h - \delta$$

(or at positions of $\tau_{12}, \tau_{11}, \tau_{22}$ maximums near the boundary $y = h$ reduced to their common point $y = h - \delta$ [Comte-Bellot, 1969]), and

$$\tau_{12} \to 0, \quad \sqrt{\tau_{11}} \to 0.3, \quad \tau_{22} \sim \tau_{11} \quad \text{for } y \to h - 0.$$

So, the required $c_{\pm}$-charactistics $\varphi(x, y) = \text{const}$ in Fig. 11.1.1 are streamlines

$$y = h - \delta + x\kappa_{=}(0, h - \delta) \quad \text{and} \quad y = h + x\kappa_{+}(0, h),$$

with slopes

$$\kappa = \kappa_{\mp} = \frac{-\tau_{12} \mp \sqrt{\tau_{12}^2 + (1 - \tau_{11})\tau_{22}}}{1 - \tau_{11}},$$

and angles

$$c_{-} = \text{arctg}\kappa_{-}(0, h - \delta) = -\text{arctg}\frac{0.4\sqrt{2.5} + \sqrt{(12)^2 - 2.1}}{(12)^2 - 2.5} \approx -5^{\circ}$$

and

$$c_{+} = \text{arctg}\kappa_{+}(0, h) \sim \text{arctg}\frac{0.3}{\sqrt{1 - (0.3)^2}} \approx 17^{\circ},$$

shown in Fig. 11.1.1.

11.2. Turbulence wave characteristics

Now, following [Troshkin, 1989, 1990a, 1990b, 1992b, 1993, 1995], assuming the fluid to be incompressible

$$\rho = \text{const} > 0 \quad (\nabla \cdot \mathbf{u} = 0)$$

and neglecting both molecular viscosity $\mu = 0$ and turbulent diffusion $\vec{\mathbf{D}}_{\mu\vec{\tau}'} = \vec{\mathbf{0}}$, let us investigate Equations (10.1.4) and (10.2.4), with (10.2.6) and (10.2.7),

$$(\rho\bar{\mathbf{u}})_t + \nabla \cdot (\rho\bar{\mathbf{u}}\bar{\mathbf{u}} + \bar{p}\vec{\mathbf{e}} + \rho\vec{\mathbf{c}}) = \rho\mathbf{g} \quad \text{for } \mu = 0, \text{ hence, } \bar{P} = \bar{p},$$

$$(\rho\vec{\mathbf{c}})_t + \nabla \cdot \rho\bar{u}\vec{\mathbf{c}} + \rho(\vec{\mathbf{c}} \cdot \nabla\bar{\mathbf{u}} + \bar{\mathbf{u}}_{\mathbf{r}} \cdot \vec{\mathbf{c}}) = \rho\vec{\mathbf{R}}_{p'\mathbf{u}'} = -\frac{\rho c^2}{\gamma}(\mathfrak{y} - c^2\vec{\mathbf{e}}),$$

or the corresponding *quasilinear system* in partial differential derivatives of the first order

$$\bar{\mathbf{u}}_{(t)} + \nabla \cdot \vec{\mathbf{c}} + \nabla\frac{\bar{p}}{\rho} = \mathbf{g}, \quad \nabla \cdot \bar{\mathbf{u}} = 0,$$

$$\vec{\mathbf{c}}_{(t)} + \vec{\mathbf{c}} \cdot \nabla\bar{u} + \bar{u}_{\mathbf{r}} \cdot \vec{\mathbf{c}} = -4\pi\sigma c(\vec{\mathbf{c}} - c^2\vec{\mathbf{e}}),$$

$$\text{where } \partial_{(t)} = \frac{d}{dt} = \partial_t + \bar{u} \cdot \nabla, \quad \sigma = \frac{c}{4\pi\gamma} \quad \text{and} \quad 3c^2 = Sp(\vec{\mathbf{c}}), \tag{11.2.1}$$

on the subject of *wave* characteristic equation (11.1.2).

After substituting small disturbances

$$\bar{\mathbf{u}} + \xi, \quad \vec{\mathbf{c}} + \vec{\eta} \quad \text{and} \quad \bar{p} + \rho\zeta \quad \text{of } \bar{\mathbf{u}}, \vec{\mathbf{c}} \quad \text{and}$$
$$\bar{p} \quad \text{for } \vec{\eta} = \vec{\eta}^*, \tag{11.2.2}$$

with *deviations* ξ, $\vec{\eta}$ and $\rho\zeta$ (or ζ, respectively) of (11.2.2) from the *equilibrium state* of $\bar{\mathbf{u}}$, $\vec{\mathbf{c}}$ *and* $\bar{p}$ that satisfies (11.2.1) and neglecting quadratical members in (11.2.1) as

$$\xi \cdot \nabla\xi \to \mathbf{0}, \quad \xi \cdot \nabla\vec{\eta}, \vec{\eta} \cdot \nabla\xi, \xi_{\mathbf{r}} \cdot \vec{\eta} \to \vec{\mathbf{0}} \quad \text{and}$$
$$c^2(\vec{\mathbf{c}} + \vec{\eta})(\vec{\mathbf{c}} - c^2(\vec{\mathbf{c}} + \vec{\eta})\vec{\mathbf{e}}) - c^2(\vec{\mathbf{c}})$$
$$\times(\vec{\mathbf{c}} - c^2(\vec{\mathbf{c}})\vec{\mathbf{e}}) \to c^2(\vec{\eta} - \eta\vec{\mathbf{e}}) + \eta(\vec{\mathbf{c}} - c^2\vec{\mathbf{e}}),$$

for the *turbulent energy deviation*

$$\eta = Sp(\vec{\eta})/3, \tag{11.2.3}$$

we come to the *linear system* of partial differential equations

$$\begin{aligned}
&\xi_{(t)} + \xi \cdot \nabla \bar{u} + \nabla \cdot \vec{\eta} + \nabla \zeta = \mathbf{0}, \quad \nabla \cdot \xi = 0 \quad \text{and} \\
&\vec{\eta}_{(t)} + \xi \cdot \nabla \vec{\mathbf{c}} + \vec{\eta} \cdot \nabla \bar{u} + \vec{\mathbf{c}} \cdot \nabla \xi + \xi_{\mathbf{r}} \cdot \vec{\mathbf{c}} \\
&\quad = -4\pi\sigma c(\vec{\eta} - \eta \vec{\mathbf{e}}) - 4\pi\sigma c\eta(\vec{\mathbf{c}} - c^2 \vec{\mathbf{e}}),
\end{aligned} \tag{11.2.4}$$

for infinitely small *deviations* ξ, η and $\rho\zeta$ for the averaged *flow* $\bar{\mathbf{u}}$, *turbulence* $\vec{\mathbf{c}}$, and *pressure* $\bar{p}$, respectively.

Then, following the standard scheme of reasoning [Petrovsky, 1954; Vladimirov, 1971] and leaving *only derivatives of* $\xi, \vec{\eta}, \zeta$ in (11.2.4),

$$\begin{aligned}
&\xi_{(t)} + \nabla \cdot \eta + \nabla \zeta = \mathbf{0}, \quad \nabla \cdot \xi = 0 \quad \text{and} \\
&\vec{\eta}_{(t)} + \vec{\mathbf{c}} \cdot \nabla \xi + \xi_{\mathbf{r}} \cdot \vec{\mathbf{c}} = \vec{\mathbf{0}},
\end{aligned} \tag{11.2.5}$$

we will be interested in a nontrivial solution

$$\xi = \xi(\varphi), \quad \vec{\eta} = \vec{\eta}(\varphi), \quad \zeta = \zeta(\varphi) \quad \text{for } \varphi_t |\varphi_{\mathbf{r}}||\xi_\varphi| \neq 0 \tag{11.2.6}$$

of (11.2.5) depending only on a smooth scalar function, or a *phase* $\varphi = \varphi(t, \mathbf{r})$, or on a corresponding level surface, or a *characteristic* $\varphi(t, \mathbf{r}) = \text{const}$.

In this connection, accounting for identities

$$\begin{aligned}
&\varphi_{\mathbf{r}} = \varphi_x \mathbf{i} + \cdots = \mathbf{i}\varphi_x + \cdots = \nabla\varphi \quad \text{with} \\
&\nabla\xi = \mathbf{i}\xi_x + \cdots = \mathbf{i}\varphi_x \xi_\varphi + \cdots = (\nabla\varphi)\xi_\varphi \quad \text{and} \\
&\xi_{\mathbf{r}} = \xi_x \mathbf{i} + \cdots = \xi_\varphi \varphi_x \mathbf{i} + \cdots = \xi_\varphi \varphi_{\mathbf{r}},
\end{aligned}$$

and introducing the vector

$$\eta^\varphi = \varphi_{\mathbf{r}} \cdot \vec{\eta}_\varphi = \vec{\eta}_\varphi \cdot \nabla\varphi(\eta_\varphi = \eta^*_\varphi),$$

we have the following.

Theorem 11.2.1. *The system* (11.2.5) *admits nontrivial solutions of the form* (11.2.6) *only with the specific pressure deviation* $\zeta = \text{const}$ *and characteristic equation* (11.1.2) *fulfilled.*

Proof. Really, substituting (11.2.6) into (11.2.5), we come to

$$\varphi_{(t)}\xi_\varphi + \eta^\varphi + \zeta_\varphi\, \varphi_{\mathbf{r}} = \mathbf{0} \quad \text{and}$$
$$\varphi_{(t)}\vec{\eta}_\varphi + (\vec{\mathbf{c}}\cdot\nabla\varphi)\,\xi_\varphi + \xi_\varphi\varphi_{\mathbf{r}}\cdot\vec{\mathbf{c}} = \vec{\mathbf{0}},$$

with

$$\varphi_{\mathbf{r}}\cdot\xi_\varphi = \xi_\varphi\cdot\nabla\varphi = 0, \text{ hence, } \varphi_{(t)}\eta^\varphi + \xi_\varphi\varphi_{\mathbf{r}}\cdot\vec{\mathbf{c}}\cdot\nabla\varphi = \mathbf{0},$$

so

$$(\varphi_{(t)}^2 - \varphi_{\mathbf{r}}\cdot\vec{\mathbf{c}}\cdot\nabla\varphi)\xi_\varphi + \varphi_{(t)}\zeta_\varphi\varphi_{\mathbf{r}} = \mathbf{0}$$

provided that

$$\varphi_{(t)}\zeta_\varphi|\nabla\varphi|^2 = 0, \quad \text{or} \quad \zeta_\varphi = 0 \quad (\text{since } \varphi_{(t)}|\nabla\varphi| \neq 0),$$

and

$$(\varphi_{(t)}^2 - \varphi_{\mathbf{r}}\cdot\vec{\mathbf{c}}\cdot\nabla\varphi)|\xi_\varphi|^2 = 0, \quad \text{or } \varphi_{(t)}^2 = \varphi_{\mathbf{r}}\cdot\vec{\mathbf{c}}\cdot\nabla\varphi \quad (\text{since } |\xi_\varphi| \neq 0),$$

which concludes the proof. □

11.3. Electromagnetic structure

Following Troshkin [1989, 1990a, 1990b, 1992b, 1993, 1995] again, let us consider the linear system (11.2.4) for infinitesimal deviations $\xi, \vec{\eta}, \zeta$ of *disturbed averaged fields*

$$\bar{\mathbf{u}} + \xi, \quad \vec{\mathbf{c}} + \vec{\eta} \quad \text{and} \quad \bar{p} + \rho\zeta \quad \text{of } \bar{\mathbf{u}},\ \vec{\mathbf{c}} \quad \text{and} \quad \bar{p} \quad \text{for } \vec{\eta} = \vec{\eta}^*,$$

in a *kernel* (or a domain) of a *homogeneous and isotropic turbulence at rest* in which *basic* (or undisturbed) fields are

$$\bar{\mathbf{u}} = \mathbf{0}, \quad \vec{\mathbf{c}} = c^2\vec{\mathbf{e}}, \quad \bar{p} = \text{const}$$
$$\text{for } c, \sigma = \text{const} > 0 \quad \text{and} \quad c = \sqrt{\frac{Sp(\vec{\mathbf{c}})}{3}} \qquad (11.3.1)$$

(to satisfy the initial quasilinear equations (11.2.1)).

With definitions (11.2.3) and (11.3.1) and evident identities

$$\vec{\mathbf{e}} \cdot \nabla \xi = \nabla \xi \quad \text{and} \quad \xi_{\mathbf{r}} \cdot \vec{\mathbf{e}} = \xi_{\mathbf{r}},$$

the system (11.2.4) is reduced to the equations

$$\xi_t = -\nabla \cdot (\vec{\eta} + \zeta \vec{\mathbf{e}}), \quad \nabla \cdot \xi = 0 \quad \text{and}$$
$$\vec{\eta}_t + c^2(\nabla \xi + \xi_{\mathbf{r}}) + 4\pi\sigma c(\vec{\eta} - \eta\vec{\mathbf{e}}) = \vec{\mathbf{0}}$$

or, with identities

$$\nabla \cdot \eta\vec{\mathbf{e}} = \nabla \eta, \quad \nabla \cdot \zeta\vec{\mathbf{e}} = \nabla \zeta$$

and the fact that the symmetry deviation η is *stationary* as

$$\eta_t = \frac{1}{3} Sp(\vec{\eta}_t) = -\frac{c^2}{3} Sp(\nabla \xi + \xi_{\mathbf{r}}) - \frac{4\pi\sigma c}{3}(Sp(\vec{\eta}) - 3\eta)$$
$$= -\frac{2c^2}{3} \nabla \cdot \xi = 0,$$

to the relations

$$\xi_t + \nabla \cdot (\vec{\eta} - \eta\vec{\mathbf{e}}) + \nabla(\eta + \zeta) = \mathbf{0} \ (\xi_t = -\nabla \cdot (\vec{\eta} + \zeta\vec{\mathbf{e}})), \quad \nabla \cdot \xi = 0$$
$$\text{and} \quad \frac{1}{c}(\vec{\eta} - \eta\vec{\mathbf{e}})_t + c(\nabla \xi + \xi_{\mathbf{r}}) + 4\pi\sigma(\vec{\eta} - \eta\vec{\mathbf{e}}) = \vec{\mathbf{0}}. \tag{11.3.2}$$

Then, putting c in (11.3.1) to be the *speed of light in vacuum*,

$$c = 299\ 793\ 000\,\text{m/s},$$

that is determined only by the pulsating velocity component, and hence, proves to be an *absolute quantity*

$$c((\mathbf{u} + \mathbf{a})') = c(\mathbf{u}') = c = \sqrt{\frac{Sp(\vec{\mathbf{c}})}{3}} \quad \text{for any} \quad \mathbf{a} = \bar{\mathbf{a}}(\mathbf{a}' = \mathbf{0}),$$

as required by the *partial relativity principle* [Einstein, 1905], and introducing, further, the *electrical equivalent of force*, κ, to be the reciprocal of the *charge-to-mass ratio*

$$\frac{1}{\kappa} = \frac{e}{m_e} = 1.75882 \cdot 10^{11} C/\mathrm{kg}\ (C = 1\ \text{Coulomb}),$$

and related *scalar,vector* and *divector* (matrix) *potentials*, with *electrical charge density*,

$$\varphi = \kappa(\eta + \zeta), \quad \mathbf{A} = c\kappa\boldsymbol{\xi} \quad \text{and} \quad \vec{\mathbf{b}} = \kappa(\vec{\boldsymbol{\eta}} - \eta\vec{\mathbf{e}}),$$
$$\text{with} \quad \delta = \frac{\nabla \cdot \nabla \cdot \vec{\mathbf{b}}}{4\pi}, \tag{11.3.3}$$

we may rewrite *mechanical* equations (11.3.2) in an *electromagnetic form*

$$\frac{1}{c}\mathbf{A}_t + \nabla \cdot \vec{\mathbf{b}} + \nabla\varphi = \mathbf{0}, \quad \nabla \cdot \mathbf{A} = 0 \quad \text{and}$$
$$\frac{1}{c}\vec{\mathbf{b}}_t + \nabla\mathbf{A} + \mathbf{A}_\mathbf{r} + 4\pi\sigma\vec{\mathbf{b}} = \vec{\mathbf{0}}$$

with relevant *magnetic* and *electrical* fields $\mathbf{H}$ and $\mathbf{E}$ of kernel (11.3.1), for corresponding *electrical current* and *charge* densities $\mathbf{j}$ and δ in proper *Ohm* and *Coulomb laws*,

$$\mathbf{H} = \nabla \times \mathbf{A} \quad \text{and} \quad \mathbf{E} = \nabla \cdot \vec{\mathbf{b}} \quad \text{for}\ \mathbf{j} = \sigma\mathbf{E} \quad \text{and} \quad -\Delta\varphi = 4\pi\delta, \tag{11.3.4}$$

respectively, which, with identities

$$\nabla \cdot \mathbf{A}_\mathbf{r} = \nabla \cdot \mathbf{A}_x\mathbf{i} + \cdots = (\nabla \cdot \mathbf{A})_\mathbf{r} = \mathbf{0} \quad \text{and}$$
$$\nabla \cdot (\nabla\mathbf{A} + \mathbf{A}_\mathbf{r}) = \nabla \cdot \nabla\mathbf{A} = -\nabla \times (\nabla \times \mathbf{A}),$$

leads to the required *Maxwell equations*

$$\frac{1}{c}\mathbf{H}_t + \nabla \times \mathbf{E} = \mathbf{0} \quad \text{and} \quad \frac{1}{c}\mathbf{E}_t - \nabla \times \mathbf{H} + 4\pi\mathbf{j} = \mathbf{0} \tag{11.3.5}$$

for small disturbances (11.3.3) and (11.3.4) of turbulent kernel (11.3.1) to propagate with absolute velocity $c = c(\mathbf{u}')$ as shown in Fig. 11.3.1.

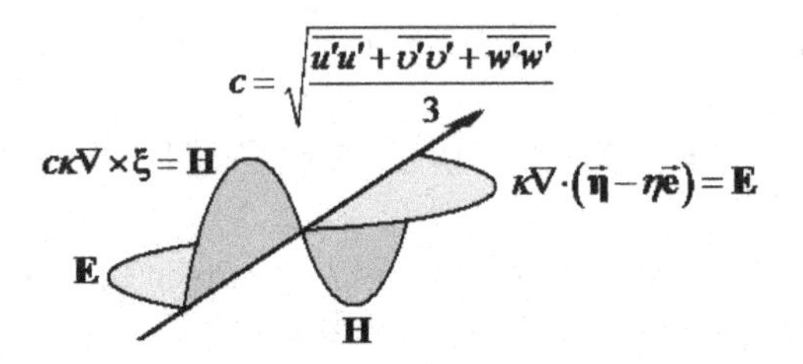

Figure 11.3.1. A wave of small disturbances ξ, $\vec{\eta}$ of turbulent kernel (11.3.1).

11.4. Dark medium source, sink, and light pulsations

Meanwhile, as has happened often in physics, just after Michelson–Morley's experiment [Michelson and Morley, 1887] and the related Einstein's relativity principle on the absoluteness of speed of light in a vacuum supposed to surround the material world [Einstein, 1905], both first analysis and observations of running off, rotation, and gravitational lensing of galaxies [Chwolson, 1924; Hubble, 1929; Zwicky, 1937a, 1937b; Rubin, 1970] evidenced that all the detecting *light part* of the universe does not dwell in the void, but is submerged in the invisible *dark medium* assumed to complement the former up to a *smooth substance* filling up all the surrounding space.

In what follows, the latter is assumed to be mobile, namely, to be described by vector fields of either *instant velocity* $\mathbf{u} = \mathbf{u}(t, \mathbf{r})$ or its *time average* $\bar{\mathbf{u}} = \lim_{T\to\infty} \frac{1}{T} \int_0^T \mathbf{u}(t+s, \mathbf{r})ds$, when supplied with matrix fields of *strain* $\mathbf{u_r}$ or *turbulence* $\overline{\mathbf{u}'\mathbf{u}'}$ (of velocity pulsations $\mathbf{u}' = \mathbf{u} - \bar{\mathbf{u}}$) taken to be prevailing in its dark or light parts, respectively.

In any dark part, the smooth medium is invisible because it is *laminar*, i.e. it makes no turbulence: $\mathbf{u} = \bar{\mathbf{u}}$. It does not obligate even to obey here the classical laws of mass conservation, dynamic equilibrium, and heat balance. Instead of them, an alternative principle of least deformation (in Theorem 2.4.1) is assumed to be valid in the noted part. Then, when free of forces and masses, as a complement to habitual translations and rotations (Theorem 2.4.1), a central expansion of the smooth medium appears with Hubble constant H [Hubble, 1929] and two stagnation points of its velocity field $\mathbf{u}$ (where $\mathbf{u} = \mathbf{0}$) to yield the required source and sink of the universe.

Now, following [Troshkin, 2017b], let us consider a pattern of trajectories for the *stationary central expansion*

$$\mathbf{u} = \mathbf{w}^{(\mathrm{iii})} = H\mathbf{r} - \frac{1}{2}\mathbf{r}\cdot\mathbf{r}\boldsymbol{\Theta}\mathbf{r}\cdot\boldsymbol{\Theta}\mathbf{r}$$

from Section 2.4 of Chapter 2 when

$$H = \mathrm{const} > 0 \quad \text{and} \quad \boldsymbol{\Theta} = \mathbf{const} \neq \mathbf{0}, \quad \text{or} \quad \Theta = \sqrt{\boldsymbol{\Theta}\cdot\boldsymbol{\Theta}} > 0.$$

In terms of dimensionless variables

$$s = Ht, \quad \mathbf{R} = \frac{\Theta\mathbf{r}}{H} = X\mathbf{i} + Y\mathbf{j} + Z\mathbf{k} \quad \text{and}$$
$$\mathbf{U} = \frac{\Theta\mathbf{u}}{H^2}, \quad \text{for } \frac{\boldsymbol{\Theta}}{\Theta} = -\mathbf{j},$$

we have that

$$\mathbf{R}_s = \mathbf{U} = \mathbf{R} + \frac{1}{2}\mathbf{R}\cdot\mathbf{R}\mathbf{j} - \mathbf{R}\cdot\mathbf{j}\mathbf{R} \quad (\mathbf{r}_t = \mathbf{u}),$$

or

$$X_s = (1-Y)X, \quad Y_s = Y + \frac{1}{2}(X^2 - Y^2 + Z^2) \quad \text{and}$$
$$Z_s = (1-Y)Z.$$

In cylindrical coordinates

$$R = \sqrt{Z^2 + X^2} \quad \text{and} \quad \theta = \operatorname{arctg}\frac{X}{Z}$$
$$(\text{or } Z = R\cos\theta \text{ and } X = R\sin\theta),$$

we obtain the equations

$$R_s = \bar{U} = -R\bar{Y}, \quad \bar{Y}_s = \bar{V} = \frac{R^2 + 1 - \bar{Y}^2}{2},$$
$$\bar{Y} = Y - 1, \quad \text{and} \quad \theta_s = 0$$

excluding the rotation around the axis $R = 0$.

Accounting for the identity

$$\left(\frac{\bar{U}}{R^2}\right)_R + \left(\frac{\bar{V}}{R^2}\right)_{\bar{Y}} = \left(\frac{-\bar{Y}}{R}\right)_R + \left(\frac{R^2+1-\bar{Y}^2}{2R^2}\right)_{\bar{Y}}$$
$$= \frac{\bar{Y}}{R^2} - \frac{\bar{Y}}{R^2} = 0,$$

trajectories of the two remaining equations reduce then to *level lines* $\psi(R,\bar{Y}) = c$ for the *first integral* of the form

$$\psi(R,\bar{Y}) = -\frac{R}{2} + \frac{1-\bar{Y}^2}{2R} \quad \text{(such that } \bar{U} = R^2\psi_{\bar{Y}} \text{ and } \bar{V} = -R^2\psi_R)$$

to form the required c-family of circular arcs

$$(R+c)^2 + \bar{Y}^2 = c^2 + 1 \text{ (or } \psi(R,\bar{Y}) = c), \quad -\infty < c = \text{const} < \infty,$$

on the $(R,\bar{Y})$-plane that come from the above-mentioned *source* $(0,-1)$ and end in the *sink* $(0,1)$ as shown in Fig. 11.4.1.

In the light part, the medium in hand is visible because it is *turbulent*, i.e. it finely pulsates to create the turbulent kernel (11.3.1), with the detected fields of small disturbances (11.3.3) and (11.3.4) propagating in space with the absolute speed of light $c = c(\mathbf{u}')$.

As this takes place, while moving with a relative velocity $\boldsymbol{v} = \boldsymbol{v}(t)$ through the cosmic medium disturbed with a small bulk rate ξ of

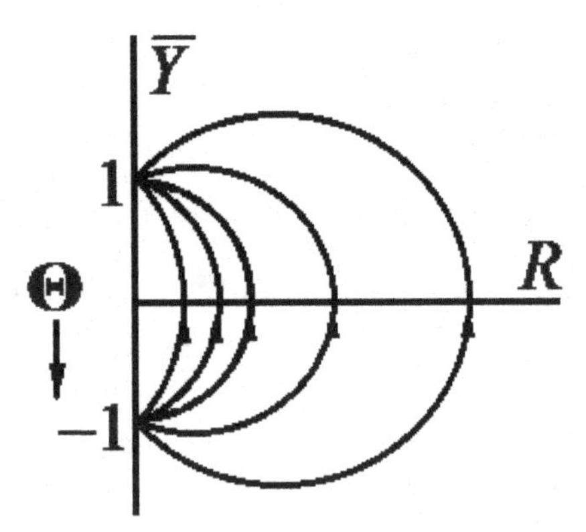

Figure 11.4.1. A pattern of a central expansion with constant $H > 0$ and the braking vector $\boldsymbol{\Theta} = -\Theta\mathbf{j}$, $\Theta > 0$, in the $(R,\bar{Y})$-plane, with $R = \sqrt{Z^2+X^2}$ and $\bar{Y} = Y - 1$ of dimensionless coordinates $X = \Theta x/H$, $Y = \Theta y/H$ and $Z = \Theta z/H$.

negligible scalar square $|\xi|^2 << c^2$ and supplied in each point $\mathbf{r}$ with *potential energy* $\eta+\zeta$ *per mass* ρdV delivered by *turbulent energy* and *average pressure* deviations, $\eta = \eta(t, \mathbf{r})$ and $\zeta = \zeta(t, \mathbf{r})$, respectively, every *particle* $\mathbf{r}$ of *mass* $m = q\kappa$ or *charge* $q = m/\kappa$ possesses a *Lagrangian*

$$L = ml(t, \mathbf{r}, \boldsymbol{v}), \quad \text{or} \quad l = L/m,$$

for

$$l = \frac{1}{2}|\boldsymbol{v} + \xi|^2 - \eta - \zeta, \quad \text{or} \quad = \frac{1}{2}|\boldsymbol{v}|^2 + \boldsymbol{v} \cdot \xi - \eta - \zeta \text{ for } |\xi|^2 \to 0,$$

including its *momentum*, *relative acceleration*, and *force* per unit mass

$$l_{\boldsymbol{v}} = \boldsymbol{v} + \xi,$$

$$\boldsymbol{v} \cdot \nabla l_{\boldsymbol{v}} = \boldsymbol{v} \cdot \nabla \xi, \quad \text{and} \quad l_{\mathbf{r}} = \nabla l = \nabla(\boldsymbol{v} \cdot \xi - \eta + \zeta),$$

respectively, and satisfying *Newton's dynamic law*

$$\frac{d}{dt} L_{\boldsymbol{v}} = L_{\mathbf{r}}, \quad \text{or} \quad \partial_t L_{\boldsymbol{v}} + \boldsymbol{v} \cdot \nabla L_{\boldsymbol{v}} = L_{\mathbf{r}}, \quad \text{or} \quad (l_{\boldsymbol{v}})_t + \boldsymbol{v} \cdot \nabla l_{\boldsymbol{v}} = l_{\mathbf{r}},$$

or

$$\boldsymbol{v}_t + \xi_t + \boldsymbol{v} \cdot \nabla \xi = \nabla \boldsymbol{v} \cdot \xi - \nabla(\eta + \zeta),$$

or, accounting for the first equation in (11.3.2) and the double cross-product identity,

$$\xi_t = -\nabla \cdot \vec{\eta} - \nabla \zeta \quad \text{and} \quad \nabla \boldsymbol{v} \cdot \xi - v \cdot \nabla \xi = \boldsymbol{v} \times (\nabla \times \xi),$$

subjected to the following *point mass motion* equation in a turbulent kernel,

$$m\boldsymbol{v}_t = m\nabla \cdot (\vec{\eta} - \eta\vec{e}) + m\boldsymbol{v} \times (\nabla \times \xi), \tag{11.4.1}$$

with the*Coriolis force* $m\boldsymbol{v} \times (\nabla \times \xi)$ (as in Section 2.6 of Chapter 2).

With the fields in (11.3.3) and (11.3.4), we may conclude, thus, that

Theorem 11.4.1. *For* $m = q\kappa$ *(or* $q = m/\kappa$*), the point mass motion equation* (11.4.1) *is equivalent to its point charge analog*

$$m\boldsymbol{v}_t = q\mathbf{E} + \frac{q}{c}\upsilon \times \mathbf{H}, \quad \frac{q}{m} = \frac{1}{\kappa},$$

with the Lorentz force $q\boldsymbol{v} \times \mathbf{H}/c$.

The theorem follows immediately from (11.4.1), (11.3.3), and (11.3.4).

PART 5

Plane Channel Flow Stability

Chapter 12

Flows as Infinitely Dimensional Rotations

12.1. Plane eddy lifetime

Alternatively to vortices in an ideal fluid [Fraenkel and Berger, 1973; Belotserkovskii *et al.*, 2014], the velocity pulsations $\mathbf{u}'$ that develop the *turbulence* $\overline{\mathbf{u}'\mathbf{u}'} \neq \vec{\mathbf{0}}$ (in Part 4) are no more than the *flow instabilities* usually caused by *eddies* spontaneously originating in a liquid of sufficiently small *kinematic viscosity* $\nu > 0$ [Reynolds, 1883].

However, when taken *alone* to be initiated by a *circulation*

$$\Gamma_0 = \int_{-\infty}^{\infty} \int_{-\infty}^{\infty} \omega dx dy,$$

in a *constant flow* $u_\infty \mathbf{i}$, each such eddy proves to be similar to a *laminar* (not turbulent) *flow* $\mathbf{u}$ ($\overline{\mathbf{u}'\mathbf{u}'} = \vec{\mathbf{0}}$) of a *vortex tube* (as in Fig. 12.1.1) to be in stock for only a finite *lifetime* $t_{\max} < t_*$ (as everything in Nature), during a definite interval $0 < t < t_{\max}$ of *real time* t bounded above by the value

$$t_* = \frac{h^2}{16\pi^2\nu}, \quad h = \frac{\Gamma_0}{u_\infty}, \quad u_\infty, \Gamma_0 = \text{const} > 0 \quad (\pi = 3.14159\ldots).$$

Indeed, let a velocity field

$$\mathbf{u} = u\mathbf{i} + v\mathbf{j} = \psi_y \mathbf{i} - \psi_x \mathbf{j}$$

of stream function

$$\psi = \psi(t, 0, 0) + u_\infty y + \Gamma_0 \varphi(t, \bar{r}), \quad \bar{r} = \sqrt{\bar{x}^2 + y^2}, \quad \bar{x} = x - u_\infty t,$$

be supported by a constant flow at infinity,

$$\mathbf{u} \to u_\infty \mathbf{i} \quad \text{for } r = \sqrt{x^2 + y^2} \to \infty,$$

when initiated by the *circulation*

$$\Gamma = \Gamma(t) = \int_{-\infty}^{\infty} \int_{-\infty}^{\infty} \omega dx dy, \quad t > 0, \quad \Gamma|_{t=+0} = \lim_{t \to 0} \Gamma(t) = \Gamma_0,$$

of *point vorticity*

$$\omega = \Gamma_0 \delta, \quad \delta = \frac{\exp(-\zeta^2)}{4\pi\sigma^2}, \quad \sigma = \sigma(t) = \sqrt{\nu t},$$

$$\zeta = \sqrt{\xi^2 + \eta^2}, \quad \xi = \frac{\bar{x}}{2\sigma}, \quad \eta = \frac{y}{2\sigma},$$

proving to be constant as

$$\Gamma = \Gamma_0 \int_0^{\infty} \int_0^{2\pi} \delta \bar{r} d\bar{r} d\theta = 2\pi\Gamma_0 \int_0^{\infty} \delta \bar{r} d\bar{r} = \Gamma_0 \int_0^{\infty} \exp(-\xi) d\xi = \Gamma_0,$$

in which the *source*

$$\omega_t - \nu\Delta\omega = \left(\frac{\bar{r}^2}{4\nu t} - 1 + \frac{u_\infty \bar{x}}{2\nu}\right)\frac{\omega}{t} - \left(\frac{\bar{r}^2}{4\nu t} - 1\right)\frac{\omega}{t}$$

$$= \frac{u_\infty \bar{x}}{2\nu}\frac{\omega}{t} = -u_\infty \omega_x$$

of derivatives

$$\omega_t = -\frac{\omega}{t} + \bar{x}_t \omega_x = -\frac{\omega}{t} - u_\infty \omega_x, \quad \omega_x = \bar{x}_x \bar{r}_{\bar{x}} \omega_{\bar{r}} = -\frac{\bar{x}\omega}{2\nu t},$$

$$\omega_y = \bar{r}_y \omega_{\bar{r}} = -\frac{y\omega}{2\nu t}$$

$$\text{and} \quad \Delta\omega = \omega_{xx} + \omega_{yy} = -\frac{\bar{x}\omega_x + y\omega_y + 2\omega}{2\nu t} = \left(\frac{\bar{r}^2}{4\nu t} - 1\right)\frac{\omega}{\nu t}$$

when reinforced with *convective transfer*

$$u\omega_x + v\omega_y = \psi \times \omega = u_\infty \omega_x + \Gamma_0 \varphi \times \omega = u_\infty \omega_x$$

$$(\varphi \times \omega = (\bar{r}_y \bar{r}_x - \bar{r}_x \bar{r}_y)\varphi_{\bar{r}} \omega_{\bar{r}} = 0)$$

provides the dynamic equilibrium

$$\omega_t - \nu\Delta\omega + \psi \times \omega = 0, \quad -\Delta\psi = \omega, \quad t > 0,$$

for the *circulation stream function*

$$\varphi = \varphi(t, \bar{r}) = -\frac{1}{4\pi}\int_0^{\zeta^2} \Phi(s)ds, \quad \zeta = \frac{\bar{r}}{2\sigma} = \frac{\bar{r}}{2\sqrt{\nu t}},$$

$$\Phi(s) = \frac{1 - \exp(-s)}{s} > 0,$$

satisfying the required identity

$$-\Delta\psi = \frac{\Gamma_0}{\bar{r}}(-\bar{r}\varphi_{\bar{r}})_{\bar{r}} = \frac{\Gamma_0}{2\pi\bar{r}}(\zeta^2\Phi(\zeta^2))_{\bar{r}} = \frac{\Gamma_0\zeta\zeta_{\bar{r}}}{\pi\bar{r}}\exp(-\zeta^2) = \Gamma_0\delta = \omega.$$

Then, in terms of *characteristic* length h, time χ, and Reynolds number Re determined for a dimensionless *decreasing scale* $\eta_\dagger$ of real time t,

$$h = \frac{\Gamma_0}{u_\infty}, \quad \chi = \frac{h^2}{\nu} \quad \text{and} \quad \mathrm{Re} = \frac{u_\infty h}{\nu} = \frac{\Gamma_0}{\nu} \quad \text{for}$$

$$\eta_\dagger = \frac{h}{2\sigma} = \frac{1}{2}\sqrt{\frac{\chi}{t}} \quad \text{of} \quad t > 0,$$

respectively, we have the following dimensionless velocity components:

$$u/u_\infty = 1 + \frac{\Gamma_0}{u_\infty}\varphi_y = 1 - \frac{\eta_\dagger\eta}{2\pi}\Phi(\zeta^2) \quad \text{and}$$

$$v/u_\infty = -\frac{\Gamma_0}{u_\infty}\varphi_x = \frac{\eta_\dagger\xi}{2\pi}\Phi(\zeta^2),$$

with flow patterns in Fig. 12.1.1 of level lines $\psi(t, x, y) = \text{const}$ at instants

$$\nu t/h^2 = t/\chi = 1, \ 35, \ 51.1.$$

As a result, an *isolated plane vortex*, as a flow structure consisting of *stagnation* points of $\mathbf{u}$ where, by definition,

$$v = 0 \quad \text{and} \quad u = 0, \quad \text{or} \quad \xi = 0 \quad \text{and} \quad 1 - \frac{\eta_\dagger\eta}{2\pi}\Phi(\eta^2) = 0$$

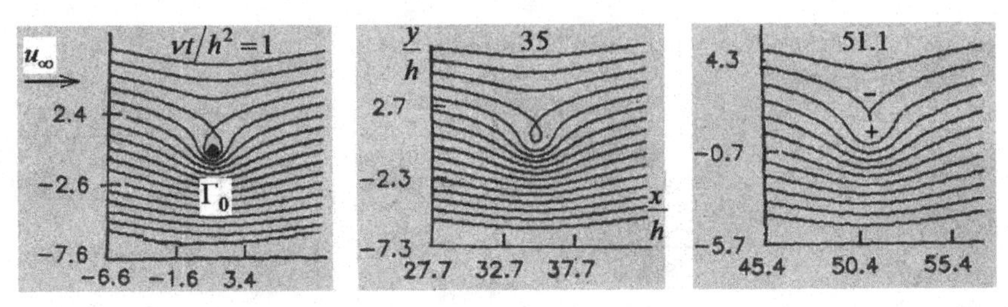

Figure 12.1.1. Plain point vortex circulation Γ_0 in a flow u_∞ [Troshkin, 1992a].

or of plane *centre* (+) and *saddle* (−) in Fig. 12.1.1 where

$$\xi = 0 \quad \text{and} \quad \eta = \eta_\pm, \quad \text{or} \quad f(\eta) = \eta\Phi(\eta^2) = \frac{1-\exp(-\eta^2)}{\eta}$$

$$= \frac{4\pi u_\infty}{\Gamma_0}\sqrt{\nu t} = \frac{2\pi}{\eta_\dagger}$$

and

$$\pm \frac{16\pi^2\sigma^2\eta}{u_\infty^2\eta_\dagger^2 f(\eta)} \left|\begin{vmatrix} u_x & u_y \\ v_x & v_y \end{vmatrix}\right|_{\xi=0} = \pm f'(\eta) > 0 \quad (\eta = \eta_\pm),$$

does exist for only a finite duration $0 < t < t_{\max}$ of *eddy lifetime*

$$t_{\max} = \frac{1}{\nu}\left(\frac{hf_{\max}}{4\pi}\right)^2 < \frac{1}{\nu}\left(\frac{h}{4\pi}\right)^2 = t_*, \quad \text{since } f_{\max} = \max_{\eta>0} f < 1,$$

because

$$f < \frac{1}{\eta} \quad \text{and} \quad f < \eta \quad (\text{hence, } f^2 < \frac{1}{\eta}\eta = 1),$$

as illustrated in Fig. 12.1.2.

So,

$$t_{\max} = \frac{1}{\nu}\left(\frac{hf_{\max}}{4\pi}\right)^2 \quad \text{and} \quad f_{\max} = f(\eta_*) \simeq 0.64$$

$$\text{for } \eta_+ = \eta_- = \eta_* \simeq 1.1.$$

In spite of the lifetime boundness for every particle of turbulence, or eddy, it is seems, however, to be natural to wonder about the

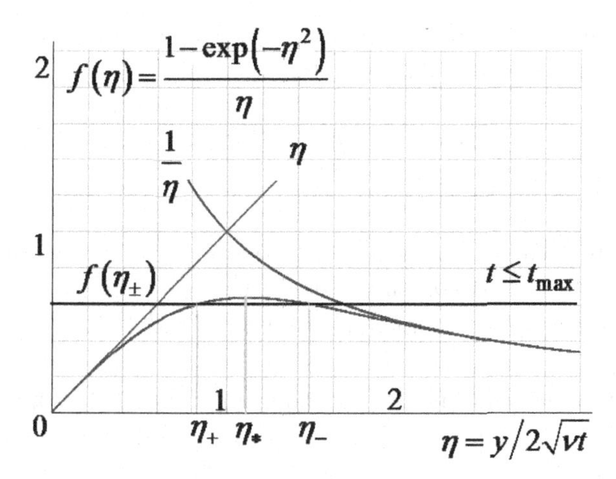

Figure 12.1.2. An eddy lives until its center η_+ merges with saddle point η_- in η_* at $t_{\max}$.

stability of a laminar *stationary flow* visible, and hence, living *infinitely long* by definition.

12.2. Vortex stability

In what follows, we will deal with two-dimensional *solenoidal* velocity fields

$$\mathbf{u} = u\mathbf{i} + v\mathbf{j} = \mathbf{u}(t, x, y), \ \nabla \cdot \mathbf{u} = u_x + v_y = 0, \ \partial_z = w = 0, \quad t \geq 0,$$

determined in such a bounded flow region V as a *rectangular* $V_0 = \{0 < x < l, 0 < y < h\}$ of *aspect ratio*

$$\alpha = h/l,$$

or the same rectangular as a *periodic cell* V_K or V_N of an unbounded horizontal *channel*

$$K : u, v(t, x, y) = u, v(t, x + l, y), \quad -\infty < x < \infty, \quad 0 \leq y \leq h,$$

or vertical *strip*

$$N : u, v(t, x, y) = u, v(t, x, y + h), \quad 0 \leq x \leq l, \quad -\infty < y < \infty,$$

for *smooth* (infinitely differentiable) velocity components u, v *periodic* on x or y, respectively, to form a *flow region* V, with *boundary*

$$\partial V_K = \{y = 0, h : 0 \leq x \leq l\} \quad \text{and} \quad \partial V_N = \{x = 0, l : 0 \leq y \leq h\}$$
$$\text{or } \partial V_0 = \partial V_K \cup \partial V_N$$

of one (as to ∂V_0) or two circles (to be either $y = 0, h$ for ∂V_K or $x = 0, l$ for ∂V_N) and *compact closure* $\bar{V} = V + \partial V$ equivalent (i.e. equal topologically, or *homeomorphic*) to a *disk* ($V = V_0$) or a *ring* ($V = V_{K,N}$).

At least one boundary circle of the ring in hand is supposed to be a *rigid wall* impenetrable in the sense that boundary normal velocity components

$$v|_{y=0} = v(t, x, 0) = 0(0 \leq x \leq l) \quad \text{for } V_K \quad \text{and}$$
$$u|_{x=l} = 0 \quad \text{for } V_N(t \geq 0),$$

so contour integrals,

$$\int_0^y u(t, x, y')dy' = \psi(t, x, y) - \psi(t, 0, 0) \quad \text{in} \quad V_K$$

and

$$\int_x^l v(t, x', y)dx' = \psi(t, x, y) - \psi(t, l, h) \quad \text{in} \quad V_N,$$

lead *mechanically* first to a mass *expenditure* $\rho\psi$ per unit area and time [Milne-Thomson, 1955], and then to a *specific* (as divided by a density $\rho = \text{const} > 0$) one, ψ, or to a *stream function* $\psi = \psi(t, x, y)$, which formally is either determined up to such an arbitrary additive and constant *extra expenditure* $C = C(t)$ as $C = \psi(t, 0, 0)$, $C = \psi(t, l, h)$, etc., or *point normed* as $C = 0$, so that

$$(\psi + C)_y = \psi_y = u(t, x, y) \quad \text{and} \quad (\psi + C)_x = \psi_x = -v(t, x, y)$$
$$\text{everywhere in} \quad V = V_0, V_K, V_N \text{ and on } \partial V.$$

The existence of ψ in such a *one-connected region* as a rectangular V_0 (where every simple circle is contracted to a point) is well known

[Hurwitz and Courant, 1922]. So, in any case, there is a *vector potential*

$$\boldsymbol{\psi} = \psi \mathbf{k} \quad \text{such that } \nabla \times \boldsymbol{\psi} = \mathbf{u} \quad \text{and} \quad \nabla \cdot \boldsymbol{\psi} = 0 \quad \text{for } \psi_z = 0,$$

supplied with *vector* and *scalar curls* of $\mathbf{u}$, or *vorticities*

$$\boldsymbol{\omega} = \nabla \times \mathbf{u} = -\Delta \boldsymbol{\psi} = \omega \mathbf{k} \quad \text{and}$$
$$\omega = v_x - u_y = -\Delta \psi = -\psi_{xx} - \psi_{yy},$$

respectively, and a *vector commutator*

$$\mathbf{u} \cdot \nabla \boldsymbol{\omega} - \boldsymbol{\omega} \cdot \nabla \mathbf{u} = \mathbf{u} \cdot \nabla \boldsymbol{\omega} = (\psi \times \omega)\mathbf{k} \quad (\boldsymbol{\omega} \cdot \nabla \mathbf{u} = \omega \mathbf{u}_z = \mathbf{0})$$

to give an *algebraic screw* to be a commutator of *Poisson brackets*

$$\psi \times \omega = [\psi, \omega] = \psi_y \omega_x - \psi_x \omega_y = u\omega_x + v\omega_y.$$

All the *basic* (stationary) flows $\mathbf{u}^{(0)}$ treated below are *disturbed initially* by non-stationary velocity fields $\mathbf{u}$ starting at $t = 0$ from *arbitrary* smooth stationary ones, $\mathbf{u}_0$, given in the *flow region* $V = V_0, V_K, V_N$ and on the boundary ∂V to keep boundary conditions of $\mathbf{u}^{(0)}$ for all $t \geq 0$, including both *boundary expenditure* ψ^* prescribed to be constant on a *rigid* ∂V, with *no-slip conditions* $u, -v = \psi_y, \psi_x = 0$, for a *viscous* fluid in $V = V_K$ (when $\nu > 0$) and *inflow vorticity* ω^+ fixed on the boundary points $x = 0$ where inflow velocity $\psi_y^* > 0$ for an *ideal* fluid in $V = V_0, V_N$ (when $\nu = 0$):

$$\begin{aligned}
&\mathbf{u}, \psi, \omega|_{t=0} = \mathbf{u}_0, \psi_0, \omega_0 \quad \text{on} \quad \bar{V}, \psi, \psi_0, \psi^{(0)}\Big|_{\partial V} = \psi^*, \\
&V = V_{0,K,N}, \nu \geq 0, \\
&\omega, \omega_0, \omega^{(0)}\Big|_{\partial V, x=0, \psi_y^*>0} = \omega^+ \quad \text{for } V = V_{0,N} \quad \text{and} \quad \nu = 0, \\
&u, v, u^{(0)}, v^{(0)}\Big|_{\partial V} = 0 \quad \text{for } V = V_K \quad \text{and} \nu > 0, t \geq 0. \qquad (12.2.1)
\end{aligned}$$

In doing so, both basic fields $\mathbf{u}^{(0)}$ and their *disturbances* $\mathbf{u}$ remain subjected to the *Euler–Navier equations* (7.1.8) in V for either the ideal or viscous incompressible fluid:

$$\mathbf{u}_t + \mathbf{u}\cdot\nabla\mathbf{u} - \nu\Delta\mathbf{u} + \frac{1}{\rho}\nabla p = \mathbf{u}^{(0)}\cdot\nabla\mathbf{u}^{(0)} - \nu\Delta\mathbf{u}^{(0)}$$

$$+\frac{1}{\rho}\nabla p^{(0)} = \mathbf{g}, \quad \mathbf{u}_t^{(0)} = \mathbf{0} \quad \text{and}$$

$$\nabla\cdot\mathbf{u} = \nabla\cdot\mathbf{u}^{(0)} = 0, \quad t > 0, \quad \text{in } V \quad \text{for } \rho = \text{const} > 0 \quad \text{and}$$

$$\nu = \frac{\mu}{\rho} = \text{const} \geq 0,$$

namely,

$$u_t + uu_x + vu_y - \nu\Delta u + \frac{1}{\rho}p_x = u^{(0)}u_x^{(0)} + v^{(0)}u_y^{(0)}$$

$$-\nu\Delta u^{(0)} + \frac{1}{\rho}p_x^{(0)} = g^x,$$

$$v_t + uv_x + vv_y - \nu\Delta v + \frac{1}{\rho}p_y = u^{(0)}v_x^{(0)} + v^{(0)}v_y^{(0)}$$

$$-\nu\Delta v^{(0)} + \frac{1}{\rho}p_y^{(0)} = g^y,$$

$$u_x + v_y = u_x^{(0)} + v_y^{(0)} = 0, \quad t > 0, \quad \text{for } \nu = 0 \text{ in } V_0, V_N \quad \text{and}$$

$$\nu > 0 \text{ in } V_K, \tag{12.2.2}$$

including the necessary *vorticity equation* (7.1.9) that for $\boldsymbol{\omega} = \omega\mathbf{k}$ takes the form:

$$\omega_t - \nu\Delta\omega + \psi\times\omega = -\nu\Delta\omega^{(0)} + \psi^{(0)}\times\omega^{(0)} = \nu f,$$

$$-\Delta\psi = \omega, \quad t > 0,$$

$$\text{and} \quad -\Delta\psi^{(0)} = \omega^{(0)} \quad \text{in } V,$$

$$\text{for } f = \frac{g_x^y - g_y^x}{\nu} \quad \text{when } \nu > 0. \tag{12.2.3}$$

Then, turning to *variations*

$$\varphi, \zeta, v = \psi - \psi^{(0)}, \omega - \omega^{(0)}, \mathbf{u} - \mathbf{u}^{(0)} = \varphi, -\Delta\varphi, \varphi_y\mathbf{i} - \varphi_x\mathbf{j}, \quad t \geq 0,$$

$$\varphi, \zeta, v|_{t=0} = \varphi_0, \zeta_0, v_0 = \psi_0 - \psi^{(0)}, \omega_0 - \omega^{(0)}, \mathbf{u}_0 - \mathbf{u}^{(0)},$$

or *deviations* of *disturbances* $\psi, \omega, \mathbf{u}$ from *basic* $\psi^{(0)}, \omega^{(0)}, \mathbf{u}^{(0)}$, with conditions

$$\begin{aligned} \varphi|_{\partial V} &= 0 \quad \text{for } V = V_{0,K,N}, \nu \geq 0, \\ \zeta|_{x=0,\psi_y^*>0} &= 0 \quad \text{for } V = V_{0,N}, \\ \nu = 0,\ \varphi_y|_{\partial V} &= 0 \quad \text{for } V = V_K, \quad \nu > 0, \quad t \geq 0, \end{aligned} \tag{12.2.4}$$

equivalent to (12.2.1), and norms

$$\|\varphi\| = \sqrt{\int_V \varphi^2 dV} \quad \text{and} \quad \|v\| = \sqrt{\int_V \left(\varphi_x^2 + \varphi_y^2\right) dV} = \int_V \varphi\zeta dV,$$
$$\zeta = -\Delta\varphi(\varphi|_{\partial V} = 0),$$

and verifying (as below) that the velocity is to be *subjected* to the vorticity as

$$\|v\| \leq \text{const}\|\zeta\|,$$

one may assume the basic flows $\mathbf{u}^{(0)}$ are *stable by Lyapunov* [1892] and [Pontryagin, 1962], which means that the following set of *stability points*(**i**)–(**iv**) are valid for some *control instant* $\tau = \tau(\varepsilon) \geq 0$ and *majorant* $\delta = \delta(\varepsilon) > 0$ for $\varepsilon > 0$, $\lim_{\varepsilon \to +0} \delta = 0$:

(i) (***the stationary uniqueness***) The basic flow $\mathbf{u}^{(0)}$ is uniquely determined by (12.2.1)–(12.2.3);

(ii) (***the non-stationary existence***) Whatever $\varepsilon > 0$, every initial deviation $v_0 = \mathbf{u}_0 - \mathbf{u}^{(0)}$ develops uniquely to a *non-stationary* one, $v = \mathbf{u} - \mathbf{u}^{(0)}$, $t \geq 0$, $v|_{t=0} = v_0$, in (12.2.1)–(12.2.3);

(iii) (***the nonlinear stability***) The flow $\mathbf{u}^{(0)}$ is *non-linearly vortex* stable in the sense that, whatever $\varepsilon > 0$, the condition $\|\zeta_0\| \leq \delta = \delta(\varepsilon)$ implies $\|\zeta\| \leq \varepsilon$ for all $t \geq \tau = \tau(\varepsilon)$ and $\nu \geq 0$;

(iv) (***the viscous stability***) For $\nu > 0$, the flow $\mathbf{u}^{(0)}$ is *asymptotically vortex* stable in the sense that $\lim_{t\to\infty} \|\zeta\| = 0$ (hence, $\lim_{t\to\infty} \|v\| = 0$).

The point (**ii**), or the unique resolution of *non-stationary* boundary value problems (12.2.1)–(12.2.3) in the *class*

$$M = C^\infty = C^\infty(\bar{V}, t \geq 0) = \{\varphi, \psi, \chi, \ldots\}$$

of smooth *real-valued* expenditures ψ of flows $\mathbf{u}$, is well known for any incompressible fluid, be it either *ideal* ($\nu = 0$, $V = V_{0,N}$) [Yudovich, 1966] or *viscous* ($\nu > 0$, $V = V_K$) [Ladyzhenskaya, 1969].

On the whole, the *vortex stability* (**i**)–(**iv**) will be proved in M as in a linear set, or a vector space supplied first with a *metric* and a proper *norm*,

$$\varphi \cdot \psi = \langle \varphi\psi \rangle = \int_V \varphi\psi dV = \int_0^l \int_0^h \varphi\psi dx dy \quad \text{and}$$
$$\|\psi\| = \sqrt{\psi \cdot \psi}, \quad \psi \in M, \tag{12.2.5}$$

to become a *pre-Hilbert* space M to be completed to $\bar{M}$ (to a Hilbert space) [Kantorovich and Akilov, 1982], secondly, with a subset

$$M_0 = \{\varphi \in M : \varphi|_{\partial V} = 0\}$$

dense in Min the sense that

$$\bar{M_0} = \bar{M}, \quad \text{or} \quad \|\psi - \psi_\varepsilon\| < \varepsilon \text{ for any } \psi \in M,$$
$$\varepsilon > 0 \quad \text{and proper } \psi_\varepsilon \in M_0$$

[Sobolev, 1963; Ladyzhenskaya, 1969], and, thirdly, with a *skew-symmetric screw* as a *commutator* of Poisson brackets

$$\varphi \times \psi = [\varphi, \psi] = \varphi_y \psi_x - \varphi_x \psi_y = -\psi \times \varphi, \quad \varphi, \psi \in M, \tag{12.2.6}$$

acting by parts, or subjected to the *Jacobi identity*

$$\varphi \times (\psi \times \chi) = (\varphi \times \psi) \times \chi + \psi \times (\varphi \times \chi), \quad \varphi, \psi, \chi \in M,$$

to be at that a *quasicompact Lie algebra M* [Troshkin, 1988a, 1995], or a pre-Hilbert space possessing a *dense invariant volume* $\varphi \cdot (\psi \times \chi)$,

$\varphi \in M_0$, $\psi, \chi \in M$, satisfying the required *volume invariance identity*

$$\varphi \cdot (\psi \times \chi) = \chi \cdot (\varphi \times \psi)$$
$$\text{for any } \varphi \in M_0 = \{\varphi \in M : \varphi|_{\partial V} = 0\} \quad \text{and} \quad \psi, \chi \in M, \tag{12.2.7}$$

verified immediately with evident equalities

$$\varphi\psi \times \chi - \chi\varphi \times \psi = \psi \times \varphi\chi = (\varphi\chi\psi_y)_x - (\varphi\chi\psi_x)_y$$

and the *Gauss–Stokes formula*

$$\int_V \Big((\varphi\chi\psi_y)_x - (\varphi\chi\psi_x)_y \Big) dxdy = \int_{\partial V} \varphi\chi\psi_x dx + \varphi\chi\psi_y dy = 0,$$
$$\varphi|_{\partial V} = 0 (\varphi \in M_0).$$

12.3. Non-commutative inertia and dissipation

Together with symmetric and definite *inertia*

$$A : M \to M : \psi \to A\psi = -\Delta\psi, \quad \Delta\psi = \psi_{xx} + \psi_{xx}, \quad \psi \in M,$$
$$\varphi \cdot A\xi = \zeta \cdot \xi \quad \text{and} \quad \varphi \cdot \zeta = \|\varphi\|_1^2 = \langle \varphi_x^2 + \varphi_y^2 \rangle \geq \gamma^2 \|\varphi\|^2,$$
$$\zeta = A\varphi, \ \varphi, \ \xi \in M_0 = \{\varphi \in M : \varphi|_{\partial V} = 0\}, \quad \gamma = \text{const} > 0, \tag{12.3.1}$$

or

$$\varphi \cdot A\xi - A\varphi \cdot \xi = \langle (\varphi_x\xi - \varphi\xi_x)_x - (\varphi\xi_y - \varphi_y\xi)_y \rangle$$
$$= \int_{\partial V} (\varphi\xi_y - \varphi_y\xi)\, dx + (\varphi_x\xi - \varphi\xi_x)\, dy = 0, \quad \varphi, \xi|_{\partial V} = 0,$$

and *isometric dissipation* B to be a *square* $A^2 = AA$ *of inertia* A,

$$B : M \to M : \psi \to B\psi = AA\psi$$
$$= \psi_{xxxx} + 2\psi_{xxyy} + \psi_{yyyy}, \quad \psi \in M,$$
$$\varphi \cdot B\xi = A\varphi \cdot A\xi \text{ for } \xi \in M,$$
$$\varphi \in M_{00} = \{\varphi \in M_0 : \varphi_x, \varphi_y|_{\partial V} = 0\},$$
$$\text{hence, } B\varphi \cdot \varphi' = \varphi \cdot B\varphi' \quad \text{and} \quad \varphi \cdot B\varphi = \|A\varphi\|^2 = \|\varphi\|_2^2$$
$$\text{for } \varphi, \varphi' \in M_{00}, \tag{12.3.2}$$

since

$$\begin{aligned}
&\varphi\cdot B\xi - A\varphi\cdot A\xi = \langle\varphi AA\xi - (A\varphi)\,A\xi\rangle = \langle\varphi\Delta\Delta\xi - (\Delta\varphi)\,\Delta\xi\rangle\\
&\quad= \langle\varphi\Delta\xi_{xx} - \varphi_{xx}\Delta\xi + \varphi\Delta\xi_{yy} - \varphi_{yy}\Delta\xi\rangle\\
&\quad= \langle(\varphi\Delta\xi_x - \varphi_x\Delta\xi)_x + (\varphi\Delta\xi_y - \varphi_y\Delta\xi)_y\rangle\\
&\quad= \int_{\partial V}(\varphi_y\Delta\xi - \varphi\Delta\xi_y)dx + (\varphi\Delta\xi_x - \varphi_x\Delta\xi)dy = 0 \quad \text{when}\\
&\qquad \varphi,\varphi_x,\varphi_y|_{\partial V} = 0,
\end{aligned}$$

and at the same time $B = A^2$ *not permutable* with A,

$$A\varphi\cdot B\xi \neq B\varphi\cdot A\xi \quad \text{for some } \varphi,\xi\in M_{00},$$

due to the *no-slip conditions* $\varphi,\varphi_x,\varphi_y|_{\partial V} = 0$ and $\varphi',\varphi'_x,\varphi'_y|_{\partial V} = 0$ for *shear stresses* $\Delta\varphi,\Delta\varphi'|_{\partial V} \neq 0$ such that

$$\begin{aligned}
&A\varphi\cdot B\varphi' - B\varphi\cdot A\varphi'\\
&\quad= \langle(\Delta\varphi_x\Delta\varphi' - (\Delta\varphi)\,\Delta\varphi'_x)_x - ((\Delta\varphi)\,\Delta\varphi'_y - \Delta\varphi_y\Delta\varphi')_y\rangle\\
&\quad= \int_{\partial V}((\Delta\varphi)\,\Delta\varphi'_y - \Delta\varphi_y\Delta\varphi')\,dx\\
&\qquad + (\Delta\varphi_x\Delta\varphi' - (\Delta\varphi)\,\Delta\varphi'_x)\,dy \neq 0, \quad \varphi,\varphi'\in M_{00},
\end{aligned}$$

the Pre-Hilbert space M with dense invariant volume (12.2.7), or the quasi–compact Lie algebra M acquires two Rayleigh functions with lemma 14.6.1:

Statement 12.3.1. *Symmetric and definite inertia* (12.3.1) *and isometric dissipation* (12.3.2) *make up two weak–strong metric pairs*

$$(\varphi,\xi)_w = \varphi\cdot\xi \quad \text{and} \quad (\varphi,\xi)_s = \varphi\cdot A\xi = A\varphi\cdot\xi \quad \text{in } L = M_0$$

and

$$(\varphi,\xi)_w = \varphi\cdot A\xi \quad \text{and} \quad (\varphi,\xi)_s = A\varphi\cdot A\xi \quad \text{in } L = M_{00},$$

to satisfy conditions of Lemma 14.6.1 *and thus to supply* M *with two full sets of proper eigenvalues and eigenfunctions referred further to as principal and adjoint moments,* $\lambda = \lambda_n \leq \lambda_{n+1}$ *and* $\mu = \mu_n \leq$

μ_{n+1} *including the least* $\lambda_{\min} = \lambda_1$ *and* $\mu_{\min} = \mu_1$, *with relevant rotations*

$$\varphi_n^\lambda = \varphi_n^{\lambda_n} \in M_0 \quad \text{and} \quad \varphi_n^\mu = \varphi_n^{\mu_n} \in M_{00}, \quad \varphi_n^{\lambda,\mu} \neq 0, \quad n = 1, 2, \ldots,$$

and their Fourier sums

$$\varphi_\lambda^{(N)} = \sum_{n=1}^{N} \frac{\varphi_n^\lambda \cdot \varphi}{\varphi_n^\lambda \cdot \varphi_n^\lambda} \varphi_n^\lambda \quad \text{and} \quad \varphi_\mu^{(N)} = \sum_{n=1}^{N} \frac{A\varphi_n^\mu \cdot \varphi}{A\varphi_n^\mu \cdot \varphi_n^\mu}$$

$$\varphi_n^\mu = \sum_{n=1}^{N} \frac{A\varphi_n^\mu \cdot A\varphi}{A\varphi_n^\mu \cdot A\varphi_n^\mu} \varphi_n^\mu, \quad N = 1, 2, \ldots,$$

such that

$$\begin{aligned}
&A\varphi \cdot \xi = \varphi \cdot A\xi = \lambda\varphi \cdot \xi, \quad \xi \in M_0, \quad \text{or} \quad A\varphi = \lambda\varphi, \\
&\quad \text{or} \quad \Delta\varphi + \lambda\varphi = 0 \quad \text{in } V, \quad \varphi|_{\partial V} = 0, \\
&\quad \mathit{for}\ \zeta' \cdot \varphi = \zeta' \cdot \zeta = 0, \quad \zeta = A\varphi, \quad \zeta' = A\varphi', \\
&\quad \varphi = \varphi_n^\lambda, \quad \varphi' = \varphi_{n'}^{\lambda'} \in M_0 \\
&\text{and } n' \neq n, n', n = 1, 2, \ldots,
\end{aligned}$$

and

$$\begin{aligned}
&\varphi \cdot B\xi = A\varphi \cdot A\xi = B\varphi \cdot \xi = \mu A\varphi \cdot \xi = \mu\varphi \cdot A\xi, \\
&\quad \xi \in M_{00}, \quad \text{or } B\varphi = \mu A\varphi, \\
&\text{or } \Delta\Delta\varphi + \mu\Delta\varphi = 0 \text{ in } V, \varphi, \varphi_x, \varphi_y|_{\partial V} = 0, \quad \text{for} \\
&\tilde{\zeta} \cdot \varphi = \tilde{\zeta} \cdot \zeta = \tilde{\zeta} \cdot B\varphi = 0, \zeta = A\varphi, \tilde{\zeta} = A\tilde{\varphi}, \varphi = \varphi_n^\mu, \tilde{\varphi} = \varphi_{\tilde{n}}^{\tilde{\mu}} \in M_{00} \\
&\text{and } \tilde{n} \neq n, \tilde{n}, n = 1, 2, \ldots,
\end{aligned} \tag{12.3.3}$$

and, whatever $\varphi \in M$,

$$\begin{aligned}
&\lambda_{N+1}^2 \|\varphi - \varphi_\lambda^{(N)}\|^2 \leq (\varphi - \varphi_\lambda^{(N)}) \cdot A(\varphi - \varphi_\lambda^{(N)}) = \|\varphi - \varphi_\lambda^{(N)}\|_1^2, \\
&\quad 1/\lambda_N \to 0
\end{aligned}$$

and

$$\begin{aligned}
&\mu_{N+1}\lambda_1\|\varphi-\varphi_\mu^{(N)}\|^2 \le \mu_{N+1}\|\varphi-\varphi_\mu^{(N)}\|_1^2 \\
&\quad = \mu_{N+1}(\varphi-\varphi_\mu^{(N)})\cdot A(\varphi-\varphi_\mu^{(N)}) \\
&\quad \le \|A\varphi - A\varphi_\mu^{(N)}\|^2 = \|A\varphi\|^2 - \|A\varphi_\mu^{(N)}\|^2, 1/\mu_N \to 0 \qquad (12.3.4)
\end{aligned}$$

for $N \to \infty$.

Proof. Really, the compactness of *Poincare–Steklov imbedding*

$$\|\varphi\|_1^2 = \int_V (\varphi_x^2+\varphi_y^2)\, dV \ge \gamma^2\|\varphi\|^2, \quad \gamma = \text{const} > 0, \quad \varphi|_{\partial V} = 0$$

[Sobolev, 1963; Vladimirov, 1971] provides that the statement is valid for the first weak–strong metric pair from lemma 14.6.1, §15. And owing to the latter and *Cauchy–Bunyakovsky inequality* [Vladimirov, 1971],

$$\|\varphi\|_1^2 = \varphi\cdot\zeta \le \|\varphi\|\,\|\zeta\|, \text{ hence, } \|\varphi\| \le \frac{1}{\gamma^2}\|\zeta\|, \zeta = -\Delta\varphi = A\varphi,$$

the same is true for the second pair, namely, for $\varphi = \varphi_{n+p}-\varphi_n$, $n,p = 1,2,\ldots$, both boundness $\|\zeta\| \le \text{const}$ and convergence $\|\varphi\| \to 0$ imply $\|\varphi\|_1 \to 0 (n,p\to\infty)$. The necessary *smoothness* $\varphi \in M$ [Agmon *et al.*, 1964] and Fourier convergence in Lemma 15.1 complete the proof. □

For example, the least principal moment

$$\begin{aligned}
\lambda_{\min} &= \inf_{\substack{\varphi\neq 0,\\ \varphi|_{\partial V}=0}} \frac{\|\varphi\|_1^2}{\|\varphi\|^2} \\
&= \left(\frac{\pi}{l}\right)^2, \left(\frac{\pi}{h}\right)^2, \left(\frac{\pi}{l}\right)^2 + \left(\frac{\pi}{h}\right)^2 \text{ in } V = V_{N,K,0}, \qquad (12.3.5)
\end{aligned}$$

achieved at functions

$$\varphi = \sin\frac{\pi x}{l}, \sin\frac{\pi y}{h}, \sin\frac{\pi x}{l}\sin\frac{\pi y}{h} \text{ in } V = V_{N,K,0},$$

respectively, to provide the estimates

$$\|\varphi\|_1^2 = \varphi \cdot \zeta \le \|\varphi\| \, \|\zeta\| \le \frac{\|\varphi\|_1 \, \|\zeta\|}{\sqrt{\lambda_{\min}}}, \text{ or}$$

$$\|\varphi\|_1 \le \frac{\|\zeta\|}{\sqrt{\lambda_{\min}}} \text{ for } \varphi \in M_0, \text{ or,}$$

$$\text{more precisely, } \|\varphi\|_1 \le \frac{\|\zeta\|}{\sqrt{\mu_{\min}}} \le \frac{\|\zeta\|}{\sqrt{\lambda_{\min}}}, \|\zeta\| = \|A\varphi\| = \|\varphi\|_2 \text{ for}$$

$$\varphi \in M_{00}, \tag{12.3.6}$$

with evident

$$\|\varphi\|_1^2 = \varphi \cdot \zeta \le \|\varphi\| \, \|\varphi\|_2, \quad \varphi \in M_0 \supset M_{00},$$

where it proves to be no more than the least adjoint moment

$$\mu_{\min} = \inf_{\substack{\varphi \neq 0, \\ \varphi \in M_{00}}} \frac{\|\varphi\|_2^2}{\|\varphi\|_1^2} \ge \inf_{\substack{\varphi \neq 0, \\ \varphi \in M_0}} \frac{\|\varphi\|_2^2}{\|\varphi\|_1^2} \ge \inf_{\substack{\varphi \neq 0, \\ \varphi \in M_0}} \frac{\|\varphi\|_1^2}{\|\varphi\|^2} = \lambda_{\min}.$$

With all the principal and adjoint moments, we may take the boundary value problem (12.2.1)–(12.2.3) as an *infinitely dimensional top* on M of the form

$$\begin{aligned}
&\omega_t + \nu B\psi + \psi \times \omega = \nu B\psi^{(0)} + \psi^{(0)} \times \omega^{(0)} = \nu f, \\
&\quad t > 0, \quad \omega|_{t=0} = \omega_0; \\
&\psi|_{\partial V} = \psi_0|_{\partial V} = \psi^{(0)}\Big|_{\partial V} = \psi^*, \quad \nu \ge 0; \\
&\psi_y^*\big|_{\partial V, x=0} = u_* = u_*(y), \quad \omega_0, \omega^{(0)}, \omega\Big|_{\substack{\partial V, x=0, \\ u_* > 0}} = \omega^+, \\
&\quad V = V_{0,N}, \quad \nu = 0, \psi_y, \psi_{0y}, \psi_y^{(0)}\Big|_{\partial V} = 0, \\
&\quad V = V_K, \quad \nu > 0, \quad t \ge 0,
\end{aligned} \tag{12.3.7}$$

with multidimensional inertia A (12.3.1) and dissipation B (12.3.2) for vorticities ω and $\omega^{(0)}$ as *angular momentums* and expenditures, or stream functions ψ and $\psi^{(0)}$ as*rotations*, or angular velocities.

For $\nu > 0$, due to the *no-slip conditions* $\psi - \psi^{(0)} \in M_{00}$, the top (12.3.7) proves to be *non-commutative* ($A\varphi \cdot B\xi \neq B\varphi \cdot A\xi$) to be distinguished from the commutative one [Troshkin, 1995] that generalizes stability theorems of Meshalkin and Sinai [1961] and Yudovich [1965, 1966] and resembles an ordinary dissipative top (5.1.1). As we shall see below, the stability theorems of the latter remain valid for the former.

After substituting *disturbances*

$$\psi = \psi^{(0)} + \varphi \quad \text{and} \quad \omega = \omega^{(0)} + \zeta, \quad \varphi \in M_0(\varphi \in M_{00})$$
$$\text{for} \quad \nu = 0(\nu > 0),$$

into relations (12.3.7), we come to an *equivalent* boundary value problem as to a *top equation in variations* φ, ζ,

$$\begin{aligned}
&\zeta_t + \nu B\varphi + \psi^{(0)} \times \zeta + \varphi \times \omega^{(0)} + \varphi \times \zeta = 0, \quad t > 0,\\
&\quad \zeta|_{t=0} = \zeta_0, \quad \zeta = A\varphi,\\
&\varphi_0, \varphi|_{\partial V} = 0, \quad \nu \geq 0, \quad \zeta_0, \zeta\Big|_{\substack{\partial V, x=0,\\ \psi_y^{(0)}>0}} = 0, \quad V = V_{0,N}, \quad \nu = 0,\\
&\varphi_y, \varphi_{0y}, \varphi_y^{(0)}\Big|_{\partial V} = 0, \quad V = V_K, \quad \nu > 0, \quad t \geq 0,
\end{aligned} \tag{12.3.8}$$

to be equivalent evidently to the top (12.3.7) and to remain further the main object of the stability considerations concerning the following stationary undisturbed, or *basic* flows $\mathbf{u}^{(0)}$.

12.4. Direct and reverse basic flows in problems D and E

For a stationary *direct flow* of an ideal incompressible and free of body forces ($g^{x,y} = \nu = 0$) fluid coming in a *duct* V_0 of length l, width h, and aspect ratio $\alpha = h/l$, with constant normal velocity component $U > 0$, point normed expenditure $\psi^{(0)}(0,0) = 0$, and inflow vorticity $\omega^+ = \Omega$ of a *level*

$$-\infty < \beta = \Omega h/U < \infty$$

a proper stationary top (12.3.7) becomes the following non-linear boundary value *problem* **D**,

$$\begin{aligned}
&-\Delta\psi^{(0)} = \omega^{(0)} \quad \text{and} \quad \psi^{(0)} \times \omega^{(0)} = 0 \text{ in} \\
&\quad V = V_0 : 0 < x < l, 0 < y < h, \\
&\text{for } \psi^{(0)}\Big|_{\partial V} = Uy \quad \text{and} \\
&\omega^{(0)}\Big|_{x=0} = \Omega, U, \Omega = \text{const}, U > 0(\mathbf{D}),
\end{aligned} \tag{12.4.1}$$

that admits a stream function with a constant vorticity,

$$\begin{aligned}
&-\Delta\psi^{(0)} = \omega^{(0)} = \Omega \text{ in } V \text{ for } \psi^{(0)}\Big|_{\partial V} = Uy, \\
&\text{or } \psi^{(0)}(x, y) = Uh\,(Y + \beta\Gamma)\,, X = \tfrac{x}{h}, Y = \tfrac{y}{h},
\end{aligned} \tag{12.4.2}$$

determined uniquely by a *vortex expenditure* Γ of the form

$$\Gamma = \Gamma(X, Y), \quad -\Gamma_{XX} - \Gamma_{YY} = 1, \quad 0 < X < \frac{1}{\alpha}, \quad 0 < Y < 1,$$

$$\Gamma|_{X=0,1/\alpha} = \Gamma|_{Y=0,1} = 0,$$

in which its symmetry

$$\Gamma(1/\alpha - X, Y) = \Gamma(X, Y) = \Gamma(X, 1 - Y)$$

and convexity

$$-\Gamma_{XX} > 0, \quad \Gamma_{YY} < 0, \quad 0 < X < \frac{1}{\alpha}, \quad 0 < Y < 1,$$

as delivered by the maximum principle [Landis, 1997] and related topological constructions [Troshkin, 1988b], lead both to normal derivative and critical value,

$$\Gamma_Y^0 = \Gamma_Y\left(\frac{1}{2}, 0\right) > 0 \quad \text{and} \quad \beta_0 = \beta_0(\alpha) = \frac{1}{\Gamma_Y^0} > 0,$$

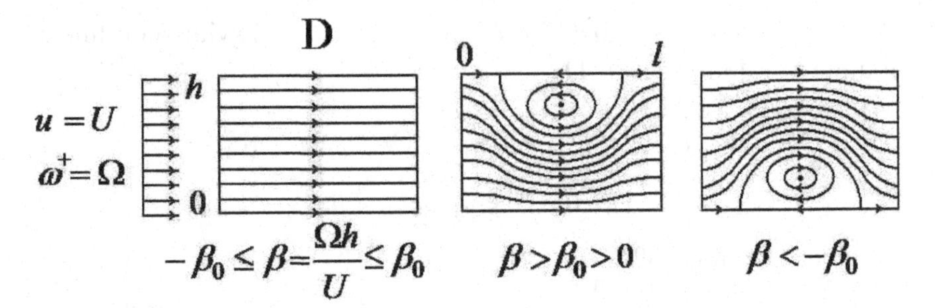

Figure 12.4.1. Plane vortices bifurcated from a direct flow at $\pm\beta_0$.

providing thereby the flow bifurcations of

$$\begin{gathered}\psi_y^{(0)}\left(l/2,0\right)=U\left(1+\beta\Gamma_Y^0\right)=U\Gamma_Y^0\left(\beta_0+\beta\right)\geq(<)\,0\\ \text{for}\quad \beta_0+\beta\geq(<)\,0,\\ \psi_y^{(0)}\left(l/2,h\right)=U\left(1-\beta\Gamma_Y^0\right)=U\Gamma_Y^0\left(\beta_0-\beta\right)\geq(<)\,0\\ \text{for } \beta_0-\beta\geq(<)\,0,\end{gathered}$$

with plane vortices in Fig. 12.4.1 corresponding to a spatial counter–flow circulation in Fig. 9.1.2.

Alternatively to a direct flow in a duct, for a stationary *reverse flow* of an ideal incompressible and free of body forces ($g^{x,y}=\nu=0$) fluid coming in a *bay* V_0 with a constant inflow vorticity $\omega^+=\Omega$ and outgoing from V_0 with an expenditure point normed as above, a top (12.3.7) takes the form of another *problem* **E**,

$$\begin{gathered}-\Delta\psi^{(0)}=\omega^{(0)},\quad \psi^{(0)}\times\omega^{(0)}=0,\\ 0<x<l,0<y<h,\quad U,\Omega=\text{const},\\ \psi^{(0)}\Big|_{x=0}=Uy\left(1-\frac{y}{h}\right),\quad \psi^{(0)}\Big|_{\substack{x=l,\\ y=0,h}}=0,\\ \omega^{(0)}\Big|_{\substack{x=0,\\ 0<y<\frac{h}{2}}}=\Omega,\quad U>0(\mathbf{E}),\end{gathered}\tag{12.4.3}$$

that admits a new stream function with old constant vorticity,

$$-\Delta\psi^{(0)} = \omega^{(0)} = \Omega \text{ in } V \text{ for } \left.\psi^{(0)}\right|_{x=0} = Uy\left(1-\frac{y}{h}\right), \left.\psi^{(0)}\right|_{\substack{x=l,\\ y=0,h}} = 0,$$

$$\text{or } \psi^{(0)} = Uh\left(\Phi+\beta\Gamma\right), \Phi, \Gamma = \Phi, \Gamma\left(X, Y\right), \quad X = \frac{x}{h}, Y = \frac{y}{h}, \tag{12.4.4}$$

and vortex flow of the former vortex expenditure Γ imposed additively and proportionally on an irrotational flow of a potential Φ,

$$\Phi_{XX} + \Phi_{YY} = 0 \quad 0 < X < \frac{1}{\alpha}, \quad 0 < Y < 1, \quad \Phi|_{X=0} = Y\left(1-Y\right),$$
$$\Phi|_{X=1/\alpha} = \Phi|_{Y=0,1} = 0.$$

Due to the same reasons as before [Landis, 1997; Troshkin, 1988b], the latter proves to be symmetric and *saddle* as

$$\Phi(X,Y) = \Phi(X, 1-Y) \quad \text{and} \quad \Phi_{XX}, -\Phi_{YY} > 0,$$

respectively, to lead to the normal derivatives

$$\Phi_X^{\mp} < 0, \quad \Gamma_X^{-} = -\Gamma_X^{+} > 0, \quad \Phi_X^{\mp}, \Gamma_X^{\mp} = \Phi_X, \Gamma_X\left(X^{\mp}, \frac{1}{2}\right),$$
$$X^{-} = 0, \quad X^{+} = \frac{1}{\alpha},$$

related *critical points*

$$\beta_{+} = -\frac{\Phi_X^{-}}{\Gamma_X^{-}} > 0 \quad \text{and} \quad \beta_{-} = \frac{\Phi_X^{+}}{\Gamma_X^{-}} < 0,$$

and flow bifurcations of

$$-\psi_x^{(0)}\left(0, h/2\right) = -U\left(\Phi_X^{-} + \beta\Gamma_X^{-}\right) = U\Gamma_X^{-}\left(\beta_{+} - \beta\right) \geq (<) 0$$
$$\text{for } \beta_{+} - \beta \geq (<) 0,$$
$$-\psi_x^{(0)}\left(1, h/2\right) = -U\left(\Phi_X^{+} + \beta\Gamma_X^{+}\right) = U\Gamma_X^{-}\left(\beta_{-} - \beta\right) \leq (>) 0$$
$$\text{for } \beta_{-} - \beta \leq (>) 0,$$

depicted in Fig. 12.4.2.

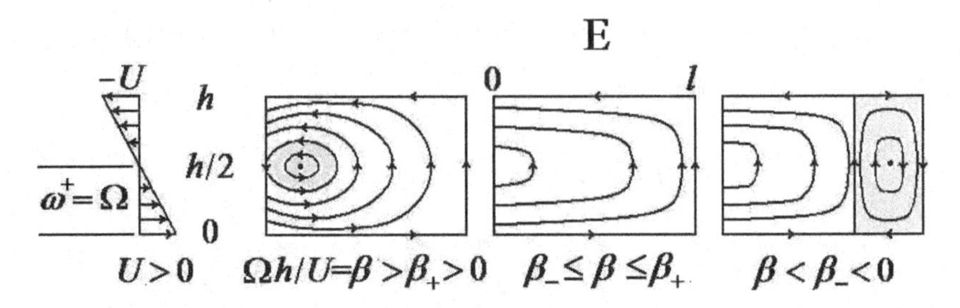

Figure 12.4.2. Vortices bifurcated from a reverse flow at $\beta_\pm$.

12.5. Vortex strip in F and velocity profiles in P,S

Returning finally to the particular reason and disturbed equilibrium which hydrodynamic stability had come from (in Figs. 0.1–03, Chapter 0), let us start with the *Jacobs–John–Niderhouse experiment* [Jacobs *et al.*, 1996] that returns the gas instability in Fig. 0.2 back to liquids in Fig. 12.5.1 by means of the *water hammer* and the related phenomenon of *contact inversion* that flattens out the contact surface as follows.

When the tank hits the floor, a plane front of pressure drop originating at the bottom starts to move (at a rate of approximately 1000m/s) and get to the *nearest* points of a contact surface to turn the *falling* speeds $-u$ to the opposite, or *reactive* ones, u, as in Fig. 12.5.1. At that, pairs of contact points with opposite normal velocities $\mp u$ form the *inflow vorticity* $\omega^+ = \omega^+(y)$ to lead with (12.3.7) to the following boundary value *problem* **F** for an ideal incompressible fluid free of body forces ($g^{x,y} = \nu = 0$) treated in a flow region $V = V_N$, with a compact closure $\bar{V}$ (as above), or, equivalently, in an unbounded vertical periodic strip N:

$$-\Delta\psi^{(0)} = \omega^{(0)}, \quad \psi^{(0)} \times \omega^{(0)} = 0, \quad 0 < x < l,$$

$$-\infty < y < \infty, \quad \psi_y^{(0)}\Big|_{x=l} = 0,$$

$$\psi^{(0)}(x, y+h) = \psi^{(0)}(x, y), \quad \psi_y^{(0)}\Big|_{x=0} = u_*(y) = -U\cos\kappa y,$$

$$\kappa = \frac{2\pi}{h},$$

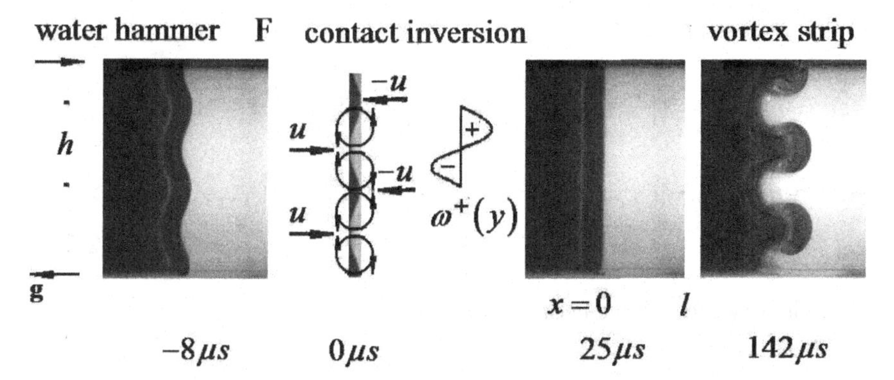

Figure 12.5.1. What happens during the first 150 microseconds (μs) after a tank filled with two liquids contacting on a sinusoidal surface of a period h falls at the gravity $\mathbf{g}$ and hits the floor [Jacobs *et al.*, 1996].

$$\omega^{(0)}\Big|_{x=0,u_0>0} = \omega^{+}(y) = \Omega \sin \kappa y = \frac{\Omega}{U} u_* \left(y - \frac{h}{4}\right),$$
$$U, \Omega = \text{const}, \quad U > 0(\mathbf{F}), \tag{12.5.1}$$

with dimensionless *aspect ratio* $\alpha = h/l$ and *inflow vorticity level*

$$\beta = \frac{\Omega}{\kappa U} = \frac{\alpha\alpha^{+}}{4}, \quad \alpha^{+} = \frac{2l\Omega}{\pi U}, \quad -\infty < \Omega < \infty.$$

The problem $\mathbf{F}$ admits a *stationary solution*

$$\omega^{(0)} = \lambda\psi^{(0)}, \quad \lambda = \beta\kappa^2, \quad \beta = \frac{\Omega}{\kappa U} = \frac{\alpha\alpha^{+}}{4}, \quad \alpha^{+} = \frac{2l\Omega}{\pi U},$$
$$\psi^{(0)} = -\frac{U}{\kappa} A(x) \sin \kappa y, \quad A(x) = A_{\beta>1,\beta<1,\beta=1}, \quad V = V_N(\mathbf{F}), \tag{12.5.2}$$

found first in [Troshkin, 1988b], with amplitudes

$$A_{\beta>1} = \frac{\sin \theta X}{\sin \theta}, \quad A_{\beta<1} = \frac{\mathrm{sh}\theta X}{\mathrm{sh}\theta} \quad \text{and} \quad A_{\beta=1} = X = 1 - \frac{x}{l} \tag{12.5.3}$$

and spectral parameter

$$\theta = \kappa l\sqrt{|\beta - 1|} = \theta(\alpha, \beta) = \frac{2\pi}{\alpha}\sqrt{|\beta - 1|}, \text{ or } |\beta - 1| = \left(\frac{\alpha\theta}{2\pi}\right)^2,$$

$$\text{and } \theta \neq \pi n, \text{ or } \beta \neq \beta_n^+ = 1 + \left(\frac{\alpha n}{2}\right)^2, n = 1, 2, \ldots, \quad \text{for } \beta > 1, \tag{12.5.4}$$

that do resolve (12.5.1), or (12.3.7).

Really, we have

$$u = \psi_y^{(0)} = -A(x)U\cos\kappa y, \quad A(0) = 1, \quad A(l) = 0,$$
$$A_{xx} - A\kappa^2 = -(\beta - 1 + 1)\kappa^2 A_{\beta>1}, (1 - \beta - 1)\kappa^2 A_{\beta<1},$$
$$-\kappa^2 A_{\beta=1} = -\beta\kappa^2 A = -\lambda A$$

and

$$\omega^{(0)} = -\Delta\psi^{(0)} = \frac{A_{xx} - A\kappa^2}{\kappa}U\sin\kappa y = -\lambda\frac{U}{\kappa}A\sin\kappa y = \lambda\psi^{(0)},$$

which proves the required resolution.

For $\beta \leq 1$, the streamlines, as level lines $\psi(x, y) = \text{const}$, delivered by the stream function (12.3.2)–(12.3.4) form the *reverse flow*, as in Fig. 12.5.2, where curves start and end on one side $x = 0$ without stagnation points inside the strip N.

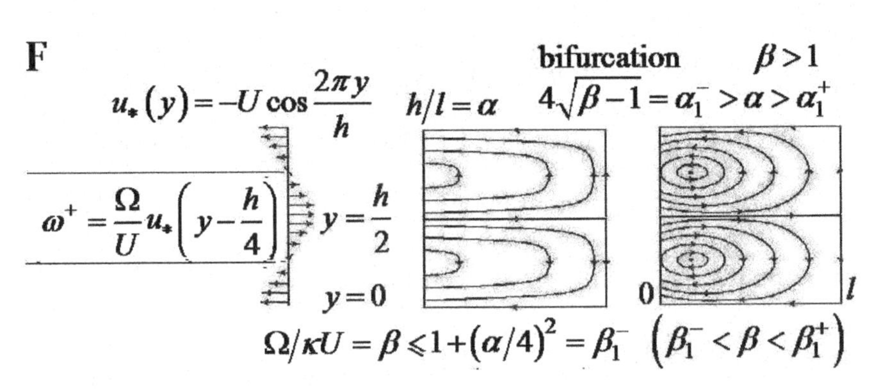

Figure 12.5.2. The first flow bifurcation in (12.5.1) at α_1^- (or β_1^-).

For $\beta > 1$, as can be verified immediately (or with [Troshkin, 1988b]), the *moments* $\theta_n^{\mp}$ of topological *change* (or violation of *homeomorphism*) are *zeros* θ_n^- and *poles* θ_n^+ of boundary value

$$v\left(0, \frac{h}{4}\right) = -\psi_x^{(0)}\left(0, \frac{h}{4}\right) = \frac{U}{\kappa} A_x^+(0) \sin\frac{\pi}{2} = -\frac{U}{\kappa l}\frac{\cos\theta}{\sin\theta} = 0, \infty,$$

$$\text{or} \quad \cos\theta \sin\theta = 0,$$

i.e. trigonometric roots

$$\theta = \theta(\alpha_n^{\mp}, \beta) = \theta(\alpha, \beta_n^{\mp}) = \theta_n^{\mp}, \text{ as } \theta_n^- = \theta_n^+ - \frac{\pi}{2} \quad \text{and}$$

$$\theta_n^+ = \pi n, \quad n = 1, 2, \ldots,$$

determined by the corresponding *critical values* of either aspect ratio,

$$\alpha_n^- = \frac{4\sqrt{\beta - 1}}{2n - 1} \quad \text{and} \quad \alpha_n^+ = \frac{2\sqrt{\beta - 1}}{n} > \alpha_n^-, \quad \text{for any } \beta > 1,$$

or inflow vorticity level

$$\beta_n^- = 1 + \left(\frac{\alpha}{2}\left(n - \frac{1}{2}\right)\right)^2 \quad \text{and} \quad \beta_n^+ = 1 + \left(\frac{\alpha n}{2}\right)^2 > \beta_n^-,$$

$$\text{for any } \alpha > 0.$$

Really, when a varying parameter, be it α diminishing or β increasing, crosses α_n^- or β_n^- (jumps over α_n^+, or β_n^+), the flow pattern (of level lines $\psi^{(0)}(x, y) = \text{const}$) does change *continuously* (*abruptly*) or *bifurcates* (*reconstructs*) into a new picture consisting of vortex centers originating in the boundary points $(0, h/4)$ and $(0, 3h/4)$, as in Fig. 12.5.2 (or far from points $(0, h/4)$ and $(0, 3h/4)$, respectively, as in Fig. 12.5.3).

The *turbulisation* of a gas behind the shock wave front shown earlier in Fig. 0.2 seems to resemble a sequence of bifurcations and reconstructions that may be treated in turn as an illustration of the short-wave character of the Richtmayer–Meshkov instability [Aleskin *et al.*, 1988; Troshkin, 2016a, 2016b].

For a periodic channel K filled with a viscous, $\nu > 0$, incompressible, $\rho = const > 0$, fluid free of body forces, $g^{x,y} = 0$, and stationary, pushed by a constant *pressure drop* of $-p_x^{(0)} = \sigma$ and $p_y^{(0)} = 0$ directed

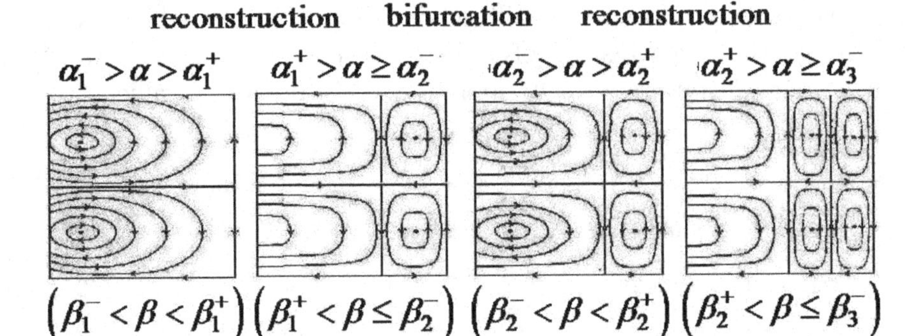

Figure 12.5.3. Firstly reconstruction at α_1^+ (or β_1^+) and secondly bifurcation at α_2^- (or β_2^-) and reconstruction at α_2^+ (orβ_2^+) [Troshkin, 1988b].

along the rigid walls $y = 0, h$, that are, possibly, shifted at constant *rates* a, b, with *no-slip* conditions, the proper stationary top (12.3.7) reduces to the first *problem* **P** from § 0:

$$
\begin{aligned}
&-\Delta\psi^{(0)} = \omega^{(0)}, \quad \nu\Delta\Delta\psi^{(0)} + \psi^{(0)} \times \omega^{(0)} = 0,\\
&\quad -\infty < x < \infty, \quad 0 < y < h,\\
&\psi^{(0)}(x + l, y) = \psi^{(0)}(x, y) = \psi^{(0)}(y),\\
&\quad \text{or} \quad -\psi_x^{(0)} = v^{(0)} = 0, \quad \psi_y^{(0)} = u^{(0)},\\
&-\psi_{yy}^{(0)} = \omega^{(0)} \quad \text{and} \quad -\nu u_{yy}^{(0)} + \frac{1}{\rho} p_x^{(0)} = 0, 0 < y < h,\\
&\quad \text{for} \ \ u^{(0)}\Big|_{y=0.h} = a, b,\\
&-p_x = \sigma, \quad p_y = 0, a, b, \sigma, \nu = \text{const}, \quad \sigma, \nu > 0, \quad V = V_K(\mathbf{P}),
\end{aligned} \tag{12.5.5}
$$

with a *parabolic* velocity profile (in Fig. 0.1, to the left) and linearly falling pressure

$$
u^{(0)} = a + \frac{(b-a)\,y}{h} + \frac{\sigma y\,(h-y)}{2\nu}
$$
$$
\text{and} \quad p^{(0)} = \text{const} + \rho\sigma\,(l - x)\,(v^{(0)} = 0),
$$

found first in [Stokes, 1845], and the corresponding basic

$$\psi^{(0)} = ay + \frac{(b-a)\,y^2}{2h} + \frac{\sigma y^2\,(3h-2y)}{12\nu}$$

$$\text{and } \omega^{(0)} = \frac{a-b}{h} + \frac{2\sigma y - \sigma h}{2\nu}, \omega_x^{(0)} = \omega_{yy}^{(0)} = 0 \quad \text{in } V = V_K(\mathbf{P}). \tag{12.5.6}$$

Finally, for the former channel K and fluid, however, with a *mixing* body force of $g^x = -g \sin \kappa y$ and $g^y = 0$ instead of the pushing pressure drop absent now, $p_{x,y}^{(0)} = 0$, the same top (12.3.7) takes the form of the second *problem* **S** from Chapter 0:

$$\begin{aligned}
&-\Delta\psi^{(0)} = \omega^{(0)}, \quad \nu\Delta\Delta\psi^{(0)} + \psi^{(0)} \times \omega^{(0)} = \nu f,\\
&\quad -\infty < x < \infty, \quad 0 < y < h,\\
&\psi^{(0)}\,(x+l, y) = \psi^{(0)}\,(x, y) = \psi^{(0)}\,(y),\\
&\quad \text{or } -\psi_x^{(0)} = v^{(0)} = 0, \psi_y^{(0)} = u^{(0)},\\
&-\psi_{yy}^{(0)} = \omega^{(0)} \quad \text{and} \quad -\nu u_{yy}^{(0)} = g^x\\
&= -g \sin \kappa y,\ u^{(0)}\Big|_{y=0,h} = 0, \quad \kappa = \frac{2\pi}{h},\\
&\nu f = -g_y^x = \kappa g \cos \kappa y . g, \nu = \text{const} > 0, \quad V = V_K(\mathbf{S}),
\end{aligned} \tag{12.5.7}$$

in which the *sinusoidal* velocity profile and constant pressure (in Fig. 0.1, to the right) [Troshkin, 1988a, example 5]

$$u^{(0)} = -U \sin \kappa y, \quad U = \frac{g}{\kappa^2\nu}, \quad \text{and} \quad p^{(0)} = \text{const}\ (v^{(0)} = 0).$$

correspond to basic stream function and vorticity,

$$\begin{aligned}
&\psi^{(0)} = \frac{U}{\kappa} \cos \kappa y \quad \text{and} \quad \omega^{(0)} = \kappa^2 \psi^{(0)}\\
&\left(\kappa = \frac{2\pi}{h}, U = \frac{g}{\kappa^2\nu}\right) \quad \text{in } V = V_K \quad (\mathbf{S}),
\end{aligned} \tag{12.5.8}$$

respectively.

Chapter 13

Ideal Analytic Stability

13.1. Unconditional and long-wave stabilities in problems D–F

The next step in our investigation of problems **D** (12.4.1), **E** (12.4.2), **F** (12.5.1) and **P** (12.5.5), **S** (12.5.7) is to verify stability points (i)–(iv) of Section 12.2 for corresponding basic flows $\mathbf{u}^{(0)}$: (12.4.2) in Fig. 12.4.1, (12.4.4) in Fig. 12.4.2, ((12.5.2)–(12.5.4)) in Figs. 12.5.2 and 12.5.3 and (12.5.6) and (12.5.8) in Fig. 0.1, respectively. Omitting the first point (i) for a while and dropping the second one (ii) which has been proved already [Yudovich, 1966; Ladyzhenskaya, 1969], we may turn directly to the remaining points (iii) and (iv) to be obtained with the help of the three following energy identities derived from the equation in variations (12.3.8).

Consecutively multiplying scalarly equality (12.3.8) by 2φ and 2ζ, with the use of identities,

$$2\zeta \cdot \zeta_t = (\zeta \cdot \zeta)_t \text{ and } \varphi_t \cdot \zeta = \varphi_t \cdot A\varphi = A\varphi_t \cdot \varphi = \zeta_t \cdot \varphi = \varphi \cdot \zeta_t,$$

$$\text{hence, } 2\varphi_t \cdot \zeta = \varphi_t \cdot \zeta + \varphi \cdot \zeta_t = (\varphi \cdot \zeta)_t,$$

skew symmetry (12.2.6) and volume invariance (12.2.7),

$$\varphi \cdot \varphi \times \omega^{(0)} = \omega^{(0)} \cdot \varphi \times \varphi = 0,$$

$$\varphi \cdot \varphi \times \zeta = \zeta \cdot \varphi \times \varphi = 0 \quad (\varphi \times \varphi = 0),$$

$$\varphi \cdot \psi^{(0)} \times \zeta = \zeta \cdot \varphi \times \psi^{(0)} \quad \text{and}$$

$$\zeta \cdot \varphi \times \zeta = \varphi \cdot \zeta \times \zeta = 0 \quad (\varphi \in M_0),$$

we come to the necessary *first*,

$$(\varphi \cdot \zeta)_t + 2v\varphi \cdot B\varphi + 2\zeta \cdot \varphi \times \psi^{(0)} = 0, \tag{13.1.1}$$

second,

$$(\zeta \cdot \zeta)_t + 2v\zeta \cdot B\varphi + 2\zeta \cdot \psi^{(0)} \times \zeta + 2\zeta \cdot \varphi \times \omega^{(0)} = 0, \tag{13.1.2}$$

and, hence, *third*,

$$\begin{aligned}((\zeta - \lambda\varphi) \cdot \zeta)_t + 2v(\zeta - \lambda\varphi) \cdot B\varphi + 2\zeta \cdot \psi^{(0)} \times \zeta \\ + 2\zeta \cdot \varphi \times (\omega^{(0)} - \lambda\psi^{(0)}) = 0 \text{ for any } \lambda = \text{const},\end{aligned} \tag{13.1.3}$$

energy identities (13.2.1)–(13.2.3).

In so doing, we have in (13.1.1)–(13.1.3) for **D** (12.4.1), **E** (12.4.2) and **F**(12.5.1) that

$$\begin{aligned} v^{(0)}\Big|_{y=0,h} = 0,\ u^{(0)}\Big|_{x=0,l} > 0 \text{ and} \\ \zeta|_{x=0} = \zeta|_{x=0,u}\,{}^{(0)}{}_{>0} = 0 \quad \text{in } \mathbf{D}(V = V_0), \\ v^{(0)}\Big|_{y=0,h} = u^{(0)}\Big|_{x=l} = \zeta|_{x=0,u^{(0)}>0} = 0 \\ \text{in } \mathbf{E}(V = V_0) \text{ and } \mathbf{F}(V = V_N), \end{aligned}$$

so the vortex *relaxation*

$$2\zeta \cdot \psi^{(0)} \times \zeta = \langle (u^{(0)}\zeta^2)_x + (v^{(0)}\zeta^2)_y \rangle \text{ in } \mathbf{D}, \mathbf{E}, \mathbf{F},$$

proves to be non-negative:

$$\begin{aligned} 2\zeta \cdot \psi^{(0)} \times \zeta = \int_0^h u^{(0)}\zeta^2\Big|_{x=l} dy \geq 0 \text{ in } \mathbf{D} \text{ and} \\ 2\zeta \cdot \psi^{(0)} \times \zeta = -\int_0^h u^{(0)}\zeta^2\Big|_{x=0,t^{(0)}\leq 0} dy \geq 0 \quad \text{in } \mathbf{E}, \mathbf{F}, \end{aligned} \tag{13.1.4}$$

Besides, the vortex *generation*

$$\begin{aligned} \zeta \cdot \varphi \times \omega^{(0)} = 0 \quad (\omega^{(0)} = \text{const}) \text{ in } \mathbf{D}, \mathbf{E} \\ \text{and } \zeta \cdot \varphi \times \omega^{(0)} = \zeta \cdot \varphi \times \lambda\psi^{(0)} (\omega^{(0)} = \lambda\psi^{(0)}) \text{ in } \mathbf{F}. \end{aligned} \tag{13.1.5}$$

As a consequence, we have two theorems on stability:

Theorem 13.1.1 (on unconditional stability in D, E). *Whatever aspect ratio, $\alpha > 0$, ideal fluid ($v = 0$) basic flows of expenditures* (12.4.2) *and* (12.4.4) *in problems **D*** (12.4.1) *and **E*** (12.4.3), *respectively, are stable, in the sense of point* (*iii*) *in Section* 12.1 *of Chapter* 12, *or unconditionally vortex stable.*

Proof. Really, for $v = 0$, the third energy identity (13.1.3) of non-negative relaxation (13.1.4) and zero generation (13.1.5) implies that

$$(\zeta \cdot \zeta)_t \leq 0, \text{ so } \|\zeta\| \leq \|\zeta_0\|, t \geq 0, \text{ in } \mathbf{D}, \mathbf{E},$$

which with evident control instant $\tau(\varepsilon) = 0$ and majorant $\delta(\varepsilon) = \varepsilon$ leads to (iii), concluding the proof. □

Theorem 13.1.2 (on long-wave stability in F). *The ideal fluid* ($v = 0$) *basic flow of expenditure* (12.5.2)–(12.5.4) *in problem **F*** (12.5.1) *is long-wave stable or stable in the sense of point* (*iii*) *in Section* 12.1 *of Chapter* 12 *if*

$$\frac{h}{l} = \alpha > \alpha^{+} = \frac{2\Omega l}{\pi U}, \beta = \frac{\alpha\alpha^{+}}{4}, \kappa = \frac{2\pi}{h},$$

$$\text{or } \lambda = \beta\kappa^2 = \alpha\alpha^{+}\left(\frac{\pi}{h}\right)^2 < \alpha^2\left(\frac{\pi}{h}\right)^2 = \left(\frac{\pi}{l}\right)^2 = \lambda_{\min} \text{ from } (12.3.5), \tag{13.1.6}$$

particularly, stable unconditionally (*for any aspect ratio $\alpha > 0$*) *if $\alpha^{+} \leq 0$* (*or $\Omega \leq 0$*), *as in* **Fig. 12.5.2**, *to the left, or limitedly if*

$$\alpha > \alpha^{+} \geq 2, \quad \alpha > \frac{4}{\sqrt{3}} \text{ and } 2\alpha^{+} - \sqrt{\alpha^{+2} - 4} \leq \alpha \leq 2\alpha^{+},$$

$$\text{or } \alpha > \alpha^{+} \geq 2 \text{ and } 2\alpha^{+} < \alpha < 2\alpha^{+} + \sqrt{\alpha^{+2} - 4}, \tag{13.1.7}$$

as in **Fig. 12.5.2**, *to the right, or as vortex mushrooms in* **Fig. 12.5.1**, $142\mu s$.

Proof. Really, for $v = 0$, the second energy identity (13.1.2) of relaxation (13.1.4) and generation (13.1.5) implies that

$$((\zeta - \lambda\varphi) \cdot \zeta)_t \leq 0$$

and means that, with (12.3.6),

$$\frac{\lambda_{\min}-\lambda}{\lambda_{\min}}\|\zeta\|^2 \le \|\zeta\|^2 - \lambda\|\varphi\|_1^2 = (\zeta-\lambda\varphi)\cdot\zeta \le (\zeta_0-\lambda\varphi_0)\cdot\zeta_0 \le \|\zeta_0\|^2, \tag{13.1.8}$$

which leads to (iii) with

$$\tau(\varepsilon) = 0 \text{ and } \delta(\varepsilon) = \varepsilon\sqrt{\frac{\lambda_{\min}}{\lambda_{\min}-\lambda}}.$$

In the remaining case of **Fig. 12.5.2**, to the right, the required inequalities,

$$\begin{aligned}\beta_1^- = 1 + \left(\frac{\alpha}{4}\right)^2 < \frac{\alpha\alpha^+}{4} < \frac{\alpha\alpha}{4} = 4\left(\frac{\alpha}{4}\right)^2 = \left(\frac{\alpha}{2}\right)^2 \\ = \beta_1^+ - 1 < \beta_1^+ \quad \text{for } \alpha > \alpha^+,\end{aligned}$$

reduce to conditions (13.1.7), which concludes the proof. □

13.2. Analytical uniqueness

Thus, the validity of the third stability point (iii) in Section 12.2 of Chapter 12 for ideal fluid basic flows in problems **D**–**F** is provided (in Theorems 13.1.1 and 13.1.2) by traditional functional analysis remedies such as compact embeddings (Lemma 14.6.1), related spectrum (12.3.5), estimates (12.3.6) and energy identities (13.1.1)–(13.1.3). As for the remaining, the first point (i) needed to complete the proper stability considerations, and this one has to show that functional analysis really serves fluid mechanics, as it can [Ladyzhenskaya, 1969; Temam, 1983], though doing it sometimes in a manner resembling the well-known *Sunset Express* that stopped at a tank to take on water and acquired some other things that were not good for it (O. Henry, *The Roads We Take*).

Indeed, all the functional constructions [Sobolev, 1963; Dezin, 1987; Ladyzhenskaya, 1969; Temam, 1983] would be impossible without Sobolev's *cut-offs* [Soboleff, 1938] or Friedrichs's *mollifiers*

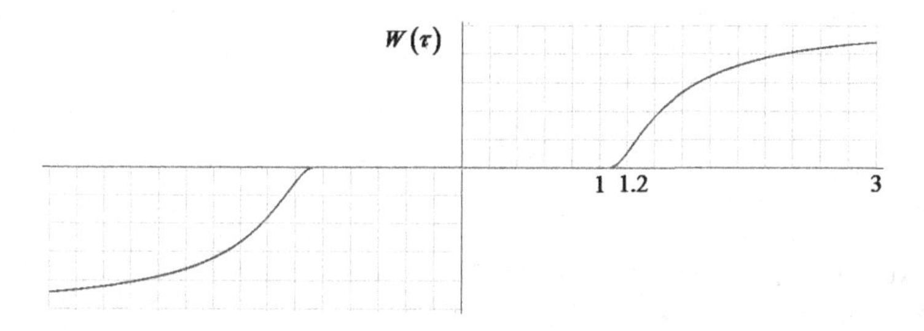

Figure 13.2.1. Infinitely smooth stair (13.2.1).

[Friedrichs, 1944] similar to a smooth skew-symmetric *stair*

$$W(\tau) = \exp\frac{1}{1-\tau^2} < 1 \text{ for } \tau > 1 \text{ and } W(\tau) = 0 \text{ for } 0 \le \tau \le 1$$
$$\text{when } W(\tau) = -W(-\tau) \text{ for } \tau \le 0 \tag{13.2.1}$$

(similar to that used in Troshkin [1989]), as in **Fig. 13.2.1**, approximated, hence, *computed* evidently to an arbitrary *finite order* $n < \infty$ in the vicinities of corresponding *angle points* $\tau = \mp 1$ by polynomials P

$$|W - P| < \text{const}|\tau \pm 1|^n, \quad |\tau \pm 1| < \varepsilon,$$
$$P = a_0 + a_1\tau + \cdots + a_{n-1}\tau^{n-1},$$
$$a_0, a_1, \ldots, a_{n-1} = \text{const, when } \varepsilon \to +0 \text{ as } n \to \infty,$$

while turning out not to be a habitual *analytic function* due to both trivial derivatives,

$$W^{(n-1)}(\mp 1) = 0, \quad n = 1, 2, \ldots, \text{ for } |\tau| \le 1,$$

and non-trivial ones,

$$0 < W_\tau = \frac{2\tau W(\tau)}{(1-\tau^2)^2} \le W'_+ = W_\tau(\tau_+),$$

$$\tau > 1, \tau_\pm = \sqrt{\frac{\sqrt{7} \pm 2}{3}}, 0 < \tau_+ - 1.244421 < 10^{-7},$$

$$W_{\tau\tau} = \frac{6\left(\tau_+^2 - \tau^2\right)\left(\tau_-^2 + \tau^2\right)W(\tau)}{\left(1-\tau^2\right)^4} \; 0 \text{ for}$$

$$\tau \pounds \tau_+ \left(0 < \tau_- - 0.46395 < 10^{-6}\right),$$

and so on.

At the same time, with the help of stair (13.2.1), the set $C^* \subset C^\infty$ of stationary *analytic* stream functions,

$$\psi^{(0)} = \psi^{(0)}(t,x,y) = \psi^{(0)}(0,x,y) = \psi^{(0)}(x,y) \in C^\infty$$
$$= C^\infty(\bar{V}, t \geq 0) \text{ on } \bar{V},$$

determined up to an additive constant (so that both $\psi^{(0)}$ and $\psi^{(0)}+$ const are expenditures for the same flow $\mathbf{u}^{(0)}$) and represented by a convergent Taylor series

$$\psi^{(0)}(x,y) = \sum_{m+n=0}^{\infty} \frac{\partial_x^m \partial_y^n \psi^{(0)}}{m!n!}(x_0,y_0)\left(x-x_0\right)^m\left(y-y_0\right)^n$$

in a neighbourhood $\sqrt{\left(x-x_0\right)^2+\left(y-y_0\right)^2} < \varepsilon =$ const of *every* point $(x_0,y_0) \in \overline{V}$, like the basic expenditures (12.4.2), (12.4.4) and (12.5.2)–(12.5.4) in problems **D–F**, is complemented in C^∞ by the set $C^\# \subset C^\infty$ of stationary *non-analytic* expenditures as follows.

As we shall see in the following equation, every *analytical* solution $\psi^{(0)}, \omega^{(0)} \in C^*$ of a *nonlinear* boundary value problem **D** (12.4.1), **E** (12.4.3), or **F** (12.5.1), or

$$-\Delta\psi^{(0)} = \omega^{(0)}, \quad \psi^{(0)} \times \omega^{(0)} = 0 \text{ in } V = V_{0,N},$$

$$\psi^{(0)}\Big|_{\partial V} = \psi^*, \quad u_* = \psi_y^*\big|_{x=0}, \quad \omega^{(0)}\Big|_{\substack{x=0,\\ u_*>0}} = \Omega, \text{ in } V = V_0, \mathbf{D}, \mathbf{E},$$

$$\omega^{(0)}\Big|_{\substack{x=0,\\ u_*>0}} = \lambda\psi^* (\text{in } V = V_N, \mathbf{F}). \tag{13.2.2}$$

as taken uniquely from the corresponding *linear* one, (12.4.2), (12.4.4), or (12.5.2)–(12.5.4), or

$$-\Delta\psi^{(0)} = \omega^{(0)}, \omega^{(0)} = \Omega\left(\omega^{(0)} = \lambda\psi^{(0)}\right) \text{ in } \mathbf{D}, \mathbf{E} \text{ (in } \mathbf{F}),$$

$$\psi^{(0)}\Big|_{\partial V} = \psi^*, \Omega, \lambda = \text{ const}, \tag{13.2.3}$$

even when stable in Theorems 13.1.1 and 13.1.2, may not be unique in C^∞ because, *whatever small* $\varepsilon = \text{const} > 0$, it can be *stationary disturbed* by a smooth flow of expenditure and vorticity, $\psi = \tilde{\psi}_0^{(0)}$ and $\omega = \tilde{\omega}^{(0)}$, satisfying the same boundary value problem (13.2.2) as $\psi^{(0)} \in C^*$, namely,

$$-\Delta\psi = \omega, \quad \psi \times \omega = 0, \quad \text{in } V \text{ and } \psi|_{\partial V} = \psi^{(0)}\Big|_{\partial V},$$
$$\psi = \psi^{(0)} + \varphi, \quad \omega|_{\substack{x=0,\\ u>0}} = \omega^{(0)}\Big|_{\substack{x=0,\\ u_0>0}} \quad \text{in } V = V_{0,N}\,(\mathbf{D}, \mathbf{E}, \mathbf{F}) \tag{13.2.4}$$

for either

$$\omega = \Omega + \varepsilon\eta(\psi) \text{ in } \mathbf{D}_\pm = \mathbf{D}|_{\beta>\beta_0, \beta<-\beta_0}, \quad \mathbf{E}_\pm = \mathbf{E}|_{\beta>\beta_+, \beta<\beta_-} \tag{13.2.5}$$

or

$$\omega = \lambda\psi + \varepsilon\eta(\psi) \text{ in } \mathbf{F}_+ = \mathbf{F}|_{\alpha>\alpha_*} \text{ where } \alpha_* = \max\left\{\alpha^+, \frac{4}{\sqrt{3}}\right\}$$
$$\text{with } \alpha^+ = \frac{2l\Omega}{\pi U} \leq 0 (> 0) \text{ for } \Omega \leq 0 (> 0), \tag{13.2.6}$$

and

$$\eta = \eta(\psi) = W\left(\tau = \frac{2\psi - b - a}{b - a}\right),$$
$$\eta|_{\substack{a \leq \psi \leq b\\ (|\tau| \leq 1)}} = 0, \quad 0 < -\eta|_{\substack{\psi<a\\ (\tau<-1)}} = \eta|_{\substack{\psi>b\\ (\tau>1)}} < 1,$$
$$0 < \eta_\psi|_{\psi<a} = \eta_\psi|_{\psi>b} \leq \frac{2W'_+}{b-a}, \quad \text{for } a = \min_{\partial V}\psi^{(0)} < b = \max_{\partial V}\psi^{(0)}, \tag{13.2.7}$$

with the corresponding *deviation* $\varphi = \psi - \psi^{(0)}$, satisfying a *weakly nonlinear* elliptic *homogeneous* (or with zero conditions on ∂v) boundary value problem of the form,

$$-\Delta\varphi = f(\varphi) \text{ in } V \text{ and } \varphi|_{\partial V} = 0 \text{ in } \mathbf{D}_\pm, \mathbf{E}_\pm, \mathbf{F}_+ \text{ for}$$
$$f(\varphi) = \varepsilon\eta\left(\psi^{(0)} + \varphi\right) \text{ in } \mathbf{D}_\pm, \mathbf{E}_\pm,$$

$$f(\varphi) = \lambda\varphi + \varepsilon\eta\left(\psi^{(0)} + \varphi\right) \text{ in } \mathbf{F}_+ \text{ and } \varepsilon = \text{const} > 0, \tag{13.2.8}$$

to make up a proper stationary disturbance $\psi = \psi^{(0)} + \varphi$ of analytical expenditure $|\psi^{(0)}$ in (13.2.4), or in each of the five *vortex basic flows* in **Fig. 12.4.1** ($\mathbf{D}_\pm$), **Fig. 12.4.2** ($\mathbf{E}_\pm$) and **Fig. 12.5.2** ($\mathbf{F}_\pm$), in which the *maximum principle* $a \leq \psi^{(0)} \leq b$ is *violated*, i.e. when

$$\max_{\bar{V}} \psi^{(0)} > \max_{\partial V} \psi^{(0)} = b \text{ or } \min_{\bar{V}} \psi^{(0)} < \min_{\partial V} \psi^{(0)} = a$$

$$\text{for } \pm\beta > \beta_0 \text{ in } \mathbf{D}_\pm, \pm\beta > \pm\beta_\pm > 0 \text{ in } \mathbf{E}_\pm \text{ and } \alpha > \alpha_* \text{ in } \mathbf{F}_+, \tag{13.2.9}$$

or (13.1.7) to provide

$$\beta_1^- = 1 + \left(\frac{\alpha}{4}\right)^2 < \beta = \frac{\alpha\alpha^+}{4} < \frac{\alpha\alpha}{4} = 4\left(\frac{\alpha}{4}\right)^2$$
$$= \left(\frac{\alpha}{2}\right)^2 = \beta_1^+ - 1 < \beta_1^+,$$
$$\text{hence, } \lambda = \beta\kappa^2 = \alpha\alpha^+\left(\frac{\pi}{h}\right)^2 < \alpha^2\left(\frac{\pi}{h}\right)^2 = \left(\frac{\pi}{l}\right)^2 = \lambda_{\min} \text{ in } \mathbf{F}_+$$

(with (12.3.5)).

However, all the disturbances ψ prove to be trivial ($\phi = 0$) in C^*.

Theorem 13.2.1 (on analytic uniqueness). *The analytical solution* $\psi^{(0)}, \omega^{(0)} \in C^*$ *of the stationary boundary value problem* (13.2.2) *is unique, or every smooth non-trivial solution* $\varphi \neq 0$ *of problem* (13.2.8) *is non-analytic:* $\varphi \in C^\# = C^\infty \backslash C^*$.

Proof. Really, following Troshkin [1988b], we find from the inflow vorticity condition in (13.2.2) that the analytic function $\omega^{(0)} - \Omega = 0$ in **D, E** ($\omega^{(0)} - \lambda\psi^{(0)} = 0$ in **F**) on a neighborhood $N(x_0, 0)$ of a boundary point $(x_0, 0) \in \partial V$. Then, due to the *uniqueness of analytic extension* from $N(x_0, 0)$ to V [Hurwitz and Courant, 1922], $\omega^{(0)} - \Omega = 0$ in **D, E** ($\omega^{(0)} - \lambda\psi^{(0)} = 0$ in **F**) everywhere in V, and the nonlinear boundary value problem (13.2.2) reduces to the linear one, (13.2.3), uniquely resolved in C^∞ [Vladimirov, 1971]. On the contrary, any problem from (13.2.8) evidently reduces to (13.2.2). The proof is complete. □

In other words, to satisfy the point (i), every basic flow $\mathbf{u}^{(0)}$ during vortex generation has to find oneself in an *analytic medium*, with the expenditure $\psi^{(0)}$ of the class C^*. This cannot be said of the velocity fields of a general ideal continuous medium of class C^∞. So, we may proceed on finding $\varphi \neq 0$ for (13.2.8) in $C^\#$ or among expenditures in the *dark medium* of the subspace $C^\#$.

13.3. Vortex phantoms

For the purpose of finding $\varphi \neq 0$ in $C^\#$ for (13.2.8), let us return to stair (13.2.1) and basic expenditures $\psi^{(0)}$ in problems (13.2.2) generating *vortex chambers* or sub-regions

$$V_\pm = \left\{(x,y) \in V : \psi^{(0)}(x,y)^{>b}_{<a}, \text{ or } \eta(\psi^{(0)})^{>0}_{<0}\right\} \text{ in } \mathbf{D}_\pm, \mathbf{E}_\pm, \mathbf{F}_+, \tag{13.3.1}$$

where the maximum principle is violated as in (13.2.9).

With evidently regular partially smooth and even analytic *contact boundary*

$$\partial V_\pm = \left\{(x,y) \in \bar{V} : \psi^{(0)}(x,y)^{=b}_{=a}, \eta(\psi^{(0)}) = 0\right\},$$

every vortex camera, V_+ (in $\mathbf{D}_+$, $\mathbf{E}_+$, $\mathbf{F}_+$) or V_- (in $\mathbf{D}_-$, $\mathbf{E}_-$, $\mathbf{F}_-$), is separated from its *non-vortex complement*

$$\begin{aligned} V^*_\pm &= \left\{(x,y) \in V : a < \psi^{(0)}(x,y) < b, \eta(\psi^{(0)}) = 0\right\}, \\ \partial V^*_\pm &= \partial V_\pm \cup \partial V, \end{aligned} \tag{13.3.2}$$

having got such *beak boundary points* P_* as

$$\left(\frac{l}{2}, h\right) \text{ in } \mathbf{D}_+, \left(\frac{l}{2}, 0\right) \text{ in } \mathbf{D}_-, \left(0, \frac{h}{2}\right) \text{ in } \mathbf{E}_+$$
$$\text{or } \left(0, \frac{h}{4}\right), \left(0, \frac{3h}{4}\right) \text{ in } \mathbf{F}_+$$

to be tangent for $\partial V_\pm \cap \partial V$, as illustrated in **Fig. 12.4.1** ($\mathbf{D}_\pm$), **Fig. 12.4.2** ($\mathbf{E}_\pm$) and **Fig. 12.5.2** ($\mathbf{F}_\pm$).

Nevertheless, even with boundary *beaks* P_*, one can state as follows.

Theorem 13.3.1 (on vortex phantoms). *Whatever* $\varepsilon = \text{const} > 0$, *every classical (or smooth) solution* ϕ *of problem* (13.2.8) *for* (13.2.9) *proves to be a function* $\phi \in C^\#$ *non-trivial,* $\phi \neq 0$, *non-analytic inside the vortex chamber* $V_\pm$, *in* $\pm\phi > 0$ *in* $V_\pm$, *and harmonic into the irrotational complement* $V_\pm^*$ *as* $\Delta\phi = 0$ *in* $V_\pm^*$, *so that* $0 < \varphi < \max_{\partial V_+} \varphi$ *in* V_+^* *or* $\min_{\partial V_-} \varphi < \varphi < 0$ *in* V_-^*, *generating a non-analytic smooth stationary vortex phantom as a flow* $\mathbf{u}$ *of stream function* $\psi = \psi^{(0)} + \varphi$.

Proof. Really, using stair (13.2.1), chambers (13.3.1), complements (13.3.2) together with ordinary, strong and harmonic *maximum principles* [Landis, 1997], we have the following:

$$-\Delta\varphi \overset{\geq 0}{\leq 0} \text{ in } V \text{ and } \varphi|_{\partial V} = 0 \text{ imply } \varphi|_V \overset{\geq 0}{\leq 0},$$

$$\max_{\partial V_+} \varphi > 0 \text{ and } \min_{\partial V_-} \varphi < 0.$$

Further,

$$-\Delta\varphi \overset{>0}{<0} \text{ in } V_\pm \text{and } \varphi|_{\partial V_\pm} \overset{\geq 0}{\leq 0} \text{ imply } \varphi|_{V_\pm} \overset{>0}{<0},$$

and, finally,

$$\Delta\varphi = 0 \text{ in } V_\pm^*, \quad \varphi|_{\partial V_\pm^* | \partial V = \partial V_\pm \leq 0} \overset{\geq 0}{\leq 0} \text{ and}$$

$$\varphi|_{\partial V} = 0 \text{ imply } \overset{0}{\min_{\partial V_-} \varphi} < \varphi|_{V_\pm} < \overset{\max_{\partial V_+} \varphi}{0},$$

respectively, which concludes the proof. □

13.4. Smooth vortex catastrophe

Vortex phantoms that are supposed to be available in Theorem 13.3.1 do exist in the *dark medium* of $C^\# \subset C^\infty$, violating the stability point (i) in Section 12.2 of Chapter 12.

Theorem 13.4.1 (on vortex catastrophe). *In addition to Theorem* 13.3.1, *with a stair* (13.2.7) *of* (13.2.1), *for any*

$$0 < \varepsilon < \frac{(b-a)\lambda_{\min}}{2W'_+} \text{ and } \lambda_{\min} = \left(\frac{\pi}{l}\right)^2 + \left(\frac{\pi}{h}\right)^2 \text{ in } \mathbf{D}_\pm, \mathbf{E}_\pm$$

$$\text{or} \quad 0 < \varepsilon < \frac{(b-a)\left(\lambda_{\min} - \lambda\right)}{2W'_+} \text{ and } \lambda_{\min} = \left(\frac{\pi}{l}\right)^2 \text{ in } \mathbf{F}_+, \text{(12.3.5)}, \tag{13.4.1}$$

there is a unique solution $\phi = \phi_\varepsilon \in C^\infty$ *of the problem* (13.2.8).

Proof. Really, with the use of identity,

$$-\langle \chi \Delta \varphi \rangle = \langle \nabla \varphi \cdot \nabla \chi \rangle,$$
$$\varphi \in M = C^\infty, \quad \chi \in M_0 = C_0^\infty = \{\varphi \in M : \varphi|_{\partial V} = 0\}$$

and *strong norm*,

$$\|\varphi\|_1 = \|\nabla \varphi\| = \sqrt{\langle \nabla \varphi \cdot \nabla \varphi \rangle} \geq \|\varphi\| \sqrt{\lambda_{\min}},$$
$$\|\varphi\| = \langle \varphi^2 \rangle = \int_V \varphi^2 dV,$$
$$\lambda_{\min} = \inf_{M_0, \varphi \neq 0} \left(\|\varphi\|_1^2 / \|\varphi\|^2\right) = \text{const} > 0 \quad \text{as in (12.3.5)},$$

proper *closure*

$$\bar{M}_1 = \Big\{\varphi = \lim_{n\to\infty} \varphi_n : \|\nabla(\varphi - \varphi_n)\| \to 0, n \to \infty,$$
$$\varphi_n \in M_0, n = 1, 2, \ldots\}$$

and *potential*

$$F(\varphi) = \int_0^\varphi f(s) ds \text{ for } f(s) = \varepsilon\eta\left(\psi^{(0)} + s\right) \text{ in } \mathbf{D}_\pm, \mathbf{E}_\pm,$$
$$f(s) = \lambda s + \varepsilon\eta\left(\psi^{(0)} + s\right) \text{ in } \mathbf{F}_+$$

of functional

$$E = E(\varphi) = \left\langle \frac{\nabla\varphi \cdot \nabla\varphi}{2} - F(\varphi) \right\rangle, \varphi \in \bar{M}_1,$$

in which the first *Frechet derivatives* [Ekeland and Temam, 1976] are linear and bilinear forms,

$$\begin{aligned}(E'(\varphi),\chi) &= \langle \nabla\varphi\cdot\nabla\chi - f(\varphi)\chi\rangle \text{ and } (E''(\varphi),\chi,\xi)\\ &= \langle \nabla\chi\cdot\nabla\xi - f_\varphi\chi\xi\rangle\ \varphi,\chi,\xi \in \bar{M}_1,\end{aligned}$$

respectively, one may conclude that every *classical solution* $\varphi \in C^\infty$ of (13.2.8) proves to be a *critical point* $\varphi \in \bar{M}_1$ of E, such that

$$(E'(\varphi),\chi) = 0 \quad \text{for any } \chi \in \bar{M}_1, \tag{13.4.2}$$

and *vice versa*, $\varphi \in \bar{M}_1$ and (13.2.2) implies that $\varphi \in C^\infty$ [Agmon *et al.*, 1964] and (13.2.8).

For sufficiently small ε from (13.4.1) and proper derivative,

$$0 \le f_\varphi = \varepsilon\psi_\varphi\eta_\psi = \varepsilon\eta_\psi < \frac{2W'_+\varepsilon}{b-a} < \lambda_{\min} \text{ in } \mathbf{D}_\pm, \mathbf{E}_\pm,$$

$$0 \le f_\varphi = \lambda + \varepsilon\eta_\psi < \lambda + \frac{2W'_+\varepsilon}{h-a} < \lambda_{\min}, \text{ in } \mathbf{F}_+,$$

from (13.2.7), the functional E becomes *definite*,

$$\begin{aligned}\frac{(E''(\varphi),\chi,\chi)}{\|\nabla\chi\|^2} &= 1 - f_\varphi\frac{\|\chi\|^2}{\|\nabla\chi\|^2} \ge 1 - \frac{f_\varphi}{\lambda_{\min}} > 1 - \frac{1}{\lambda_{\min}}\frac{2W'_+\varepsilon}{b-a}\\ &= \text{const } > 0 \text{ in } \mathbf{D}_\pm, \mathbf{E}_\pm,\\ \frac{(E''(\varphi),\chi,\chi)}{\|\nabla\chi\|^2} &\ge 1 - \frac{f_\varphi}{\lambda_{\min}} > 1 - \frac{1}{\lambda_{\min}}\frac{2W'_+\varepsilon}{b-a} > \frac{\lambda}{\lambda_{\min}}\\ &= \text{const} > 0 \text{ in } \mathbf{F}_+, \quad \|\nabla\chi\| \ne 0,\end{aligned}$$

possessing a necessary unique *minimum point* (or a minimizing function) φ [Ekeland and Temam, 1976],

$$E(\chi) > E(\varphi), \varphi = \varphi_\varepsilon \in \bar{M}_1, \quad \forall\chi \in \bar{M}_1, \quad \chi \ne \varphi,$$

necessarily critical for E in the sense of (13.4.2) [Ekeland and Temam, 1976], hence, smooth [Agmon *et al.*, 1964] and satisfying (13.2.8), as mentioned above.

The proof is complete. □

As a consequence of Theorems 13.3.1 and 13.4.1, to be stable by Lyapunov (points (i)–(iii) in Section 12.1 of Chapter 12), stationary flows of an ideal fluid in a plane region have to be analytic as those of problems **D**–**F**.

Chapter 14

Viscous Asymptotic Stability

14.1. Unconditional stability in P

The method for proving stability by Lyapunov in the two remaining problems **P** (12.5.5) and **S** (12.3.7) for parabolic and sinusoidal velocity profiles in Fig. 0.1 (Chapter 0) of expenditures (12.5.6) and (12.5.8), respectively, in a viscous fluid ($\nu > 0$) filling the *ring* $V = V_K$ of a plane horizontal channel K of a period $l > 0$, with *no-slipping walls* $y = 0, h$, that had been initially stated by Reynolds in his *particular case* (in § 0), repeating generally the preceding one used for an ideal fluid (in Chapter 13), consists again of the verification of stability points (**i**)–(**iv**) of § 2.2 reducing now to (**i**) and (**iv**), since (**iii**) follows evidently from (**iv**), and (**ii**) is already known [Ladyzhenskaya, 1969].

Contrary to an ideal fluid (theorems 13.1.1–2 on the one hand and theorem 13.2.1 on the other), for a basic flow $\mathbf{u}^{(0)}$ of a viscous fluid, both *stationary uniqueness* (**i**) and *asymptotic stability* (**iv**) will follow from the same energy identities (13.1.1)–(13.1.3) to be stationary for (**i**): firstly,

$$\nu\varphi \cdot B\varphi + \zeta \cdot \varphi \times \psi^{(0)} = 0 \quad (\partial_t = 0), \tag{14.1.1}$$

secondly,

$$\nu\zeta \cdot B\varphi + \zeta \cdot \psi^{(0)} \times \zeta + \zeta \cdot \varphi \times \omega^{(0)} = 0, \tag{14.1.2}$$

and hence, thirdly,

$$\nu(\zeta - \lambda\varphi) \cdot B\varphi + \zeta \cdot \psi^{(0)} \times \zeta + \zeta \cdot \varphi \times (\omega^{(0)} - \lambda\psi^{(0)}) = 0,$$
$$\lambda = \text{const}, \tag{14.1.3}$$

respectively (as was noted first in [Troshkin, 1988a]).

As this takes place, for both (13.1.1–3) and (14.1.1–3), vortex *relaxation*

$$2\zeta \cdot \psi^{(0)} \times \zeta = \left\langle \begin{matrix} (u^{(0)}\zeta^2)_x \\ +(v^{(0)}\zeta^2)_y \end{matrix} \right\rangle = \int_0^l v^{(0)}\zeta^2 \Big|_{x=0, v^{(0)}=0} dx = 0 \quad \text{in } \mathbf{P}, \mathbf{S}, \tag{14.1.4}$$

and *generation*

$$\zeta \cdot \varphi \times \omega^{(0)} = \frac{1}{2}\langle ((\varphi_x^2 - \varphi_y^2)\omega_y^{(0)})_x \rangle + \langle (\varphi_y \varphi_x \omega_y^{(0)})_y \rangle$$
$$- \langle \varphi_y \varphi_x \omega_{yy}^{(0)} \rangle = 0 \text{ in } \mathbf{P}, \tag{14.1.5}$$

because

$$\omega_x^{(0)} = \omega_{yy}^{(0)} = 0, \quad \langle ((\varphi_x^2 - \varphi_y^2)\omega_y^{(0)})_x \rangle$$
$$= \int_0^h (\varphi_x^2 - \varphi_y^2)\omega_y^{(0)} \Big|_{x=0}^{x=l} dy = 0,$$
$$\langle (\varphi_y \varphi_x \omega_y^{(0)})_y \rangle = \int_0^l \varphi_y \varphi_x \omega_y^{(0)} \Big|_{y=0}^{y=h} dx, \quad \varphi_y \varphi_x \omega_y^{(0)} \Big|_{y=0,h} = 0, \text{ in } \mathbf{P}.$$

Besides, with (12.3.4), we have a *spectral estimate*

$$\zeta \cdot B\varphi \geq \mu_{\min} \|\zeta\|^2, \quad \zeta = A\varphi, \quad \varphi \in M_{00}, \quad \mu_{\min} = \mu_1, \tag{14.1.6}$$

as a consequence of

$$\begin{aligned}(A\varphi)_\mu^{(N)} \cdot A\varphi &= \sum_{n=1}^{N} \frac{A\varphi_n^\mu \cdot A\varphi}{A\varphi_n^\mu \cdot \varphi_n^\mu} \varphi_n^\mu \cdot A\varphi \\ &= \sum_{n=1}^{N} \frac{A\varphi_n^\mu \cdot \varphi}{A\varphi_n^\mu \cdot \varphi_n^\mu} A\varphi_n^\mu \cdot A\varphi = A\varphi_\mu^{(N)} \cdot A\varphi \\ &= \sum_{n=1}^{N} \frac{(A\varphi_n^\mu \cdot A\varphi)^2}{A\varphi_n^\mu \cdot A\varphi_n^\mu} = \|A\varphi_\mu^{(N)}\|^2 \\ &= \|A\varphi\|^2 - \|A\varphi - A\varphi_\mu^{(N)}\|^2,\end{aligned}$$

and

$$\begin{aligned}(A\varphi)_\mu^{(N)} \cdot B\varphi = A(A\varphi)_\mu^{(N)} \cdot A\varphi &= \sum_{n=1}^{N} \frac{A\varphi_n^\mu \cdot A\varphi}{A\varphi_n^\mu \cdot \varphi_n^\mu} A\varphi_n^\mu \cdot A\varphi \\ &= \sum_{n=1}^{N} \mu_n \frac{(A\varphi_n^\mu \cdot A\varphi)^2}{A\varphi_n^\mu \cdot A\varphi_n^\mu} \\ &\geq \mu_1 \sum_{n=1}^{N} \frac{(A\varphi_n^\mu \cdot A\varphi)^2}{A\varphi_n^\mu \cdot A\varphi_n^\mu} = \mu_1 \|A\varphi_\mu^{(N)}\|^2,\end{aligned}$$

so

$$\begin{aligned}A\varphi \cdot B\varphi &= (A\varphi)_\mu^{(N)} \cdot B\varphi + (A\varphi - (A\varphi)_\mu^{(N)}) \cdot B\varphi \\ &\geq \mu_1 \|A\varphi\|^2 - \mu_1 \|A\varphi - A\varphi_\mu^{(N)}\|^2 \\ &\quad - \|A\varphi - (A\varphi)_\mu^{(N)}\| \|B\varphi\| \quad \text{and} \quad \|A\varphi - A\varphi_\mu^{(N)}\|, \\ &\quad \|A\varphi - (A\varphi)_\mu^{(N)}\| \to 0 \quad \text{for} \quad N \to \infty,\end{aligned}$$

which proves (14.1.6).

From (14.1.6), we come to the following.

Theorem 14.1.1 (on unconditional stability in P). *The parabolic velocity profile of expenditure* (12.5.6) *is unique, as a stationary solution of problem* ***P*** (12.5.5), *and nonlinearly, asymptotically, and* *unconditionally (for any $\alpha > 0$) vortex stable in the sense of point (**iv**) in Section 2.2 of Chapter 2, namely,*

$$\zeta = A\varphi = 0, \text{ hence, } \varphi = \psi - \psi^{(0)} = 0 \text{ for } \partial_t = 0$$
$$\text{and } \|\zeta\| \leq \|\zeta_0\| \exp(-\mu_{\min}\nu t), \ t \geq 0, \text{ in } \mathbf{P}, \qquad (14.1.7)$$
$$\text{for } \mu_{\min} = \text{const} \geq \lambda_{\min} > 0 \text{ from (12.3.5).}$$

Proof. Really, both relaxation (14.1.4), generation (14.1.5), and estimate (14.1.6) in the stationary second energy identity (14.1.2) imply the uniqueness of the corresponding solution of the problem **P** (12.5.5) (point (**i**) in Section 2.2 of Chapter 2), or the first equality in (14.1.7). In the case of the non-stationary second energy identity (13.1.2), the same three conditions provide the fulfillment of stability point (**iv**) in Section 2.2 of Chapter 3, or the non-stationary inequality in (14.1.7):

$$\exp(-2\nu\mu_{\min}t)(\|\zeta\|^2\exp(2\nu\mu_{\min}t))_t = \|\zeta\|_t^2 + 2\nu\mu_{\min}\|\zeta\|^2$$
$$= \|\zeta\|_t^2 + 2\nu\zeta\cdot B\varphi = (\zeta\cdot\zeta)_t + 2\nu\zeta\cdot B\varphi + 2\zeta\cdot\psi^{(0)}\times\zeta$$
$$+ 2\zeta\cdot\varphi\times\omega^{(0)} = 0,$$

which completes the proof. □

Note 14.1.1. When based on initial equations (12.2.2) only, with no fluid deformations, or vorticity equation (12.2.3), the classical stability considerations [Lin, 1955] in fact were confined by the above mentioned Orr–Sommerfeld equation and related neutral curves [Lin, 1955; Joseph, 1976] that commonly to the first energy identity (13.1.1) (or (14.1.1)) give no affirmative answer "Yes" or "No" to the questions of stationary uniqueness (**i**) and asymptotic stability (**iv**) in problem **P** for 2D-flows of (12.2.2) due to the alternating members such as

$$\zeta\cdot\varphi\times\psi^{(0)} = -\langle\zeta\varphi_x\psi_y^{(0)}\rangle = \langle\varphi_{xx}\varphi_x\psi_y^{(0)}\rangle + \langle\varphi_{yy}\varphi_x\psi_y^{(0)}\rangle$$
$$= \left\langle\frac{1}{2}(\varphi_x^2\psi_y^{(0)})_x + \left(\left(\varphi_y\varphi_x - \frac{\varphi_x^2}{2}\right)\psi_y^{(0)}\right)_y\right\rangle$$

$$+\left\langle\left(\frac{\varphi_x^2}{2}-\varphi_y\varphi_x\right)\psi_{yy}^{(0)}\right\rangle$$
$$=\left\langle\left(\frac{\varphi_x^2}{2}-\varphi_y\varphi_x\right)\psi_{yy}^{(0)}\right\rangle,\ \psi_{yy}^{(0)}=\frac{b-a}{h}+\frac{\sigma(3h-2y)}{6\nu},$$
$$0<y<h,\quad \text{in } (\mathbf{P})$$

for (13,14.1.1).

14.2. The adjoined spectral problem

As in the case of problem $\mathbf{F}$(12.5.1) above (Section 13.1 of Chapter 13), the first energy identity (13.1.1) (or (14.1.1)) is really used in stability analysis below only when combined with the second one (13.1.2) in the third identity (13.1.3) for the sinusoidal velocity profile of expenditure (12.5.8) in the remaining problem $\mathbf{S}$ (12.5.7) where more detailed information is required concerning basic rotations (12.3.3), or the *adjoined spectral problem*

$$\Delta(\Delta\varphi+\mu\varphi)=0,\quad \varphi(x+l,y)=\varphi(x,y),\quad \varphi,\varphi_y\big|_{y=0.h}=0, \tag{14.2.1}$$

that supplies the top (12.3.7) with *adjoint principal moments*

$$\mu=\mu_m^{n-1}=\kappa^2\gamma_m^{n-1}=\kappa^2\gamma_m(\alpha(n-1)),$$
$$\gamma_m^{n-1}=(\alpha(n-1))^2+(\beta_m^{n-1})^2=\gamma_m(\alpha(n-1)),\kappa=\frac{2\pi}{h},\alpha=\frac{h}{l}>0,$$
$$\gamma_m(\alpha)=\alpha^2+\beta_m^2(\alpha),\quad \beta_m^n=\beta_m^1(\alpha n),\ \beta_m^1=\beta_m,\ m,n=1,2,\ldots, \tag{14.2.2}$$

reduced to *spectral points* (real numbers) γ_m^{n-1}, or

$$\beta_{2m-1}^n=\beta_{2m-1}(\alpha)=m-\frac{1}{2}+\frac{1}{\pi}\text{arctg}\frac{\beta_{2m-1}(\alpha)}{\alpha\text{th}\pi\alpha}$$
$$<\beta_{2m-1}^0=\beta_{2m-1}(+0)=m<\beta_{2m}^n$$

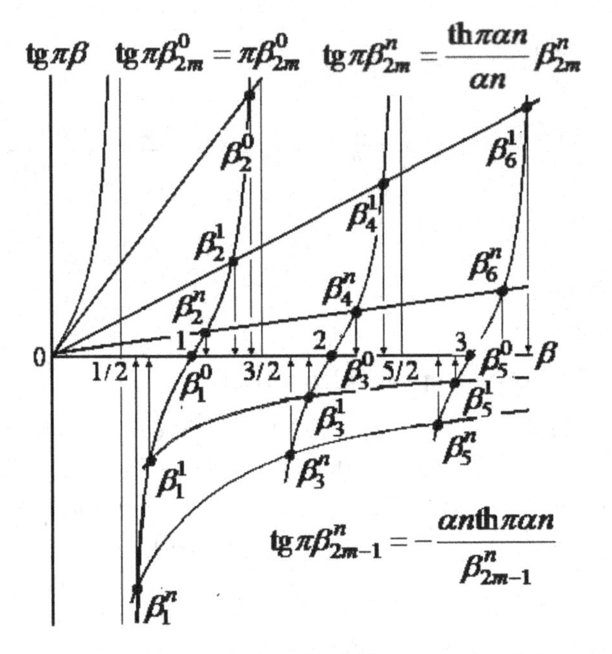

Figure 14.2.1. Spectral points β_m^n, $m, n+1 = 1, 2, \ldots$.

$$= \beta_{2m}(\alpha) = m + \frac{1}{\pi}\text{arctg}\frac{\beta_{2m}(\alpha)\text{th}\pi\alpha}{\alpha}$$

$$< \beta_{2m}^0 = \beta_{2m}(+0) = m + \frac{\text{arctg}\pi\beta_{2m}(+0)}{\pi}, \quad m = 1, 2, \ldots, \tag{14.2.3}$$

distributed as in Fig. 14.2.1 to form a double sequence

$$m - \frac{1}{2} < \beta_{2m-1}^n < \beta_{2m-1}^{n-1} \leq \beta_{2m-1}^0 = m < \beta_{2m}^n$$

$$< \beta_{2m}^{n-1} < \beta_{2m}^0 < m + \frac{1}{2}, \quad m, n = 1, 2, \ldots \tag{14.2.4}$$

while satisfying

$$\text{tg}\pi\beta_{2m-1}(\alpha) = -\frac{\alpha\text{th}\pi\alpha}{\beta_{2m-1}(\alpha)}, \quad \frac{\text{tg}\pi\beta_{2m}(\alpha)}{\pi\beta_{2m}(\alpha)} = \frac{\text{th}\pi\alpha}{\pi\alpha},$$

$$\frac{\text{tg}\pi\beta_{2m}(+0)}{\pi\beta_{2m}(+0)} = 1,$$

and being accompanied by the corresponding *adjoint basic rotations*

$$\begin{aligned}
&\varphi = \varphi_m^{n-1} = \eta_m^{n-1} \cos \kappa\alpha(n-1)x, \eta_m^{n-1} \sin \kappa\alpha(n-1)x,\\
&m, n = 1, 2, \ldots,\\
&\eta_{2m-1}^0 = 1 - \cos \kappa m y = 2 \sin \frac{\kappa m y}{2}, \quad \kappa = \frac{2\pi}{h}, \quad \alpha = \frac{h}{l},\\
&\eta_{2m}^0 = \sin(\beta_{2m}^0(\kappa y - \pi)) - \beta_{2m}^0(\kappa y - \pi) \cos \pi\beta_{2m}^0,\\
&\eta_{2m-1}^n = \cos \kappa\beta_{2m-1}^n \left(y - \frac{h}{2}\right) - \frac{\cos \pi\beta_{2m-1}^n}{\text{ch}\pi\alpha n} \text{ch}\kappa\alpha n \left(y - \frac{h}{2}\right),\\
&\eta_{2m}^n = \sin \kappa\beta_{2m}^n \left(y - \frac{h}{2}\right) - \frac{\sin \pi\beta_{2m}^n}{\text{sh}\pi\alpha n} \text{sh}\kappa\alpha n \left(y - \frac{h}{2}\right),
\end{aligned} \tag{14.2.5}$$

found by separation of variables x, y in (14.2.2) to form orthogonal *basic flows* $\boldsymbol{v} = \mathbf{i}\varphi_y - \mathbf{j}\varphi_x$ and $\boldsymbol{v}' = \mathbf{i}\varphi'_y - \mathbf{j}\varphi'_x$, or basic rotations φ and φ' *strongly orthogonal*,

$$\begin{aligned}
&\langle \boldsymbol{v} \cdot \boldsymbol{v}' \rangle = \langle \nabla\varphi \cdot \nabla\varphi' \rangle = \langle A\varphi \cdot \varphi' \rangle = \frac{\langle A\varphi \cdot A\varphi' \rangle}{\mu} \quad \text{and}\\
&\langle \boldsymbol{v} \cdot \boldsymbol{v}' \rangle = \langle \varphi \cdot A\varphi' \rangle = \frac{\langle A\varphi \cdot A\varphi' \rangle}{\mu'}\\
&\qquad \text{for } \varphi = \varphi_m^{n-1}, \varphi' = \varphi_{m'}^{n'-1}, \text{ hence,} \quad \langle \boldsymbol{v} \cdot \boldsymbol{v}' \rangle = 0 \text{ if}\\
&\qquad\qquad \mu = \mu_m^{n-1} \neq \mu' = \mu_{m'}^{n'-1}, \quad m, m', n, n' = 1, 2, \ldots,
\end{aligned}$$

due to (12.3.3).

As this takes place, the sinus of expenditure (12.5.8) corresponds to the *first basic rotation* $\varphi_1^0 = \eta_1^0 = 1 - \cos \kappa y$; namely, we have in (12.5.8) that

$$\begin{aligned}
&\psi^{(0)} = \frac{U}{\kappa}(1 - \varphi_1^0) = \frac{g}{\kappa^3\nu} \cos \kappa y, -\nu u_{yy}^{(0)} = -\nu\psi_{yyy}^{(0)}\\
&\quad = -g \sin \kappa y = g^x,\\
&\omega^{(0)} = \mu_1^0\psi^{(0)} = \kappa^2\psi^{(0)} \text{ for } \mu_1^0 = (\kappa\beta_1^0)^2 \quad \text{and}\\
&\quad \beta_1^0 = 1 \quad (\mathbf{S}).
\end{aligned} \tag{14.2.6}$$

So, like in Theorems 5.2.1 and 13.1.2 concerning such basic motions as extreme rotations of stone (5.1.1) and vortex strip of problem **F**, respectively, the angular momentum (vorticity $\omega^{(0)}$) proves to be proportional to the angular velocity (expenditure $\psi^{(0)}$), ensuring the corresponding zero vortex*generation*

$$\zeta \cdot \varphi \times (\omega^{(0)} - \lambda\psi^{(0)}) = 0(\omega^{(0)} = \lambda\psi^{(0)}), \lambda = \mu_1^0, \text{ in } \mathbf{S}.$$

in *the third energy identity* (13.1.1) (or (14.1.1)) that, with zero relaxation (14.1.4), *reduces* to

$$\begin{gathered} ((\zeta - \mu_1^0\varphi) \cdot \zeta)_t + 2\nu(\zeta - \mu_1^0\varphi) \cdot B\varphi = 0, \\ \zeta = A\varphi, \quad t > 0 \\ ((\zeta - \mu_1^0\varphi) \cdot B\varphi = 0, \partial_t = 0), \end{gathered} \tag{14.2.7}$$

and while accounting for (14.1.6), we may use the constructions of theorem 5.2.1 in problem **S** provided that, in addition to the main construction of a quasicompact algebra [Troshkin, 1988a, 1995] for the dissipative top (12.3.7), the *minimum condition*

$$\begin{aligned} &\mu_1 = \mu_{\min} = \min_{m,n \geq 1} \mu_m^{n-1} = \mu_1^0 = \kappa^2, \quad \text{or} \\ &\gamma_1^0 = 1 \leq \gamma_m^{n-1}, \quad m, n = 1, 2, \ldots. \end{aligned} \tag{14.2.8}$$

has to be fullfilled [Troshkin, 2012a, 2012b, 2013a, 2017a] for the basic flow of expenditure (14.2.6) (or (12.5.8)) in the problem **P** (12.5.7).

With the minimum condition (14.2.8), the reduced third energy identity (14.2.7) is *reformed* to

$$\begin{gathered} ((\zeta - \mu_1\varphi) \cdot \zeta)_t + 2\nu(\zeta - \mu_1\varphi) \cdot B\varphi = 0, \\ \zeta = A\varphi, \quad t > 0, \\ ((\zeta - \mu_1\varphi) \cdot B\varphi = 0, \partial_t = 0) \text{ in } \mathbf{S}. \end{gathered} \tag{14.2.9}$$

14.3. The least moment indicator

We shall start the analysis of the minimum condition (14.2.8) by comparison $\leq$, or $\leftarrow$, of first three spectral points γ_m^{n-1}, $m, n = 1, 2, \ldots,$

of two numbers,

$$\gamma_1^0 = 1 \leftarrow (\le) 2.02454 < \gamma_2^0 = \left(1 + \frac{\mathrm{arctg}\pi\sqrt{\gamma_2^0}}{\pi}\right)^2 < 2.02458$$

from (14.2.3) and the function

$$\gamma_1^1(\alpha) = \alpha^2 + \left(\frac{1}{2} + \frac{1}{\pi}\mathrm{arctg}\frac{\sqrt{\gamma_1^1(\alpha) - \alpha^2}}{\alpha \mathrm{th}\pi\alpha}\right)^2$$
$$\ge \gamma_1^1(\alpha_0) \quad \text{for all } \alpha > 0,$$
$$\text{where } \pi\alpha_0 \mathrm{th}\pi\alpha_0 = 1, \text{ or } 0.3818 < \alpha_0 < 0.3819, \quad (14.3.1)$$

of aspect ratio α from (14.2.2) and (14.2.3) in Fig. 14.3.1, to be the *least moment indicator* separating a spectrum point γ_1^{1*} from the point γ_1^0,

$$\gamma_1^0 = 1 \leftarrow (\le)\gamma_1^1(\alpha) \text{ for } \alpha \ge \alpha_1, \quad 0.3818 < \alpha_1 < 0.3819, \quad (14.3.2)$$

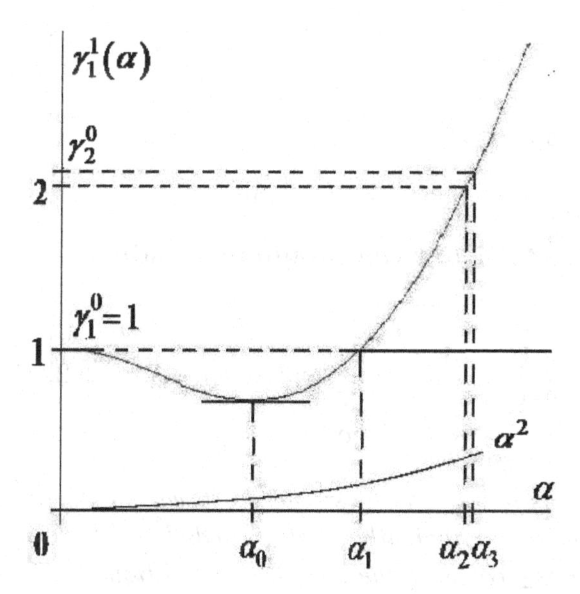

Figure 14.3.1. Least moment indicator in **S**.

the number 2,

$$\gamma_1^1(\alpha) \geq 2 \text{ for } \alpha \geq \alpha_2, \quad 1.2 < \alpha_2 < 1.3, \tag{14.3.3}$$

and the point γ_2^0,

$$2 < \gamma_2^0 \leftarrow (\leq)\gamma_1^1(\alpha) \text{ for } \alpha \geq \alpha_3, \quad 0.592 < \alpha_3 < 0.593, \tag{14.3.4}$$

to distinguish such a *spectrum structure* as

$$\begin{array}{lll}
\gamma_{2m-1}^{n-1} \leftarrow \gamma_{2m-1}^{n} \leftarrow \gamma_{2m-1}^{n+1} & & \gamma_1^0 \leftarrow \gamma_1^1 \leftarrow \gamma_1^2 \leftarrow \\
\uparrow \qquad\quad \uparrow \qquad\quad \uparrow & & \uparrow \quad \uparrow \quad \uparrow \\
\gamma_{2m}^{n-1} \leftarrow \gamma_{2m}^{n} \leftarrow \gamma_{2m}^{n+1} & \text{i.e.} & \gamma_2^0 \leftarrow \gamma_2^1 \leftarrow \gamma_2^2 \leftarrow \qquad \text{or} \\
\uparrow \qquad\quad \uparrow \qquad\quad \uparrow & & \uparrow \\
\gamma_{2m+1}^{n-1} \leftarrow \gamma_{2m+1}^{n} \leftarrow \gamma_{2m+1}^{n+1} & & \\
m, n = 1, 2, \ldots, \alpha \geq \alpha_1 & & \alpha_1 < \alpha_2 \leq \alpha < \alpha_3
\end{array}$$

$$\begin{array}{l}
\gamma_1^0 \leftarrow \gamma_1^1 \leftarrow \gamma_1^2 \leftarrow \\
\uparrow \quad \uparrow \quad \uparrow \\
\gamma_2^0 \leftarrow \gamma_2^1 \leftarrow \gamma_2^2 \leftarrow \\
\uparrow \\
\\
\alpha \geq \alpha_3 > \alpha_2 > \alpha_1
\end{array} \tag{14.3.5}$$

in (14.2.2–4), with the necessary details:

Statement 14.3.1 (on the minimum condition). *The function* (14.3.1) *is such that*

$$\frac{d\gamma_1^1}{d\alpha}(\alpha > \alpha_0) > 0 \quad \text{and} \quad \gamma_1^1(\alpha_{1,2,3}) = 1, 2, \gamma_2^0$$
$$\text{for } \alpha_0 < \alpha_1 < \alpha_2 < \alpha_3,$$

respectively, as in Fig. 14.3.1, *so the short-wave disturbances* (14.2.5) *as determined by sufficiently small periods $l \leq h/\alpha_1$, or for a large aspect ratio $\alpha \geq \alpha_1$, keep the minimum condition* (14.2.8) *valued for the least moment $\mu_1 = \kappa^2$ of the basic flow* (12.5.8). *Moreover, in the* *case of*

$$
\begin{gathered}
\mu_1 = \kappa^2 < \mu_1' = \mu_1^1 = \kappa^2\gamma_1^1(\alpha) < 2\kappa^2 \text{ for } \alpha_1 < \alpha < \alpha_2, \\
2\kappa^2 = 2\mu_1 < \mu_1' < \kappa^2\gamma_2^0 \text{ for } \alpha_2 < \alpha < \alpha_3 \quad \text{and} \\
\kappa^2\gamma_2^0 < \mu_1' \text{ for } \alpha_3 < \alpha,
\end{gathered} \tag{14.3.6}
$$

the next least moment $\mu_1' = \min_{m,n\geq 1}\{\mu_m^{n-1} > \mu_1\} = \kappa^2\gamma_1^1(\alpha)$ *is supplied with the structure* (14.3.5) *in addition to the least one* (14.2.8).

Proof. The statement follows from the relationships

$$
\begin{aligned}
&\pi^2\gamma_1^1(\alpha) = \gamma = a^2 + b^2, \quad \pi\alpha = a, \quad \pi\beta_1^1 = b, \\
&a\text{th}a = c, \quad b = \frac{\pi}{2} + \text{arctg}\frac{b}{c}, \\
&a^2 - c^2 + c = a^2 - a^2\text{th}^2 a + c \geq c > 0, \\
&b^2 + c^2 - c \geq b^2 - \frac{1}{4} > \frac{\pi^2 - 1}{4} > 0
\end{aligned}
$$

and identities

$$
\begin{aligned}
&a^2 - c^2 + c = a\frac{dc}{da}, \quad b^2 + c^2 - c = -b\frac{dc}{db} \quad \text{and} \\
&\frac{d\gamma}{dc} = \frac{2c(c-1)\gamma}{(a^2 - c^2 + c)(b^2 + c^2 - c)},
\end{aligned}
$$

which concludes the proof. □

14.4. Refined spectral estimates

For the next least moment $\mu_1' > \mu_1$ from (14.3.6) and corresponding orthogonal complements

$$
\begin{aligned}
&\varphi' = \varphi - \varphi^{\mu_1}, \quad \varphi' \cdot A\varphi^{\mu_1} = A\varphi' \cdot \varphi^{\mu_1} = 0, \\
&\zeta^{\mu_1} = A\varphi^{\mu_1} = \frac{1}{\mu_1}B\varphi^{\mu_1}
\end{aligned}
$$

and $\zeta' = A\varphi' = \zeta - \zeta^{\mu_1}, \quad \zeta' \cdot \zeta^{\mu_1} = A\varphi' \cdot A\varphi^{\mu_1} = \mu_1\varphi' \cdot A\varphi^{\mu_1} = 0,$

of φ^{μ_1} and ζ^{μ_1} to φ and $\zeta = A\varphi$, respectively, in M_{00}, both the properties (12.3.2) and the last identities in (12.3.3) provide such *refinements of estimate* (14.1.6) as

$$(\zeta - \mu_1\varphi)\cdot\zeta = (\zeta' - \mu_1\varphi')\cdot\zeta' \geq (1 - \mu_1/\mu_1')\|\zeta'\|^2, \zeta' = A\varphi' \quad \text{and}$$
$$(\zeta - \mu_1\varphi)\cdot B\varphi = (\zeta' - \mu_1\varphi')\cdot B\varphi' \geq (\mu_1' - \mu_1)\|\zeta'\|^2,$$
$$\varphi' = \varphi - \varphi^{\mu_1}, \quad \varphi \in M_{00}, \tag{14.4.1}$$

to be valued.

Really, from (12.3.2) and (12.3.3), we have that

$$(\zeta - \mu_1\varphi)\cdot\zeta - (\zeta' - \mu_1\varphi')\cdot\zeta'$$
$$= (\zeta' - \mu_1\varphi' + \zeta^{\mu_1} - \mu_1\varphi^{\mu_1})\cdot\zeta - (\zeta' - \mu_1\varphi')\cdot\zeta'$$
$$= (\zeta^{\mu_1} - \mu_1\varphi^{\mu_1})\cdot\zeta = (\zeta' - \mu_1\varphi')\cdot\zeta^{\mu_1} = 0$$

and

$$(\zeta - \mu_1\varphi)\cdot B\varphi - (\zeta' - \mu_1\varphi')\cdot B\varphi'$$
$$= 2\zeta^{\mu_1}\cdot(B\varphi^{\mu_1} - \mu_1\zeta^{\mu_1}) + \zeta^{\mu_1}\cdot B\varphi' - \mu_1\zeta^{\mu_1}\cdot\zeta'$$
$$= \zeta^{\mu_1}\cdot B\varphi' = \zeta^{\mu_1}\cdot(B\varphi')^{(N)}_{\mu} + \zeta^{\mu_1}\cdot(B\varphi' - (B\varphi')^{(N)}_{\mu})$$

where, in terms of Lemma 14.6.1 and Statement 12.3.1, with a *multiplicity* r_1 of μ_1,

$$\zeta^{\mu_1}\cdot(B\varphi')^{(N)}_{\mu} = \sum_{n\geq r_1+1}^{N} \frac{A\varphi_n^{\mu_n}\cdot B\varphi'}{A\varphi_n^{\mu_n}\cdot\varphi_n^{\mu_n}} A\varphi^{\mu_1}\cdot\varphi_n^{\mu_n} = 0,$$
$$\|B\varphi' - (B\varphi')^{(N)}_{\mu}\| \to 0, \quad N \to \infty,$$

which implies both equalities in (14.4.1).

Further, the first inequality in (14.4.1) follows immediately from embeddings of Lemma 14.6.1,

$$\mu_1' = \mu_{r_1+1} = \inf_{\substack{\varphi'\in M_{00},\varphi'\neq 0,\\ \varphi'\cdot A\varphi_1=\cdots=\varphi'\cdot A\varphi_{r_1}=0}} \frac{\|A\varphi'\|^2}{\varphi'\cdot A\varphi'}, \quad \text{or}$$

$$\mu_1'\varphi'\cdot\zeta' \leq \|\zeta'\|^2, \quad \zeta' = A\varphi', \tag{14.4.2}$$

the second one is provided by the proper Fourier decomposition,

$$(A\varphi')_{\mu}^{(N)} \cdot B\varphi' = A(A\varphi')_{\mu}^{(N)} \cdot A\varphi' = \sum_{n=r_1+1}^{N} \frac{A\varphi_n^{\mu_n} \cdot A\varphi'}{A\varphi_n^{\mu_n} \cdot \varphi_n^{\mu_n}} A\varphi_n^{\mu_n} \cdot A\varphi'$$

$$= \sum_{n=r_1+1}^{N} \mu_n \frac{(A\varphi_n^{\mu_n} \cdot A\varphi')^2}{A\varphi_n^{\mu_n} \cdot A\varphi_n^{\mu_n}} \geq \mu_{r_1+1} \sum_{n=r_1+1}^{N} \frac{(A\varphi_n^{\mu} \cdot A\varphi')^2}{A\varphi_n^{\mu} \cdot A\varphi_{n^{\mu}}}$$

$$= \mu'_1 \|A\varphi'_{\mu'}(N)\|^2,$$

so,

$$A\varphi' \cdot B\varphi' = (A\varphi')_{\mu}^{(N)} \cdot B\varphi' + (A\varphi' - (A\varphi')_{\mu}^{(N)}) \cdot B\varphi'$$

$$\geq \mu'_1 \|A\varphi'\|^2 - \mu'_1 \|A\varphi' - A\varphi'^{(N)}_{\mu}\|^2$$

$$-\|A\varphi' - (A\varphi')_{\mu}^{(N)}\| \|B\varphi'\| \quad \text{and} \quad \|A\varphi' - A\varphi'^{(N)}_{\mu}\|,$$

$$\|A\varphi' - (A\varphi')_{\mu}^{(N)}\| \to 0 \quad \text{for } N \to \infty,$$

which concludes the verification of (14.4.1).

14.5. Short-wave stability in S

Identity (14.2.9), Statement 14.3.1, and estimates (14.4.1–2) imply

$$\exp(-2\nu(\mu'_1 - \mu_1)t)(((\zeta' - \mu_1\varphi') \cdot \zeta' \exp(2\nu(\mu'_1 - \mu_1)t))_t$$

$$= ((\zeta' - \mu_1\varphi') \cdot \zeta')_t + 2\nu(\mu'_1 - \mu_1)(\zeta' - \mu_1\varphi') \cdot \zeta'$$

$$\leq ((\zeta' - \mu_1\varphi') \cdot \zeta')_t + 2\nu(\mu'_1 - \mu_1)\zeta' \cdot \zeta'$$

$$\leq ((\zeta' - \mu_1\varphi') \cdot \zeta')_t + 2\nu(\zeta' \cdot B\varphi' - \mu_1\zeta' \cdot \zeta')$$

$$= ((\zeta - \mu_1\varphi) \cdot \zeta)_t + 2\nu(\zeta - \mu_1\varphi) \cdot B\varphi = 0, \quad t > 0,$$

so

$$\left(1 - \frac{\mu_1}{\mu'_1}\right) \|\zeta'\|^2 \leq (\zeta' - \mu_1\varphi') \cdot \zeta' \leq (\zeta'_0 - \mu_1\varphi'_0) \cdot \zeta'_0 e^{-2\nu(\mu'_1 - \mu_1)t}$$

$$\leq \|\zeta'_0\|^2 e^{-2\nu(\mu'_1 - \mu_1)t},$$

or

$$\|\zeta'\|^2 = \|\zeta'\|^2(t) \le \frac{\mu_1' \|\zeta_0'\|^2}{\mu_1' - \mu_1} e^{-2\nu(\mu_1'-\mu_1)t}, \quad t \ge 0, \tag{14.5.1}$$

and, accounting for modes $\varphi^{\mu_1} = \varphi(y)$ of the least moment $\mu_1 = \mu_1^0$ depending not on the variable x in (14.2.5), as well as $\psi^{(0)} = \psi^{(0)}(y)$ and $\omega^{(0)} = \omega^{(0)}(y)$ in (14.2.6),

$$(\mu_1' - \mu_1)\|\zeta'\|^2 \le (\zeta' - \mu_1\varphi') \cdot B\varphi' = (\zeta - \mu_1\varphi) \cdot B\varphi = 0,$$

$$\text{or } \zeta' = 0, \text{ i.e. } \zeta = \zeta^{\mu_1} = \zeta(y), \text{ hence, } \varphi = \varphi^{\mu_1} = \varphi(y) \text{ for } \partial_t = 0,$$

which with (14.1.1) means that

$$\nu\|A\varphi\|^2 = \nu\varphi \cdot B\varphi = -\zeta \cdot \varphi \times \psi^{(0)} = 0, \text{ or } \varphi = 0, \quad \nu > 0$$

$$(\varphi = \varphi^{\mu_1} = \varphi(y), \psi^{(0)} = \psi^{(0)}(y)) \text{ for } \partial_t = 0. \tag{14.5.2}$$

Theorem 14.5.1 (on short-wave stability in S). *With the least and next least moments* $\mu_1 = \kappa^2 < \mu_1' = \kappa^2\gamma_1^1(\alpha)$ *from Statement* 14.3.1, *the sinusoidal velocity profile of expenditure* (12.5.8) *is unique, as a stationary solution of problem* $\boldsymbol{S}$ (12.5.7) (*in the sense of point* (**i**), *Section* 2.2 *of Chapter* 2), *and nonlinearly, asymptotically, and short-wave vortex stable for* $\alpha > \alpha_1$, *or for short periods* $l < h/\alpha_1$ *in the sense of point* (**iv**) *in Section* 2.2 *of Chapter* 2, *namely,*

$$\|\zeta\|^2 - \|\zeta_0\|^2 e^{-2\nu\kappa^2 t} \le \frac{2g\|\zeta_0'\|^2\sqrt{\gamma_1^1(\alpha)_1}}{\nu\kappa(\gamma_1^1(\alpha) - 1)} F(t) \quad \textit{in } \mathbf{S} \quad \textit{for}$$

$$F(t) = e^{-2\nu\kappa^2 t} \int_0^t e^{2\nu\kappa^2(2-\gamma_1^1(\alpha))t'} dt' < \frac{e^{-2\nu\kappa^2(\gamma_1^1(\alpha)-1)t}}{2\nu\kappa^2(2 - \gamma_1^1(\alpha))}$$

$$\textit{if } 1 < \gamma_1^1(\alpha) < 2,$$

$$F(t) = te^{-2\nu\kappa^2 t} \quad \textit{if } \gamma_1^1(\alpha) = 2, \quad F(t) < \frac{e^{-2\nu\kappa^2 t}}{2\nu\kappa^2(\gamma_1^1(\alpha) - 2)}$$

$$\textit{if } \gamma_1^1(\alpha) > 2. \tag{14.5.3}$$

Proof. Really, the stationary uniqueness (**i**) (Section 2.2 of Chapter 2) is provided by (14.5.2). The second energy identity

(13.1.2) with zero relaxation (14.1.4),

$$(\zeta \cdot \zeta)_t + 2\nu\zeta \cdot B\varphi + 2\zeta \cdot \varphi \times \omega^{(0)} = 0,$$

and generation

$$\zeta \cdot \varphi \times \omega^{(0)} = (\zeta^{\mu_1} + \zeta') \cdot (\varphi^{\mu_1} + \varphi') \times \omega^{(0)}$$
$$= (\zeta^{\mu_1} + \zeta') \cdot \varphi' \times \omega^{(0)} = \zeta' \cdot \varphi' \times \omega^{(0)}$$

without μ_1-modes,

$$\varphi^{\mu_1} \times \omega^{(0)} = 0 \quad \text{and} \quad \zeta^{\mu_1} \cdot \varphi' \times \omega^{(0)} = \varphi' \cdot \omega^{(0)} \times \zeta^{\mu_1} = 0, \varphi' \in M_0,$$
$$\text{for } \varphi^{\mu_1}, \zeta^{\mu_1}, \omega^{(0)} = \varphi^{\mu_1}, \zeta^{\mu_1}, \omega^{(0)}(y)$$

(where the volume identity (12.2.7) is used) takes the form

$$(\zeta \cdot \zeta)_t + 2\nu\zeta \cdot B\varphi = -2\zeta' \cdot \varphi' \times \omega^{(0)}$$

and, with an estimate similar to the third inequality in (12.3.6) (for μ_1' and φ' instead of $\mu_1 = \mu_{\min}$ and φ) and formulae (14.2.6), implies

$$e^{-2\nu\mu_1 t}(e^{2\nu\mu_1 t}\|\zeta\|^2)_t = \|\zeta\|_t^2 + 2\nu\mu_1\|\zeta\|^2$$
$$\leq (\zeta \cdot \zeta)_t + 2\nu\zeta \cdot B\varphi \leq 2|\zeta' \cdot \varphi' \times \omega^{(0)}|$$
$$\text{and } |\zeta' \cdot \varphi' \times \omega^{(0)}| \leq \|\zeta'\|\|\varphi_x'\omega_y^{(0)}\| \leq \|\zeta'\|\|\varphi'\|_1 \max_{0\leq y\leq h} |\omega_y^{(0)}|$$
$$\leq \frac{g\|\zeta'\|^2}{\nu\sqrt{\mu_1'}} \quad (g, \nu = \text{const} > 0),$$

or

$$\|\zeta\|^2 \leq \|\zeta_0\|^2 e^{-2\nu\mu_1 t} + \frac{2g}{\nu\sqrt{\mu_1'}} e^{-2\nu\mu_1 t} \int_0^t \|\zeta'\|^2(t') e^{2\nu\mu_1 t'} dt',$$

which, accounting for (14.5.1), leads to (14.5.3) and completes the proof. □

14.6. The spectrum of compact imbedding

Like computer processing based on a robust code required to cope with those or other input errors, the scheme of reasoning used above leans on the following statement on coping robustly with both initial and boundary disturbances [Rellich, 1930; Sobolev, 1963; Dezin, 1987; Troshkin, 1995]:

Lemma 14.6.1 (on the Rayleigh function). *Let a real linear set L of elements $\varphi, \xi, \eta, \ldots$ (or a vector space taken over the field of real numbers as constants, or coefficients for $\varphi, \xi, \eta, \ldots$) be a pre-Hilbert space supplied with a pair of metrics (point, or scalar products), $(\xi, \eta)_s$ and $(\xi, \eta)_w$, which are relatively strong, or the upper norm $\|\varphi\|_s = \sqrt{(\varphi, \varphi)_s}$ is embedded into the remaining weak one, $\|\varphi\|_w = \sqrt{(\varphi, \varphi)_w}$, or $\|\varphi\|_s \geq \gamma\|\varphi\|_w$ for any $\varphi \in L$ and some* $\gamma = \text{const} > 0$, *and, what is more, is embedded compactly in the sense that every sequence $\{\varphi_n\} \subset L$ strongly unites as $\|\varphi_n\|_s = 1, n = 1, 2, \ldots,$ proves to be partially weakly fundamental by Cauchy, or $\|\varphi_{n_{k+p}} - \varphi_{n_k}\|_w \to 0$ for $n_k, n_{k+p} \to \infty$, or $k, p \to \infty (k, p = 1, 2, \ldots)$, and, hence, partially convergent in the weak norm $\|\varphi\|_w$ to a limit element $\varphi^1 = \lim_{k\to\infty} \varphi_{n_k}, \|\varphi^1 - \varphi_{n_k}\|_w \to 0, n_k \to \infty$, of proper closure $\bar{L}_w$ (as the Hilbert space) of L. In such an event, the square rate of the embedding $\|\varphi\|_s \geq \gamma\|\varphi\|_w$, or the Rayleigh function*

$$R(\varphi) = \|\varphi\|_s^2/\|\varphi\|_w^2 \geq \gamma^2, \quad \varphi \in L, \quad \varphi \neq 0,$$

when taken on the strongly united sphere $S^0 = \{\varphi \in L : |\varphi\|_s = 1\}$, has got its least value

$$\Lambda_{\text{min}} = \Lambda_1 = \inf_{S^0} R(\varphi) = 1/\sup_{S^0} \|\varphi\|_w^2 > 0,$$
$$S^0 = \{\varphi \in L : \|\varphi\|_s = 1\},$$

achieved at an element $\varphi^1 = \lim_{k\to\infty} \varphi_{n_k}, \|\varphi^1\|_s = 1\|\varphi^1 - \varphi_{n_k}\|_s \to 0 (n_k \to \infty)$, of the stronger Hilbert space $\bar{L}_s$ (the closure of L in the strong norm $\|\varphi\|_s$), so that

$$R(\varphi^1) = \Lambda_1 \quad \text{and} \quad (\varphi^1, \xi)_s = \Lambda_1(\varphi^1, \xi)_w \quad \text{for any } \xi \in L.$$

The same is true for $R(\varphi)$ restricted further to the orthogonal complement $S^1 = \{\varphi \in S^0 : (\varphi, \varphi^1)_s = 0\}$ of φ^1 in S^0 while leading *to the next* $\Lambda_2 \geq \Lambda_1$ *and* $\varphi^2 \in \bar{L}_s \subset \bar{L}_w$, *such that*

$$\Lambda_2 = R(\varphi^2) = \inf_{S^1} R(\varphi), \quad \|\varphi^2\|_s = 1, \quad (\varphi^2, \varphi^1)_s = 0 \quad \textit{and}$$
$$(\varphi^2, \xi)_s = \mu_2(\varphi^2, \xi)_w, \xi \in L,$$

and so on. As a result, the obtained non-decreasing sequence of spectrum moments $\Lambda_n \leq \Lambda_{n+1}, n = 1, 2, \ldots,$ *of embedding* $R(\varphi) \geq \gamma^2, \varphi \in L$, *has no finite accumulation points proving unbounded and admitting only finite multiplicities*

$$1 \leq r_n < \infty, \quad 0 < \Lambda_n = \cdots = \Lambda_{n+r_n-1} < \Lambda_{n+r_n},$$
$$n = 1, 2, \ldots, \quad \lim_{n\to\infty} 1/\Lambda_n = 0.$$

At that, the set of corresponding modes as non-trivial proportions

$$\varphi_n = C_n\varphi^n \in \bar{L}_s \subset \bar{L}_w, \quad C_n = \text{const} \neq 0, \quad \|\varphi^n\|_s = 1,$$
$$(\varphi_{n+1}, \varphi_k)_s = 0, \quad k = 1, \ldots, n,$$
$$\Lambda_1 = R(\varphi_1) = R(\varphi^1) = \inf_{\varphi\in M, \varphi\neq 0} R(\varphi) = \inf_{\varphi\in S^0} R(\varphi),$$
$$\Lambda_{n+1} = R(\varphi^{n+1}) = R(\varphi_{n+1}) = \inf_{\substack{\varphi\in L, \varphi\neq 0, \\ (\varphi,\varphi_1)_s=\cdots=(\varphi,\varphi_n)_s=0}} R(\varphi)$$
$$= \inf_{\varphi\in S^n} R(\varphi), \quad n = 1, 2, \ldots,$$

as eigenfunctions of the spectrum problem

$$(\varphi_n, \xi)_s = \Lambda_n(\varphi_n, \xi)_w \quad \textit{for any } \xi \in L, \quad n = 1, 2, \ldots,$$

is full by Fourier to form a common orthogonal basis of orts φ_n *(or* φ^n*) for both* $\bar{L}_s$ *and* $\bar{L}_w$, *so that, whatever* $\varphi \in \bar{L}_s \subset \bar{L}_w$, *the strong norm*

$$\|\varphi^{(N)}\|_s^2 = \sum_{n=1}^{N} \frac{(\varphi_n, \varphi)_s^2}{(\varphi_n, \varphi_n)_s^2} = \sum_{n=1}^{N} (\varphi^n, \varphi)_s^2 \quad \textit{of}$$
$$\varphi^{(N)} = \sum_{n=1}^{N} \frac{(\varphi_n, \varphi)_s}{(\varphi_n, \varphi_n)_s} \varphi_n = \sum_{n=1}^{N} (\varphi^n, \varphi)_s \varphi^n$$

is bounded with the Bessel inequality and tends to the Perceval equality, or

$$\|\varphi^{(N)}\|_s^2 \leq \|\varphi\|_s^2 \quad \textit{and} \quad \lim_{N\to\infty} \|\varphi^{(N)}\|_s^2 = \|\varphi\|_s^2,$$

respectively, with residue $\delta^{(N)} = \varphi - \varphi^{(N)}$ *estimated as*

$$\Lambda_{N+1}^2 \|\delta^{(N)}\|_w^2 \leq \|\delta^{(N)}\|_s^2 = \|\varphi\|_s^2 - \|\varphi^{(N)}\|_s^2, \quad N = 1, 2, \ldots.$$

Proof. Really, since $R(\varphi) \geq \gamma^2$, $\varphi \in L$, there is $\Lambda_1 = \inf_{\varphi\in L,\varphi\neq 0} R(\varphi) = \inf_{\varphi\in S^0} R(\varphi)$ such that $\Lambda_1 \leq R(\varphi^n) = 1/\|\varphi^n\|_w^2 = \Lambda_1 + \varepsilon_n/2$, for $\varepsilon_n \to +0$, and $\|\varphi^m\|_s = 1$, $m, n = 1, 2, \ldots$, hence, owing to the compactness of imbedding $R(\varphi) \geq \gamma^2$ and with no loss of generality, we come to the weak convergence $\|\varphi^m - \varphi^n\|_w \to 0$ as $m, n \to \infty (\varphi^n = \varphi^{n_k})$. By parallelogram identity and definitions of Λ_1,

$$\|\varphi^m - \varphi^n\|_{s,w}^2 + \|\varphi^m + \varphi^n\|_{s,w}^2 = 2\|\varphi^m\|_{s,w}^2 + 2\|\varphi^n\|_{s,w}^2 \quad \text{and}$$
$$R(\varphi^m + \varphi^n) \geq \Lambda_1,$$

we derive the strong convergence as

$$\begin{aligned}
\|\varphi^m - \varphi^n\|_s^2 &= \|\varphi^m\|_s^2 + \|\varphi^n\|_s^2 - \|\varphi^m + \varphi^n\|_s^2 \\
&= 2R(\varphi^m)\|\varphi^m\|_w^2 + 2R(\varphi^n)\|\varphi^n\|_w^2 - R(\varphi^m + \varphi^n)\|\varphi^m + \varphi^n\|_w^2 \\
&\leq \varepsilon_m\|\varphi^m\|_w^2 + \varepsilon_n\|\varphi^n\|_w^2 + \Lambda_1(2\|\varphi^m\|_w^2 + 2\|\varphi^n\|_w^2 - \|\varphi^m + \varphi^n\|_w^2) \\
&\leq \frac{\varepsilon_m + \varepsilon_n}{\Lambda_1} + \Lambda_1\|\varphi^m - \varphi^n\|_w^2 \to 0 \quad \text{for } m, n \to \infty,
\end{aligned}$$

which leads to the existence of the required $\varphi^1 = \lim_{n\to\infty} \varphi_n \in \bar{L}_s \subset \bar{L}_w$, $R(\varphi^1) = \Lambda_1 = \inf_{L^0} R(\varphi)$, and so on.

In so doing, both the same compactness and the supposed infinite multiplicity r_n, when taken together, would be impossible because the assumption

$$\|\varphi^{n+p-1}\|_s^2 = 1 \quad \text{for } p = 1, \ldots, r_n \quad \text{and} \quad r_n = \infty$$

implies the convergence

$$\|\varphi^{n+p_k-1} - \varphi^{n+p_m-1}\|_w^2 \to 0, \quad \text{for } p_k, p_m \to \infty,$$
$$p_k \neq p_m (k \neq m, k, m = 1, 2, \ldots),$$

which contradicts the orthogonality

$$(\varphi^{n+p_k-1}, \varphi^{n+p_m-1}) = 0, \text{ hence, } \|\varphi^{n+p_k-1} - \varphi^{n+p_m-1}\|_w^2 = \frac{2}{\Lambda_n} \neq 0.$$

As this takes place, the *Freshet derivative* [Ekeland and Temam, 1976],

$$R'(\varphi;\xi) = \frac{(\varphi,\xi)_s - R(\varphi)(\varphi,\xi)_w}{\|\varphi\|_w^2/2},$$

$$R(\varphi) = \frac{\|\varphi\|_s^2}{\|\varphi\|_w^2}, \quad \text{at } \varphi \in L \quad \text{for any } \xi \in L,$$

approximating the difference

$$R(\varphi+\xi) - R(\xi) = R'(\varphi;\xi) + o(\|\xi\|_s),$$

$$o(\|\xi\|_s)/\|\xi\|_s \to 0, \quad \|\xi\|_s \to +0,$$

for

$$o(\|\xi\|_s) = \frac{\|\varphi\|_w^2\|\xi\|_s^2 - \|\xi\|_w^2\|\varphi\|_s^2}{\|\varphi+\xi\|_w^2\|\varphi\|_w^2} - R'(\varphi;\xi)\frac{2(\varphi,\xi)_w + \|\xi\|_w^2}{\|\varphi+\xi\|_w^2}$$

and

$$\|\xi\|_w < \|\varphi\|_w(\|\varphi+\xi\|_w > \|\varphi\|_w - \|\xi\|_w > 0),$$

necessarily vanishes at the next minimum point φ^n as

$$R'(\varphi^n;\xi) = 0 \quad \text{if } R(\varphi) \geq R(\varphi^n) = \mu_n = \inf_{S^{n-1}} R(\varphi),$$

$$\text{for any } \xi, \varphi, \varphi^n \in \bar{L}_s,$$

$$\varphi, \varphi^n \neq 0, \quad (\varphi,\varphi^{n-1}) = \cdots = (\varphi,\varphi^1)$$

$$= (\xi,\varphi^{n-1}) = \cdots = (\xi,\varphi^1) = 0,$$

hence,

$$(\varphi^n,\xi)_s = \Lambda_n(\varphi^n,\xi)_w \quad \text{for any } \xi \in \bar{L}_s$$

$$(\text{since } (\varphi^n,\varphi^{n-1}) = \cdots = (\varphi^n,\varphi^1) = 0).$$

The sequence of residues $\delta^{(N)} = \varphi - \varphi^{(N)}$, $N = 1, 2, \ldots$, is more strongly bounded,

$$\|\delta^{(N)}\|_s^2 = (\varphi - \varphi^{(N)}, \varphi - \varphi^{(N)})_s = \|\varphi\|_s^2 - \|\varphi^{(N)}\|_s^2 \leq \|\varphi\|_s^2,$$

and hence, is weakly degenerated in $\bar{L}_w$, as

$$\|\delta^{(N)}\|_w = \sqrt{\sum_{n>N} (\varphi^n, \varphi)_w^2} = \sqrt{\sum_{n>N} \frac{(\varphi^n, \varphi)_s^2}{\Lambda_n^2}}$$

$$\leq \frac{\|\varphi\|_s}{\Lambda_{N+1}} \to 0, \quad N \to \infty,$$

and, at the same time, convergent strongly (although partially) to some $\delta^* = \lim_{k\to\infty} \delta^{(N_k)} \in \bar{L}_s$, which is necessarily zero in $\bar{L}_w$ and therefore orthogonal to every ort φ^n in $\bar{L}_w$, and hence, $(\delta^*, \xi)_s = \Lambda_n(\delta^*, \xi)_w = 0$, $\xi \in \bar{L}_s$, $n = 1, 2, \ldots$, which is evidently possible only when $\delta^* = 0$ in $\bar{L}_s$.

The proof is complete. □

References

Abarzhi S.I. Review of theoretical modeling approaches of Rayleigh–Taylor instabilities and turbulent mixing. *Phil. Trans. R. Soc. A.* 2010, 368, 1809–1828.

Abramovich, G.N.*Applied Gas Dynamics.* — 3rd edn. Amazon.com, 1973, 1139 p.

Agmon S., Douglis A., Nirenberg L. Estimates near the boundary for solutions of elliptic partial differential equations satisfying general boundary conditions. I, II. *Comm. Pure Appl. Math.* 1959, 12(4), 623–727. 1964, 17(1), 35–92.

Aleshin A.N., Gamalii E.G., Zaitsev S.G., Lazareva E.V., Lebo I.G., Rozanov V.B. Investigation of nonlinear and intermediate stages in the development of Richmyer–Meshkov instability. *Sov. Tech. Phys. Lett.* 1988, 14, 466–470.

Arnold V.I. *Sur la geometrie differentielle des groupes de Lie de dimension infinie et ses applications a l'hydrodynamique de fluids parfaits. Ann. Inst. Fourier (Grenoble).* 1966, 16, 319–361.

Arnold V.I. *Mathematical Methods of Classical Mechanics.* Springer–Verlag, 1989, 512 p.

Batchelor G.K. *An Introduction to Fluid Dynamics.* Cambridge Univ. Press, 1967, 615 p.

Beavers G.S., Sparrow E.M. Low Reynolds number turbulent flow in large aspect ratio rectangular ducts. *J. Basic Engn.* 1971, 93, 296–299.

Belotserkovskii O.M., Betelin V.B., Borisevich V.D., Denisenko V.V., Eriklintsev I.V., Kozlov S.A., Konyukhov A.V., Oparin A.M., Troshkin O.V. To the theory of counterflow in a viscous heat conducting gas. *Zh. Vychisl. Mat. Mat. Fiz.* (*Comput. Math. and Math. Phys.*) 2011, 51(2), 222–236 [in Russian].

Belotserkovskii O.M., Konyukhov A.V., Oparin A.M., Troshkin O.V., Fortova S.V. Structurization of chaos. *Comput. Math. and Math. Phys.* 2011b, 51(2), 222–234.

Belotserkovskii O.M., Belotserkovskaya M.S., Denisenko V.V., Eriklintsev I.V., Kozlov S.A., Oparina E.I., Troshkin O.V., Fortova S.V. On the development of a wake vortex in inviscid flow. *Comput. Math. Math. Phys.* 2014, 54(1), 172–176.

Belotserkovskaya M.S., Denisenko V.V., Eriklintsev I.V., Kozlov S.A., Oparina E.I., Troshkin O.V. On the short-wave nature of the Richtmayer–Meshkov instability. *Comput. Math. Math. Phys.* 2016, 56(6), 1075–1085.

Bingham E. C. *Fluidity and Plasticity.* New York, London: McGraw-Hill Book Company, Inc., 1922, 439 p.

Bizhanov A., Kurunov I., Podgorodetskyi G., Dashevskyi V., Pavlov A. Extruded briquettes — new charge component for the manganese ferroalloys production. *ISIJ Int.* 2014, 54(10), 2206–2214.

Bizhanov A., Kurunov I., Dalmia Y., Mishra B., Mishra S. Blast furnace operation with 100% extruded briquettes charge. *ISIJ Int.* 2015, 55(1), 175–182.

Brown R. A brief account of microscopical observations made in the months of June, July and August, 1827, on the particles contained in the pollen of plants; and on the general existence of active molecules in organic and inorganic bodies. *Philos. Magaz.* 1828, 4, 161–173.

Borisevich V.D., Wood H.G. Gas centrifugation. *Encyclopedia of Separation Science. Isotope Separations.* London, UK: Academic Press, 2000, 3202–3207.

Carnot S. *Réflexions sur la puissance motrice du feu et sur les machines propres à développer cette puissance.* Paris: Bachelier Libraire, 1824, 57 p.

Comte-Bellot G. *Turbulent Flow Between Two Parallel Walls.* Aeronautical Research Council, 1969, 150 p.

Chwolson O. *Über eine mögliche Form fiktiver Doppelsterne. Astronomische Nachrichten.* 1924, 221, 329–330.

Dezin A.A. *Partial Differential Equations: An Introduction to a General Theory of Linear Boundary Value Problems.* Springer, 1987, 161 p.

Dirac P.A.M. The motion in a self-fractionating centrifuge: DTA Rept M.S.D.I., May 1942; declassified in 1946 as Report BDDA 7 (Report Br-42). London: HMSO, pp. 1–7. *Collected Works of P.A.M. Dirac 1924–1948.* Cambridge: Cambridge Univ. Press, 1995, Article 1942:3, 1063–1074.

Drazin P.G. *Introduction to Hydrodynamic Stability.* Cambridge University Press, 2002, 238 p.

Einstein A. *Zur Elektrodynamik bewegter Körper. Annalen der Physik. Chemie.* 1905, 17, 891–921.

Ekeland I., Temam R. *Convex Analysis and Variational Problems.* Amsterdam and Oxford: North-Holland Publ. Company, 1976, 402 p.

Euler L. *Principes généraux de l'etat d'equilibre des fluids. Mémoires de l'Academie des Sciences de Berlin.* 1757, 11, 217–273; Euler L. *Principes généraux du mouvement des fluids. Mémoires de l'Academie des Sciences de Berlin.* 1757, 11, 274–315; Euler L. *Continuation des recherrches sur la theorie du mouvement des fluids. Mémoires de l'Academie des Sciences de Berlin.* 1757, 11, 316–361.

Euler L. *Theoria motus corporum solidorum seu rigidorum ex primis nostrae cognitionis principiis stabilita et ad omnes motus, qui in huiusmodi corpora cadere possunt, accommodata.* Rostochii et Gryphiswaldiae: Litteris et Impensis A. F. Röse, 1765, 520 p.

Fourier J. *Théorie analytique de la chaleur.* Paris: *Firmin Didot Père et Fils*, 1822, 639 p.

Fraenkel L.E., Berger M.S. A global theory of steady vortex rings in an ideal fluid. *Acta Math.* 1974, 132(1), 13–51; *Bulletin of the American Mathematical Society.* 1973, 79(3), 806–810.

Friedrichs K.O. The identity of weak and strong extensions of differential operators. *Trans. Amer. Math. Soc.* 1944, 55, 132–151.

Friedman A. *Über die Krümmung des Raumes. Zeitschrift für Physik.* 1922, 10(1), 377–386.

Frost U., Moulden T. (ed.) *Handbook of Turbulence. Volume 1. Fundamentals and Applications.* NY, London: Plenum Press, 1977, 497 p.

Heisenberg W. *The Physical Principles of the Quantum Theory.* Dover Publication, Inc., 1949 (1930), 183 p.

Helmholtz H. *Uber Integrale der hydrodynamishen Gleichungen, welche der Wirbelbewegung entsprechen. J. fur die reine und angewandte Mathematik* 1858, 55, 25–55.

Hagen, G.H.L. *Über den Bewegung des Wassers in engen cylindrischen Röhren. Poggendorfs Ann. Physik Chemie.* 1839, 46, 423–442.

Händle, Frank (ed.). *Extrusion in Ceramics.* Berlin, Heidelberg and New York: Springer, 2007, 470 p.

Händle F., Laenger F., Laenger J. Determining the forming pressures in the extrusion of ceramic bodies with the help of the Benbow–Bridgwater equation using the capillar check. *Process Eng.* 2015, 92(10–11), 1–7.

Hubble E. A relation between distance and radial velocity among extragalactic nebulae. *Proc. N.A.S.* 1929, 15(3), 168–173.

Hurwitz A., Courant R. *Vorlesungen über allgemeine Funktionentheorie und elliptische Funktionen.* Berlin: Julius Springer, 1922, 399 p.

Hussain F., Zaman K.B.M.Q. Vortex paring in a circular jet under controlled excitation — Part 2: Coherent structure dynamics. *J. Fluid Mech.* 1980, 101(03), 493–544.

Jacobs J.W., Jones M.A., Niederhaus C.E. Experimental studies of Richtvyer–Meshkov instability. In *Proceedings of the Fifth International Workshop on Compressible Turbulent Mixing*. R. Young, J. Glimm and B. Boston (eds.). World Scientific, 1996, pp. 195–202.

Jevons W.S. On the cirrous form of cloud. *Dublin Philos. Mag. J. Sci. Ser. 4*. 1857, 14, 22–35.

Joseph D. D. *Stability of Fluid Motions*, Vols. I, II. Springer, 1976, 282 p.

Joule J.P. On the mechanical equivalent of heat. *Phil. Trans.* 1850, 140(1), 61–82.

Joukowsky N.E. *Über den hydraulischen Stoss in Wasserleitungsröhren. Mémoires de I'Académie Impériale des Sciences de St.-Pétersbourg*, 1900, Ser. 8, 9, pp. 1–72.

Kantorovich L.V., Akilov G.P. *Functional Analysis in Normed Spaces.* 2nd edn. Pergamon, 1982, 684 p.

Khariton Yu. B. On gas separation by centrifuging. *J. Appl. Phys.* 1937, 7(14), 441–443.

Kofman E.B. Constructions of the contemporary ultracentrifuges. *Adv. Phys. Sci.* 1941, 25(3), 340–361.

Kolmogorov A.N. Local structure of turbulence in incompressible viscous fluid for very large Reynolds numbers. *Proc. Math. Phys. Sci.* Lond. 1991 (1941), 434(1890), 9–13.

Kurosh A. *Higher Algebra.* Moscow: Mir Publishers, 1972, 432 p.

Kurosh A. *Lectures on General Algebra.* Chelsea Pub. Co, 1965, 335 p.

Kurunov I.F., Bizhanov A.M. *Stiff Extrusion Briquetting in Metallurgy.* Springer, 2017, 169 p.

Ladyzhenskaya O.A. *The Mathematical Theory of Viscous Incompressible Flow.* NY: Gordon and Breach, 1969, 224 p.

Landau L.D., Lifshitz E.M. *Fluid Mechanics.* Pergamon Press, London, and Addison-Wesley Publishing Co., Reading, Mass., 1959.

Landis E.M. Second order equations of elliptic and parabolic type. Translation of Mathematical Monographs. *Rhode Island, Providence: Am. Math. Soc.* 1997, 171, 203 p.

Langlois W.E., Deville M.O. *Slow Viscous Flow.* Springer, 2014, 324 p.

Lewis D. J. The instability of liquid surfaces when accelerated in a direction perpendicular to their planes. II. *Proc. R. Soc. Lond. Ser. A.* 1950, 202(1068), 81–96.

Libov R.L. *Introduction to the Theory of Kinetic Equations.* NY, London, Sydney, and Toronto: John Wiley & Sons, Inc., 1969.

Lin C. C. *The Theory of Hydrodynamic Stability.* Cambridge University Press, 1955, 155 p.

Lindermann F.A., Aston, F. A. The possibility of separating isotopes. *Phil. Mag. Ser. 6.* 1919, 37, 523–534.

Lindl J. Development of the indirect-drive approach to inertial confinement fusion and the target physics basis for ignition and gain. *Phys. Plasmas.* 1995, 2, 3933–4024.

Loitsyanskii L. G., *Mechanics of Liquids and Gases.* Oxford, London, Edinburgh, New York, Toronto, and Paris, Frankfurt: Pergamon, 1966, 481 p.

Lyapunov A.M. *Stability of Motion.* New York and London: Academic Press, 1966, 202 p. (Habilitation thesis. 1892).

Mayer J.R. Bemerkungen über die Kräfte der unbelebten Natur. *Ann. Chem. Pharm.* 1842, 42, 233–240.

Meshalkin L.D., Sinai Ia.G. Investigation of the stability of a stationary solution of a system of equations for the plane movement of an incompressible viscous liquid. *J. Appl. Math. Mech.* 1961, 25(6), 1700–1705.

Meshkov E. G. Instability of the interface of two gases accelerated by a shock wave. *Soviet Fluid Dynam.* 1969, 4, 101–104.

Michelson A. A., Morley E. W. On the relative motion of the earth and the luminiferous ether. *Am. J. Sci.* 1887, 34, 333–345.

Mikhailov A.L., Nevmerzhitskii N.V., Raevskii V.A. Hydrodynamic instabilities. *Physics — Uspekhi.* 2011, 54(4), 392–397.

Milne-Thomson L.M. *Theoretical Hydrodynamics.* London: Macmillan, 1955, 743 p.

Moffatt H.K., Shimomura Y. Classical dynamics: Spinning eggs — a paradox resolved. *Nature.* 2002, 416, 35–386.

Moffatt H.K. The topology of scalar fields in 2D and 3D turbulence. In *Geometry and Statistic of Turbulence*, T. Kambe *et al.* (eds.), Kluwer, 2001, pp. 12–32.

Moffatt H.K. Helicity and singular structures in fluid dynamics. *PNAS.* 2014, 11(111), 3663–3670.

Moreau J.J. *Une methode de "cinematique fonctionnelle" en hydrodynamigue. C.R.A.S.* 1959, 249(21), 2156–2158.

Monin A.S., Yaglom A.M. *Statistical Fluid Mechanics. Volume I: Mechanics of Turbulence.* Cambridge: The MIT Press. 1st edn. 1971, 769 p.

Mulliken R.S. The separation of isotopes by thermal and pressure diffusion. *J. Am. Chem. Soc.* 1922, 44(5), 1033–1051.

Navier C.L.M.H. *Memoire sur les lois du mouvement des fluids. Memoires de I' Academic Royale des Sciences de I'Institut de France.* 1823, 6, 389–440.

Nevzglyadov V.G. The theory of anisotropic turbulence. *Dokl. Akad. Nauk SSSR.* 1960, 135(2), 283–286.

Newton I. *The Mathematical Principles of Natural Philosophy.* 1687. NY: Published by Daniel Adee, Univ. of California, 1846, 581 p.

Obukhov A.M. On the distribution of energy in the spectrum of turbulent flow. *Dokl. Akad. Nauk SSSR.* 1941, 32(1), 22–24.

Pascal B. *Récit de la grande expérience de l'équilibre des liqueurs.* Hachette Livre BNF, 1648, 22 p.

Patel V.C., Head M.R. Some observations on skin friction and velocity profiles in fully developed pipe and channel flows. *J. Fluid Mech.* 1969, 38(1), 181–201.

Petrovsky I.G. *Lectures on Partial Derivatives Equations.* NY, London: Interscience Publishers Inc., 1954, 265 p.

Poiseuille J.L.M. *Recherches expérimentales sur le mouvement des liquides dans les tubes de très-petits diamètres. I. Influence de la pression sur la quantité de Iiquide qui traverse les tubes de très-petits diameters. C.R.A.S.* 1840, 11, 961–967.

Pontryagin L.S. *Ordinary Differential Equations.* Pergamon, Elsevier, 1962, 304 p.

Rayleigh S. J. W. Investigation of the character of the equilibrium of an incompressible heavy fluid of variable density. *Proc. Lond. Math. Soc.* 1883, 14, 170–177.

Rellich F. *Ein Satz über mittlere Konvergenz. Nachrichten von der Gesellschaft der Wissenschaften zu Göttingen, Mathematisch-Physikalische Klasse.* 1930, 30–35.

Reynolds O. An experimental investigation of the circumstances which determine whether the motion of water shall be direct or sinuous, and of the law of resistance in parallel channels. *Phil. Trans. R. Soc. Lond.* 1883, 174, 935–982.

Reynolds O. On the dynamical theory of incompressible viscous fluids and the determination of the criterion. *Phil. Trans. R. Soc. Lond. Ser. A.* 1894, 186, 123–164.

Richtmyer R. D. Taylor instability in shock acceleration of compressible fluids. *Comm. Pure Appl. Math.* 1960, 13, 297–319.

Rotta J. C. *Turbullente Strömungen.* Stuttgart: Teubner, 1972, p. 267.

Rubin V.C., Ford W.K. Rotation of the Andromeda nebula from a spectroscopic survey of emission regions. *Astrophys. J.* 1970, 159, 379.

Soboleff S. Sur un thèoréme d'analyse fonctionnelle. *Rec. Math.* [*Mat. Sbornik*] *N.S.* 1938, 4(46), 471–497.

Sobolev S.L. Some applications of functional analysis in mathematical physics. Translation of Mathematical Monographs. *Rhode Island, Providence: Am. Math. Soc.*, 1963, 7, 239 p.

Stokes, G.G. On the theories of the internal friction of fluids in motion. *Trans. Cambridge Philos. Soc.* 1845, 8(22), 287–342.

Taylor G.I. The instability of liquid surfaces when accelerated in a direction perpendicular to their planes. I. Waves on fluid sheets. *Proc. R. Soc. Lond. Ser. A.* 1950. 201(1065), 192–196.

Temam R. *Navier–Stokes Equations and Nonlinear Functional Analysis.* 2nd edn. Philadelphia: SIAM, 1983.

Thomson W. On the propagation of laminar motion through a turbulently moving inviscid liquid. *Phil. Mag.* 1887, Ser. 5, 24(149), 342–353.

Troshkin O.V. Algebraic structure of two-dimensional steady-state Navier–Stokes equations and global uniqueness theorems. *Soviet Physics Doklady.* 1988a, 33, 112.

Troshkin O.V. Topological analysis of the structure of hydrodynamic flows. *Russ. Math. Surv.* 1988b, 43(4), 153–190.

Troshkin O.V. On propagation of small disturbances in an ideal turbulent medium. *Sov. Acad. Sci. Doklady.* 1989, 307(5), 1072–1076.

Troshkin O.V. On wave properties of an incompressible turbulent fluid. *Phys. A.* 1990a, 168, 881–899.

Troshkin O.V. A two-dimensional flow problem for steady-state Euler equations. *Math. USSR-Sbornik.* 1990b. 66(2), 363–382.

Troshkin O.V. A separated vortex in a flow of viscous liquid. *Comput. Math. Math. Phys.* 1992a, 32(2), 273–274.

Troshkin, O.V. Waves of turbulent media. In *Nonlinear Dispersive Wave Systems.* Debnath, Lokenath (ed.), 1992b, pp. 591–609.

Troshkin O.V. Perturbations waves of turbulent media **Волны возмущений турбулентных сред**. *Comput. Math. Math. Phys.* 1993, 33(12), 1844–1863.

Troshkin, O.V. Nontraditional methods in mathematical hydrodynamics. *Translation of Mathematical Monographs. Rhode Island, Providence: Am. Math. Soc.* 1995, 144, 197 p.

Troshkin O.V. A rotating gas tube: Heating by torsion. *Phisica Scripta.* 2010a, T142, 014051, pp. 1–5.

Troshkin O.V. On rotating gas heating. *Comput. Math. Math. Phys.* 2010b, 50(6), 1004–1112.

Troshkin O.V. A dissipative top in a weakly compact lie algebra and stability of basic flows in a plane channel. *Doklady Phys.* 2012a, 57(1), 36–41.

Troshkin O.V. Nonlinear stability of Couette, Poiseuille and Kolmogorov plane channel flows. *Doklady Math.* 2012b, 85(2), 181–185.

Troshkin O.V. Nonlinear stability of a parabolic velocity profile in a plane periodic channel. *Comput. Math. Math. Phys.* 2013, 53(11), 1729–1747.

Troshkin O.V. On the stability of plane flow vortex. *Doklady Math.* 2014, 90(2), 584–588.

Troshkin O.V. On stable plane vortex flows of an ideal fluid. *Math. Statist.* 2016a, 4(2), 47–57.

Troshkin O.V. On the stability of reverse flow vortices. *Comput. Math. Math. Phys.* 2016b, 56(12), 2092–2096.

Troshkin O.V. Stability theory for a two-dimensional channel. *Comput. Math. Math. Phys.* 2017a, 57(8), 1320–1334.

Troshkin O.V. Dark deformations and light pulsations of cosmic medium. *Global J. Sci. Fron. Res.* 2017b, 17(3). 1–4.

Yudovich V.I. Example of the birth of secondary steady-state of periodic flow at stability failure of the laminar flow of viscous incompressible fluid. *J. Appl. Math. Mech.* 1965, 29(3), 527–544.

Yudovich V.I. A two-dimensional problem of unsteady flow of an ideal incompressible fluid across a given domain. *Amer. Math. Soc. Trans.* 1966, 57, 277–304.

Urey H.C. Separation of isotopes. *Reports Prog. Phys.* 1939, 6, 48–77.

Vladimirov V.S. *Equations of Mathematical Physics.* Marcel Dekker, 1971, 424 p.

Zubarev D.N., Morozov V.G., Troshkin O.V. Turbulence as a nonequilibrium phase transition. *Theor. Math. Phys.* 1992, 92, 896–914.

Zwicky F. On the masses of nebulae and of clusters of nebulae. *The Astrophysical Journal.* 1937a, 86(3), 217–246.

Zwicky F. Nebulae as gravitational lenses. *Phys. Rev.* 1937b, 51, 290–290.

Index

D

E

F

Printed in the United States
by Baker & Taylor Publisher Services

Printed in the United States
by Baker & Taylor Publisher Services